Efficiency and Costing

Efficiency and Costing
Second Law Analysis of Processes

Richard A. Gaggioli, EDITOR
The Catholic University of America

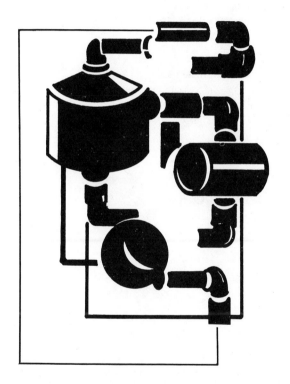

ACS SYMPOSIUM SERIES **235**

AMERICAN CHEMICAL SOCIETY
WASHINGTON, D.C.   1983

Library of Congress Cataloging in Publication Data

Efficiency and costing.
(ACS symposium series, ISSN 0097–6156; 235)

Bibliography: p.
Includes indexes.

1. Thermodynamics.
I. Gaggioli, Richard A. II. Series.

QC311.E33   1983   660.2'969   83–19707
ISBN 0–8412–0811–5

ACS Symposium Series

M. Joan Comstock, *Series Editor*

FOREWORD

The ACS Symposium Series was founded in 1974 to provide a medium for publishing symposia quickly in book form. The format of the Series parallels that of the continuing Advances in Chemistry Series except that in order to save time the papers are not typeset but are reproduced as they are submitted by the authors in camera-ready form. Papers are reviewed under the supervision of the Editors with the assistance of the Series Advisory Board and are selected to maintain the integrity of the symposia; however, verbatim reproductions of previously published papers are not accepted. Both reviews and reports of research are acceptable since symposia may embrace both types of presentation.

CONTENTS

vii

CONTENTS

EXERGY PROPERTY EVALUATION

PREFACE

OVER THE YEARS, PROMINENT THERMODYNAMICISTS have advocated second law analyses for properly evaluating energy conversion processes on the basis of available energy. The concept of available energy, now also called exergy and essergy, originated with Maxwell and Gibbs. Unfortunately, it has not taken hold in engineering practice or in managerial decision making. Many practitioners still view energy as the commodity of value, as the "potential to cause change," and therefore produce confusing and erroneous analyses. For example, we know from the first law of thermodynamics that energy is not consumed in a process; therefore, whatever energy is supplied with fuel must end up somewhere—if not in the desired product, then in some waste. Consequently, effluent wastes are grossly overestimated in value while consumptions within processes—the major inefficiencies—are overlooked completely.

The second law of thermodynamics makes a distinction between the total energy, which remains constant, and the exergy or available energy, which is consumed as it drives a process. The key to resolving inaccurate or inconsistent process efficiency analyses is simply to recognize that exergy is the proper measure. With exergy analysis, which involves the same calculational procedures as energy analysis, engineers can determine the true inefficiencies and losses, and only then make valid decisions concerning design and operation parameters.

The concept of exergy is crucial not only to efficiency studies but also to cost accounting and economic analyses. Costs should reflect value; because the value is not in energy but in exergy, assignment of cost to energy often leads to large misappropriations. Management should also use exergy content as a basis for pricing products and evaluating their profits.

An earlier book, ACS Symposium Series No. 122, "Thermodynamics: Second Law Analysis," is an introduction to the direct application of the second law of thermodynamics to (1) process efficiency analysis and (2) cost accounting in energy conversion systems and chemical/metallurgic processes. Since the publication of that volume, there has been a steady growth in the interest in applying these methods, and hence, more applications that encompass a greater realm of processes have surfaced. The purpose of this sequel is to present these new applications—in particular those that shed additional light on the theory and practice of the subject. The reader may wish to refer to the first volume for further elaboration of the fundamentals and the details of application procedures. The table of contents of the earlier book is presented on page 461.

Exergy analyses not only avoid many misconceptions resulting from energy analyses but also point out the way to economic energy conservation.

RICHARD A. GAGGIOLI
The Catholic University of America
School of Engineering and Architecture
Washington, D.C.

August 1983

INTRODUCTION AND SURVEY

Second Law Analysis for Process and Energy Engineering

RICHARD A. GAGGIOLI

School of Engineering and Architecture, Catholic University of America, Washington, DC 20064

The case for practical application of Second Law Analysis to Process and Energy Engineering is developed by (i) elucidating the 1st and 2nd Laws, and the concepts of energy and exergy, (ii) showing typical results of 2nd Law analysis, pinpointing the inefficiencies in various processes, devices, systems, industries, and sectors--including comparisons with 1st-Law analyses; (iii) presenting the conclusions of economic analyses-- of single projects; while (iv) showing that the Second Law cost accounting method is a valuable tool for optimizing the development, design and utilization of plants and facilities.

Josiah Willard Gibbs (1) and James Clerk Maxwell (2) gave form to the concept of "available energy" more than one hundred years ago; however, efforts in this century to popularize its practical use (see, for examples, the classical engineering thermodynamics texts of Goodenough (3), Kennan (4), and Dodge (5)) have met with limited acceptance.

Available energy, now called exergy, is a property which measures an object's maximum capacity to cause change, a capacity which exists because the substance is not in complete, stable equilibrium. Consequently, it is a perfectly rational basis for assigning value to a "fuel"--any commodity having the potential to drive a process. Exergy is irreversibly annihilated in any process where a potential (voltage difference, pressure difference, chemical affinity, temperature difference, etc.) is allowed to decrease without causing a fully equivalent rise in some potential elsewhere. It is a simple and understandable concept, completely consistent with our intuition and everyday perceptions. Exergy is what the layman calls "energy".

Unfortunately, another property, called energy by scientists and engineers, has become the traditional basis for assigning fuel value to materials. And because of this, process efficien-

0097–6156/83/0235–0003$12.75/0

cies have come to be defined as energy ratios. Energy efficiency is only an approximation of the true efficiency with which a
fuel resource is used, and often a poor one.

A barrier to the utilization of exergy has been the slow
historical refinement of the theory. It has been a common viewpoint until quite recently that the development of Thermodynamics
as a subject was virtually complete, and that little further investment of scientific research was warranted. It is quite clear
now that this is not the case. Thermodynamic theory is receiving
renewed interest, and deservedly so for many reasons.

The Roles of Second Law Analysis

Exergy analysis is intended to complement, not to replace, energy
analysis. Energy balances, when used in conjunction with mass
balances and other theoretical relations, are employed to design
a workable process or system. One principal role of exergy analysis is to assist in approaching optimal design or optimal operation.

One of two ways in which exergy analysis assists is by pinpointing and quantifying both the annihilations ("consumptions")
of exergy, used to drive processes, and the effluent losses of
exergy. These are the true inefficiencies, and therefore they
point the way to improvement of a system, and they stimulate creativity, leading to entirely new concepts--new technology.

Another manner in which exergy can be employed for optimization is with "Second Law Costing." Exergy, the extent to which
a material is out of equilibrium with its environment, provides
a true measure of the material's potential to cause change and/or
of the degree to which it has been processed (i.e., the degree to
which it has been driven away from stable equilibrium with the
environment). Therefore it gives a common and rational basis for
costing all the chemical flow streams, utility flow streams, heat
transports and work transfers in an energy-conversion or chemi-
cal-process system. Consequently, the traditional tradeoff between the operating and capital costs can be optimized unit by
unit within the system. Exergy costing is of value not only for
optimization, but also for cost accounting purposes.

In the role of optimization it is exergy analysis, not energy
analysis, which is appropriate, because exergy is the "common denominator." That is, all forms of exergy are equivalent to each
other as measures of departure from equilibrium, and hence as
measures of (i) a material's capacity to cause change, or (ii)
the extent to which raw materials have been processed.

I.   Thermodynamic Principles

Part I gives a simple, comprehensible presentation of (a) the
First and Second laws of Thermodynamics; (b) their associated
basic concepts of Energy and Exergy respectively; and, (c) their

practical implications on the performance of processes and equipment. It will be seen that it is exergy, not energy, which is the commodity of value and, hence, the proper measure for assessing inefficiencies and wastes.

Thermodynamics - Its Basic Implications

The basic concepts of Thermodynamics are two commodities called Energy and Exergy. The basic principles are the First Law, dealing with energy, and the Second Law, dealing with exergy.

   To illustrate the basic concepts and principles, picture a conduit carrying some commodity such as electric charge, or high-pressure water, or some chemical like hydrogen (H_2). The flow rate of any such commodity is called a current and may be expressed as

$$I_q \qquad\qquad \text{coulombs per second (amperes)}$$

$$I_v \qquad\qquad \text{gallons per minute}$$

$$I_{H_2} \qquad\qquad \text{moles per second}$$

The conduit could be a heat conductor carrying a thermal current, I_θ. Whatever the commodity might be, energy is transported concurrently with it. The rate I_E, at which energy flows is proportional to the commodity current. Thus, with charge current, I_q, the electric flow rate of energy past a cross-section of the conduit is

$$I_E \quad = \quad \phi\, I_q$$

where ϕ is the local value of the electric potential at that cross-section.

   Likewise, the hydraulic energy flow rate associated with the volumetric current, I_v, is

$$I_E \quad = \quad p I_v$$

where p is the pressure. When a material flows and carries energy not only because of its pressure but also because of its composition, the flow of energy can be called a hydro-chemical flow, and

$$I_E \quad = \quad \mu_{H_2}\, I_{H_2}$$

where μ_{H_2} is the chemical potential.

   Notice that, in each of the above examples, the proportionality factor between the commodity current and the associated energy current turns out to be the "potential" which drives the commodity through the conduit.

The driving force which causes a thermal current is a temperature difference, and the flow rate of energy with thermal current is given by

$$I_E \quad = \quad TI_\theta$$

Traditionally, in science and engineering, it is the flow rate of energy, I_E, that has been called the rate of <u>heat flow</u> (and symbolized by $\dot{Q}$). It would have been better to use the word "heat" (or "heat content") for the commodity flowing with current I_θ, but this commodity was not recognized until later, and has been named <u>entropy</u>. Obert (<u>6</u>) introduced entropy as that commodity with which heat transfers of energy are associated, with temperature T as the proportionality coefficient -- in analogy with p as the proportionality coefficient between energy and volume transfers (or ϕ as that between energy and charge transfers).

<u>Commodity Balances</u>. In analysis of energy converters, <u>balances</u> are applied for each of the relevant <u>commodities</u>; for examples, mass balances, energy balances, chemical compound balances, and so on. The amount of any given <u>commodity</u> in some <u>container</u> can in general be changed either (1) by transporting the commodity into or out of the container, or (2) by production or consumption inside. Thus, on a rate basis

The rate of change in the amount of the commodity contained	=	The sum of all the inlet rates	−	The sum of the outlet rates
+ The rate of production inside		−		the rate of consumption inside

for <u>steady</u> operation the rate of change in the amount of commodity contained within the device or system is equal to zero.

Some commodities, like charge, that cannot be produced or consumed, are said to be <u>conserved</u>.

<u>The First Law of Thermodynamics</u> can be stated:

(1) <u>Energy is conserved</u>
(2) <u>The transport of any commodity has an associated energy transport</u>.

<u>The Potential to Cause Change for Us: A Commodity</u>. When does a commodity have the capacity to cause change for us? The answer is: whenever it is not in complete, stable equilibrium with our environment (<u>8,9</u>). Then, it can be used to accomplish any kind of change we want, to some degree. Thus, charge has this capaci-

ty whenever it is at a potential different from "ground."   A
charge current carries "capacity to cause charge" to the extent
that ϕ differs from the ground value, ϕ_0:

$$P_A \quad = \quad [\phi - \phi_0] I_q$$

Similarly,

$$P_A \quad = \quad [p - p_0] I_v$$

$$P_A \quad = \quad [\mu_{H_2} - \mu_{0,H_2}] I_{H_2}$$

$$P_A \quad = \quad [T - T_0] I_\varrho$$

The charge current is represented by I_q and $I_E = \phi I_q$ represents
energy current; $P_A = [\phi - \phi_0] I_q$, the current of the commodity
called <u>exergy</u>, is the useful <u>power</u> or available power, representing the "capacity to cause change" transmitted by the charge
current.

<u>Potential To Cause Change for Us</u>: A Commodity Different from
Energy. Potential energy (exergy) does represent the capacity to
cause change for us. It is a commodity. It is distinct from
energy; it is not the same commodity. Energy cannot serve as a
measure of capacity to cause change for us; only exergy can.
      An important point is that the "capacity to cause change,"
the exergy, that a material has when it is not in equilibrium
with our environment in general is not simply equal to the difference between the energy it has, E, and the energy, E_0, it
would have were it brought to its "ground state" or "dead state,"
in equilibrium with the environment. The difference between the
exergy and $E-E_0$ stems from the fact that, while bringing the material to stable equilibrium with the environment in order to get
its exergy, it may be necessary to exchange things like volume,
"heat" and environmental <u>components</u> (7) with the environment.
The exergy content A of a material is given by (<u>1</u>,<u>8</u>,<u>9</u>)

$$A = E + p_0 V - T_0 S - \Sigma \mu_{0i} N_i$$

This equation is an important one, for calculating the exergy content of any material -- any material which could be brought to
complete stable equilibrium with the reference environment by
processes involving transports of only V, S and the components
N_i between the material and the environment.
<u>Exergy Consumption</u>. In contrast with energy and charge, exergy
is not a conserved commodity. Exergy is called "energy" in lay
terminology, and is the true measure of the potential of an object to cause change; some is irreversibly destroyed (annihilated,

consumed) in any real process. Unreal, ideal operation without
irreversible annihilation is the theoretical limit which can be
approached, but never reached in practice. Associated with real
hardware and processes, there will always be dissipations
of exergy -- consumption thereof -- used up to make the hardware
"go." These dissipations manifest themselves in "heat production;" i.e., with the production of entropy. It can be seen,
from the preceding expression for A, that the rates of exergy
consumption and entropy production are proportional:

$$\dot{A}_c = T_0 \dot{S}_p$$

This is true insofar as E, V and the N_i -- <u>components</u> (<u>7</u>)-- are
conserved quantities.

<u>The Second Law</u>. In summary, then, energy does not, in general,
represent the "capacity to cause change for us;" rather:
> <u>Exergy, which anything has when it is not in com-</u>
> <u>plete equilibrium with our environment, does repre-</u>
> <u>sent the capacity to cause change for us; it can be</u>
> <u>transferred from one thing to any other, completely</u>
> <u>in the ideal limit. In actuality, to accomplish</u>
> <u>changes for us some exergy is invariably annihi-</u>
> <u>lated, irreversibly used up because it is needed</u>
> <u>to make the changes occur.</u>

The Roles of Thermodynamics

Traditionally, Thermodynamics has served the following purposes:
1. It provided the concept of an energy balance, which has
    commonly been employed (as one of the "governing equations")
    in the mathematical modelling of phenomena.

2. It has provided mathematical formulas for evaluating
    properties (such as enthalpy and entropy).

3. It has provided the means for establishing the final
    equilibrium state of a process.

Now, with more modern formulations of Thermodynamics, it can
be used for the following purposes as well:

4. Pinpointing the inefficiencies in and losses from processes,
    devices and systems, using exergy.

5. Cost accounting with exergy. This is useful in engineering
    (design; operation of systems), and in management (pricing;
    calculating profits).

6. The governing equations for modelling nonequilibrium phenom-
    enon can be derived, by selecting the appropriate commodity
    balances (those for all commodities transported and/or pro-
    duced during the phenomenon) and utilizing the First and
    Second Laws.

The roles of primary interest in this article are those related to the direct practical application of exergy.

The material in Part I is presented in more detail in Reference 8.

II. Efficiency Analysis

Tools Used in Second Law Efficiency Analysis

In this section, the various tools employed in exergy analyses are listed and described; in the next section, their use will be illustrated by an example.
    Exergy analyses and energy analyses use the same family of tools to evaluate and compare processes:

1) <u>Balances</u> for exergy and for each independent commodity which is transported into or out of the system.

2) <u>Transport relations</u> between companion commodities.

3) <u>Kinetic relations</u> (like $\dot{Q} = UA\Delta T_m$, or $r_{AB} = k_{AB}C_A{}^a C_B{}^b$), which relate transports or productions (reaction rates) to driving forces.

4) <u>Thermostatic properties</u> specific to the materials involved.

    Of these four tools, only the first two, <u>balances</u> and <u>transport relations</u>, need further discussion here, inasmuch as they are different for exergy analyses than for energy analyses.

<u>Exergy Balances.</u> Writing a steady-state balance for exergy is just like writing a steady-state energy balance except for one major difference. While energy is conserved, exergy can be annihilated (not lost, but actually consumed), and so the balance must contain a destruction term:

Total Exergy Transported into the System	=	Total Exergy Transported out of the System	+	Exergy Destroyed within the System
$P_{A,\ in}$	=	$P_{A,\ out}$	+	$\dot{A}_c$

    When the transport rates of independent commodities are known (given or determined from kinetic relations), then the exergy transport terms can be evaluated using transport relationships like those presented in the next section. Then, the balance can be used to evaluate the one remaining quantity, the consumption, $\dot{A}_c$.

Transport Relationships. The following expressions are used to evaluate the transports of exergy, P_A.

a) Shaft Work: When energy and exergy are transported via a turning shaft--with torque, τ , which is simply a current, I_α, of angular momentum--the energy flow is $P_E = \boldsymbol{\omega} \cdot I_\alpha$ where ω is the angular velocity. This relation is usually written as $\dot{W} = \omega \cdot \tau$ since the energy flow rate is the so-called work rate, $\dot{W}$, and the flow rate of angular momentum, I_α, is the torque, τ.

The exergy current is given by

$$P_A = [\omega - \omega_0] \cdot I_\alpha = [\omega - \omega_0] \cdot \tau$$

Since the angular velocity ω_0 of the environment can generally be taken to be zero, $P_{A, shaft} = \omega \cdot \tau$ This is identical to the work rate, $\dot{W}$:

$$P_{A, shaft} = P_{E, shaft} = \dot{W}$$

As a consequence, the conclusion can be drawn, from the second law, that "the exergy is the maximum shaft work obtainable." This statement is usually used to define exergy. Unfortunately such a definition gives the impressions (i) that exergy is relevant only to "work processes," and (ii) that work is the ultimate commodity of value. Actually, exergy is the commodity of value, regardless of the form (thermal, mechanical, chemical, electrical, ...); and it is relevant to processes involving any of these forms.

b) Thermal Transports of Exergy: The energy and exergy currents associated with a thermal current at a temperature T are $I_E = TI_S$ and $P_A = [T - T_0]I_S$. By combining these two expressions, the exergy current can be written in terms of energy current as $P_A = [1 - T_0/T]I_E$. Since the energy flow rate by heat transfer is usually represented by $\dot{Q}$,

$$P_{A, thermal} = [1 - T_0/T_Q]\dot{Q}$$

If $\dot{Q}$ represents the energy supplied at a temperature T_Q to a steady-state or cyclic "heat engine", it follows from an exergy balance that the net rate of exergy flowing from the engine in the form of shaft work can at most be equal to the thermal exergy supplied to the cycle; i.e., $P_{A, shaft} \leq P_{A, thermal}$. Using the transport relationship $P_{A, shaft} = \dot{W}_{shaft}$, it follows that $\dot{W}_{max} = [1 - T_0/T_Q]\dot{Q}$. This is the classic result usually derived in a complex manner from traditional (obtuse) statements of the second law.

c) Simultaneous Thermal and Chemical Exergy Flow with Matter: The energy and exergy flows associated with bulk transports of material j are:

THERMAL: $I_E = TI_S$ and $P_A = [T-T_0]I_S$

CHEMICAL: $I_E = \mu_j I_j$ and $P_A = [\mu_j - \mu_{j0}]I_j$

where μ_{j0} is the chemical potential of material j in the dead state. The energy current for simultaneous thermal and chemical transfers associated with the flow of material j is therefore

$$P_E = \mu_j I_j + TI_S = [\mu_j + Ts_j]I_j = h_j I_j.$$

The exergy flow rate is $P_A = [h_j - T_0 s_j - \mu_{j0}]I_j.$

Evaluation of Exergy Transport Expressions. Exergy transport relations are seen to be products of thermo-static properties with
commodity currents. Given the commodity currents (say from a
process flow diagram), the exergy transports can then be evaluated by determining the thermostatic properties, using traditional thermochemical property evaluation techniques. References
(10-15) present convenient relationships for practical evaluation
of exergy flows for several important cases.

A prerequisite for the evaluation of the exergy transports
is the selection of a proper dead state (9-17).

Second Law Efficiency--The True Efficiency

In the theoretical limit, any amount of exergy contained in one
or more given feed streams (call them fuels and feedstocks) could
be completely transferred or transformed to any other commodities
(call them products). For example, the exergy transported in
with Feeds 1 and 2, $P_1 + P_2$ in Figure 1, in theory could be completely delivered--by transformation of Feeds 1 and/or 2 into
Product A, yielding P_A, and by transfer of exergy to stream B
yielding ΔP_B. That is, ideally the output would be

$$[P_A + \Delta P_B]_{ideal} = P_1 + P_2$$

In actuality, however, there will always be a consumption of
some exergy to drive the various transformation and/or transfer
processes. Furthermore, there may be effluent losses of exergy.
Then, for real operation, an exergy balance says

$$P_A + \Delta P_B = P_1 + P_2 - P_{lost} - P_{consumed}.$$

The Second Law efficiency, measuring the ratio of actual to
ideal output, is therefore given by

$$\eta_{II} = \frac{P_A + \Delta P_B}{P_1 + P_2}$$

The denominator exceeds the numerator by the amount of exergy
consumed (annihilated) by the transformation plus the amount lost
in effluents.

More generally,

$$\eta_{II} = \frac{\Sigma[\text{net product exergies}]}{\Sigma[\text{feed exergies}]}$$

For any conversion, the theoretical upper limit of η_{II} is 100%,
which corresponds to the ideal case with no dissipations. To
approach this limit in practice requires the investment of greater and greater capital and/or time. The tradeoff, then, is the
classical one: operating costs (for fuel) versus capital (for
equipment and time). The important point here is that attainment
of the optimal, economic design can be greatly facilitated by the
application of Second-Law analysis (i.e., exergy analyses) to
processes, devices, and systems.

Traditional efficiencies (here called first law efficiencies,
η_I) based on the ratio of "product" energy to "fuel" energy are
generally faulty, to a degree that depends on the kind of device
or system to which they are applied. Because energy is conserved,
the difference between the energy output in the products from a
system and the energy input with fuels--the difference which, it
is perceived, represents the inefficiency--must be the energy
lost with effluents. Exergy consumptions, which drive the various operations in the system, are neglected. Examples which illustrate these errors will follow.

(Those traditional efficiencies such as isentropic or polytropic efficiencies of turbines and compressors are 2nd law efficiencies of a sort. They do approximate η_{II} fairly well, depending on the situation.)

The Methodology of Exergy Analyses

How the tools are organized into a methodology for process evaluation via exergy is illustrated in Reference 13 with a coal-fired
boiler. It will be used to demonstrate the calculation of exergy
flows, losses and consumptions.

Application to Coal-Fired Boiler (13). Consider this problem: A
given coal-fired boiler is burning Illinois No. 6 coal while converting 298°K (77°F) water to 755°K (900°F) steam, at 5.86 MPa
(850) psia. The boiler has a first law efficiency (η_I) of 85%.
How much of the coal's exergy is destroyed? What is the second
law of efficiency (η_{II}) of the boiler? Where are the distinct
exergy consumptions within the boiler and what are their magnitudes? Where are the losses, and what are their quantities?
Figure 2 illustrates one type of flow diagram which can be drawn
for this boiler; as in energy analyses, it serves to define the
boundaries of the process being studied as well as to establish a

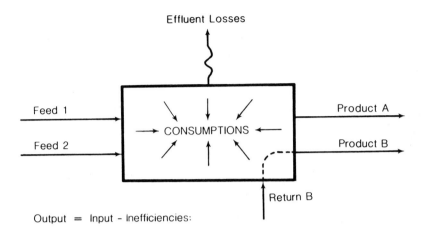

Effluent Losses

Feed 1 → CONSUMPTIONS ← → Product A

Feed 2 → → Product B

Return B

Output = Input - Inefficiencies:

$$P_A + \Delta P_B = P_1 + P_2 - P_{Lost} - P_{Consumed}$$

$$\eta = \frac{P_A + \Delta P_B}{P_1 + P_2}$$

Figure 1.   Schematic diagram of a typical energy-conversion or chemical-processing device or system.

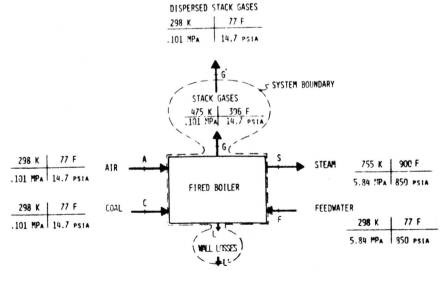

Figure 2. Flow diagram of fired boiler (with stream labels and key property data).

scheme for stream identification. Key stream properties have been included in Figure 2.

Evaluation of Transports. Using the information given on Figure 2, the composition of the coal, the composition of the stack gases, and the characteristics of the reference environment, typical thermodynamic property calculations serve to evaluate the transport terms (10,12). Thus the exergy transported with the coal is:

$$P_{coal} = 26391 \ \frac{kJ}{kg \ raw \ coal} = 11348 \ Btu/lb \ raw \ coal.$$

The combustion air, free from the environment, has zero exergy:

$$P_{air} = \text{exergy transport with combustion air} = 0$$

The transport of exergy with the feedwater is due only to its pressure since it is at T_0 and water is "free" from the environment (except for purification)

$$P_{H_2O, \ in} = 50.2 \ \frac{kJ}{kg \ coal} = 21.6 \ Btu/lb \ coal$$

The exergy in the steam can be easily evaluated using the property relations represented by the steam tables:

$$P_{H_2O, \ out} = 8963.3 \ \frac{kJ}{kg \ coal} = 3854.2 \ \frac{Btu}{lb \ coal}$$

At the location (G) the gases are at the same pressure as the environment but are not in thermal or chemical equilibrium with it. Even if cooled to T_0, the stack gases at a total pressure P_0 would still not be in complete equilibrium with the environment, because the composition is different. Using thermostatic property relations (10), the total exergy in the stack gases may be calculated:

$$P_{stack} = 1454.9 \ \frac{kJ}{kg \ coal} = 625.6 \ Btu/lb \ coal.$$

For convenience, the "system boundary" has been located far enough outside the surface of the boiler so that the exergy carried past by heat transfer is practically zero:

$$P_{wall} = [1-T_0/T_Q]\dot{Q}_L = [1-T_0/T_0]\dot{Q}_L = 0$$

Now that all the transports of exergy across the system boundary have been evaluated, the exergy consumption term can be determined:

$$\dot{A}_c = P_{in} - P_{out}$$

$$= 26391 + 0 + 50.2 - 8963.3 - 1454.9 - 0$$

$$= 16023 \; \frac{kJ}{kg \; coal} \; = 6889 \; Btu/lb \; coal$$

In answer to the second question raised earlier, the <u>second law efficiency</u> of this system is

$$\eta_{II} = \frac{P_{H_2O,out} - P_{H_2O,in}}{P_{coal}} = 0.338$$

compared to the first law efficiency of 0.85.

<u>Analysis of Sub-processes</u>. To determine the locations and magnitudes of the various different consumptions which comprise $\dot{A}_c$, one need only subdivide the system appropriately into sub-systems, and then repeat the foregoing procedure. Thus, in this problem, the consumptions within the boiler can be broken down into 1) combustion, and 2) heat transfer. Each can be analyzed for its second law efficiency and the amount of exergy it consumes.

To calculate the consumptions of exergy in the combustion process and the heat transfer process, it is supposed that the boiler may be separated into two distinct hypothetical entities: an "adiabatic combustor" and a "heat exchanger." After determining the state of the combustion products, using an energy balance, the transport of exergy from the combustion process with product gases can be determined with the same procedures as above:

P_p = exergy transport with products of combustion

$$= 19489.3 \; \frac{kJ}{kg \; coal} = 8380.4 \; Btu/lb \; coal$$

Now, having evaluated the relevant transport terms, two important consumptions of exergy within the boiler may be evaluated, by applying an exergy balance to the hypothetical "combustor" and one to the "heat exchanger"

$\dot{A}_{c, \; RXN}$ = destruction of exergy due to uncontrolled combustion of coal

$$= P_{coal} + P_{air} - P_p = 6901.7 \; \frac{kJ}{kg \; coal}$$

$$= 2967.7 \; Btu/lb \; coal$$

$\dot{A}_{c, \; HX}$ = destruction of exergy due to the heat transfer process

$$= P_p - P_{stack} + P_{H_2O,in} - P_{H_2O,out}$$

$$= 9121.3 \; \frac{kJ}{kg \; coal} = 3922.2 \; Btu/lb \; coal$$

Correspondingly, the second law efficiency for each of these internal processes may be evaluated. If the view is taken that the purpose of the combustor is to convert the (chemical) exergy of the coal into exergy of combustion products,

$$\eta_{II, RXN} = \frac{19489}{26391} = 0.738$$

In turn, the heat exchanger serves to transfer the product exergy to the H_2O

$$\eta_{II, HT} = \frac{8963-50}{19489} = 0.457$$

Here, the difference between the denominator and the numerator is the destruction by the heat transfer process plus the loss of exergy with the stack gases.

Figure 3 shows one method of presenting the results of an exergy analysis. It is similar to an energy flow diagram, with the added feature of showing consumptions of exergy as negative values within the various process blocks. Such a diagram aids in gauging the relative importance of each transport and consumption. (The so-called Sankey diagram has been used effectively in the European literature, representing each exergy flow by a band so drawn that its width is proportional to the flow.)

Discussion. The "thermal efficiency" of this boiler--that is, the net energy delivered to H_2O divided by total energy input in coal--is 85%. To cite such efficiencies--energy ratios--is misleading. As shown by the foregoing overall analysis of the boiler, the exergy of the steam--its "useful energy"--is much less than its energy content; hence, the energy efficiency, η_I, cited for the boiler is 2½ times its true efficiency, η_{II}, of 33.8%.

The detailed analysis of the different sub-processes of the boiler, as summarized in Figure 3, shows that the two largest dissipations are due to the uncontrolled kinetics of combustion (26.2% of the total exergy input), and heat transfer (34.5%) as heat passes from hot products at a high average $(1 - T_0/T)$ to liquid and gaseous H_2O with a relative low average $(1 - T_0/T)$. The stack losses, while not insignificant, represent only 5.5% of the exergy in the coal; in contrast, they represent nearly 15% of the coal's energy content.

Of course, no cost effective opportunities to reduce any consumption or loss should be overlooked. Often, the best opportunities are where the larger consumptions (and losses) occur. For example, if the steam pressure were raised, the average temperature and, therefore, the average value of $(1 - T_0/T)$ for heat addition to steam would be raised and a significant decrease in $\dot{A}_{c,HT}$ could be effected. (Design modifications may have to be made in the equipment utilizing the steam as "fuel," in order to

effectively take advantage of the steam's higher exergy content.
That is, improving η_{II} of the boiler does not necessarily imply
an improvement in the overall η_{II} for the overall process of
which the boiler is only a part. If the requirements for steam
are at low-pressure, it would be of no benefit to produce it at
high pressure and then simply throttle it to the low pressure.
On the other hand, if a turbogenerator were used in lieu of a
throttling valve, electric power could be obtained while dropping
the steam to the desired pressure--cogeneration.) In effect, the
steam's exergy content would be higher, without increasing the
exergy input (coal) to the boiler.

A detailed exergy analysis was carried out in Reference (18)
on a modern 300-MW coal-burning power plant. Although many of
the interesting details are not shown, some of the results of that
analysis are shown schematically in Figure 4; in addition to the
exergy flows and consumptions, the energy flows are shown in parentheses for the sake of comparison. It is notable that half of
the potential energy (exergy) of the incoming fuel is dissipated
immediately by the boiler (30% is used up in combustion, 15% is
consumed in the heat transfer from high-temperature products to
the steam and preheated air, and 5% is lost--primarily in chemical exergy--with the stack gases).

The corresponding energy balance implies that the boiler is
very efficient, losing only 10% of the input energy--virtually
all of it associated with stack gas thermal losses.

"Energy losses" associated with the condenser, carried into
the environment by the cooling water, are great. We hear much
about the need to utilize that energy. Actually, virtually none
(∿1%) of the resource which went into the power plant is lost in
that water. The real loss was (primarily) back in the boiler
where "heat" (entropy) was produced. Once produced, it must be
transmitted to "ground"--the environment--in order for its potential energy (i.e., exergy) to be obtained. It carries much energy with it to the environment but essentially no potential energy. Attempts to take advantage of "all that energy" and thereby improve the utilization of the fuel used by the plant are fu-
tile. On the other hand, the renewed interest in the cogeneration of electricity and "heat" is consistent with Second Law results. Cogeneration reduces the boiler inefficiencies, associated with the production of steam (or hot water) at low temperatures. (Notice that the production of hot water by cogeneration
is tantamount to raising the turbine back-pressure, so that the
condensing steam, at a high temperature, can raise the temperature of the cooling water more. Then, the cooling water becomes
useful, because it has substantial exergy.)

Summary. In illustrating the role played by exergy analysis in
design optimization, one of two methods was demonstrated. The
boiler problem used exergy balances to obtain all the pertinent

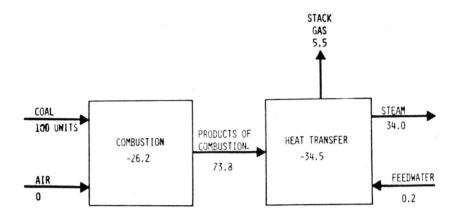

Figure 3.   Exergy flow diagram for coal-fired boiler.

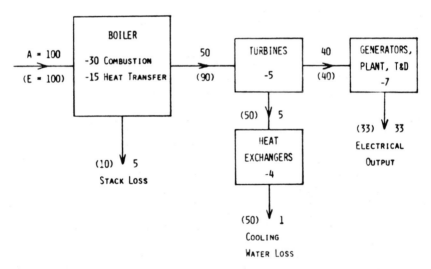

Figure 4. Exergy (and energy) flow diagram for a conventional fossil-fired steam power plant (negative numbers are exergy consumptions).

consumptions and transports of exergy. This method reveals the
relative importance of consumptions and losses with respect to
the other transports of exergy into, out of, and within the system. It also provides a consistent basis for accurate costing of
flow streams.

The error in energy efficiency analyses is that they attrib-
ute all the inefficiencies to losses, and then miscalculate those,
as was demonstrated with the coal-fired boiler.

A Survey of Typical Results of Second law Efficiency Analyses

Exergy balances can be applied at one extreme to complete sectors
of the "energy economy", such as "power generation" (the collection of electric utilities) or "industry" (the collection of process and manufacturing industries). Or, at the other extreme, to
an individual thermodynamic process within a particular device
(such as the combustion process in the boiler of the earlier example). In between such extremes there are, for example, the
various different industries (SIC's) within the industrial sector,
and in turn there are the distinct firms or plants within an SIC.
Within a plant, there are a variety of processes. Each process
consists of several units. A unit may include several devices,
and those may have several thermodynamic processes. Second Law
efficiency analysis of the specific devices, units, plants, etc.
is valuable for pinpointing inefficiencies--consumptions and los-
ses--and, in turn, the opportunities for improvement of performance. Such information is valuable when seeking to improve a
given operation, say via retro-fit modifications, or when aiming
for design improvements for a new operation.

As yet, relatively few Second-Law efficiency analyses have
been carried out. Yet, from among those which have been made
interesting patterns of information can be developed. This part
of the article will be devoted to a survey of many of the results
which have been obtained so far--for detailed processes, ..., and
for sectors. Then, important conclusions can be drawn, which are
relevant (i) to the possibilities of improving "energy" utilization, and (ii) to decisions regarding the allocation of resources
--economic and manpower--to the improved application of existing
technologies and to R&D for prospective new technologies.

The results for sectors have been obtained by "integrating"
the results for industries (while involving various approximations and assumptions), which in turn were obtained by "integrating" those for processes, ... therefore, it might seem logical
to present the detailed results first and then proceed to the integral. On the other hand, by presenting the integral results
first, the significance of the detailed results can be appreciated
in context.

Efficiency of Energy Sectors. The following information regarding
"energy" resource consumption and utilization by sector is based
almost vebatim on the work of Reistad (19, 20).

Figure 5 illustrates a typical energy flow diagram for the U.S., in 1970, of the type which was first introduced by Cook (21). In viewing such diagrams, some realized that this illustration did not tell the whole story, and in fact was quite misleading on several points. Such a diagram using exergy rather than energy would present the appropriate picture. Figure 6 presents an exergy flow diagram for the U.S. for the year 1970, from Reistad (19).

Consider a comparison of Figures 5 and 6 which are figures drawn on the same format to illustrate the difference between energy and exergy analyses. The energy flow diagram indicates that for every unit of energy that is utilized approximately one unit of energy is wasted. The exergy flow diagram of Figure 6 shows a much different picture of our technology: for each unit of exergy consumed in end uses, greater than three units of exergy is wasted. To put it another way, Figure 6 reveals that our level of technology in energy conversion and utilization is roughly one-half of that indicated by the usual energy "picture" as shown in Figure 5. From the brighter side, Figure 6 reveals that there is substantially greater room for improvement in our technology than Figure 6 would indicate to be possible.

When considering which segment of our economy is most in need of improvement, magnitude of waste is a good indication. The energy flow diagram indicates that the electrical sector and the transportation sector were the largest contributors to the wasting of energy, and that both were substantially less efficient than either the industrial or household and commercial segments. Figure 6 on the other hand, shows each of the four segments of our economy contributing roughly equal shares to the waste of exergy. Here, the transportation, and the household and commercial categories, are indicated as the least efficient with second law efficiency values roughly half that for electrical generation and industrial.

The methods, approximations and assumptions employed in constructing Figure 6, similar to those used for Figure 5, are described in some detail by Reistad (19, 20). it is clear (20, 22, 23) that the Second Law efficiencies shown in Figure 6, especially those for the Residential-Commercial and for the Industrial sectors, are conservatively high. The more recent estimates for the overall η_{II} of the Industrial sector (20, 22) are about 8% rather than the 17% given by Figure 6. The revised values are a consequence of (i) substantial amounts of additional statistical information which has become available, and (ii) the redefinition of the task (the "duty" or "load") for several specific processes. In any case,

- The energy conversion technology in the U.S. is substantially poorer than indicated by a First Law analysis, and consequently there is substantial room for improvement.
- It is not the electricity generation and auto-

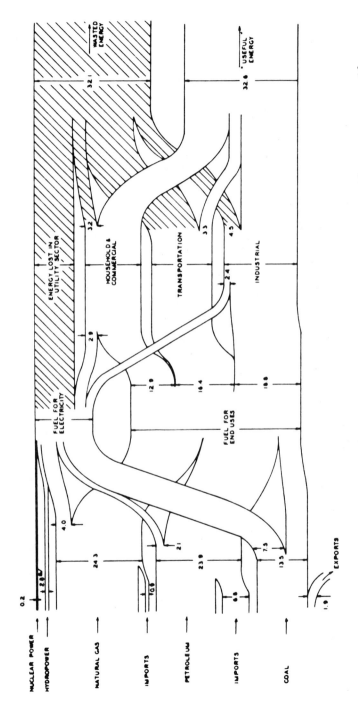

Figure 5. Flow of energy in the U.S. (21). All values reported are 10^{15} Btu [$1.055(10^{18})$J]. Footnote a: the energy value for hydropower is reported in the usual manner of coal equivalent.

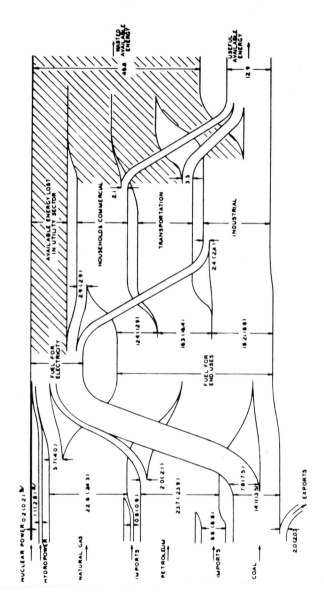

Figure 6. Flow of exergy in the U.S. (All values reported are 10^{15} Btu [$1.055(10^{18})$J] (20).) Footnote a: all values in ( ) are energy, other values reported are exergy. Footnote b: the energy value for hydropower is reported in the usual manner of coal equivalent; actual energy is 1.12×10^{15} Btu.

motive sectors alone that are the major causes
of waste as indicated by the First Law analysis. Rather, all sectors have waste of about
the same order and electricity generation is
the smallest of these.
The Household-Commercial and the Industrial
Sectors are not the most efficient, as we are
led to believe by Figure 5, but the least
efficient.
   As Reistad (20) says, in conclusion:
"The quite often used energy flow diagrams similar
to Figure 5 are quite misleading in two important
aspects. First, they imply substantial waste in
the wrong sectors, pointing the finger of blame
regarding our energy problem in the wrong direction. Secondly, they imply a technology state
of a substantially higher level than we presently
have; that is, they show "efficiencies" that are
much higher than properly evaluated efficiencies
would be for a substantial number of processes.
On the other hand, the exergy flow diagrams show
the true picture and can be used to gain useful
insight into our overall energy problems.

Exergy illustrates that our overall level of
technology is quite low, with an overall η_{II} of
less than 10% with the latest figures (22). This
point at first glance seems to be a negative aspect in our future, but in fact it is a very
positive one. Since the present efficiencies
are so low there is a lot of room for improvement in our conversion systems and consequently
a lot of room for reducing our energy consumption through improving the performance of energy
conversion and industrial processing (especially
chemical) systems."

   Interesting questions remain to be answered in the following
section: "What is the breakdown of the inefficiencies in each of
the sectors? What industries, what processes, what devices are
the sources of inefficiencies, and to what relative extent?"

   Performance of Typical Overall Systems and Processes is presented in Table I. With the nomenclature of this table, (i) $\dot{A}_s$
is "fuel" and/or "feed stock" exergies supplied, (ii) $\dot{A}_c$ is the
exergy consumed in "driving" the process, (iii) $\dot{A}_\ell$ is the exergy
lost in effluents, and (iv) the exergy delivered in the product
is $\dot{A}_p$. With the same nomenclature, Table II presents the performance of numerous devices which are common to energy-conversion
and process systems.

Table I. Performance Of Typical Overall Systems and Processes*

	$\dot{A}_c/\dot{A}_s$	$\dot{A}_\ell/\dot{A}_s$ ($\dot{E}_\ell/\dot{E}_s$)	$\eta_{II}=\dot{A}_p/\dot{A}_s$ ($\eta_I=\dot{E}_p/\dot{E}_s$)
Fossil-fired Power Plant (18)	0.55	0.06,(0.59)	0.4,(0.41)
- Boiler	0.45	0.05,(0.09)	
- Turbines	0.05		
- Condenser & heaters	0.05	0.01,(0.5)	
Co-generating Plant (24) (410 kw elec, 1130 kw steam)	0.65	0.05	0.3,(0.75)
Co-generating Plant (24) (10,000 kw elec, 17,000 st)	0.62	0.05	0.33,(0.75)
Equivalent, conventional	0.62	0.10	0.28
- 10,000 kw elec.pwr	18/30	2/30	10/30
- 17,000 kw 50 psig boiler	44/70	8/70	18/70,(55/70)
Electric Total Energy (18)	0.65	0.06,(0.67)	0.28,(0.33)
- Power prod'n & trans.	0.55	0.06,(0.67)	0.33
- Heat pump	0.05		
Fossil Total Energy(24, 26)	0.42	0.3,(0.4)	0.28,(0.6)
- Engine	0.37	0.3	
- Heating & cooling	0.05		
Equiv. conven. (18,20)	0.60	0.19	0.21,(0.5)
- Electricity	30/55	3/55	18/55
- Heating & cooling	30/45	11/45	3/45,(30/45)
Heating and Air-cond'g (23)			
- Air-conditioning	0.85	0.1	0.04(>1.0)
- Refrigerating unit	0.5	0.06	
- Compressor	0.15		
- Condensor	0.15	0.06	
- Expansion valve	0.06		
- Evaporator	0.15		
- Air-handling unit	0.3	0.04	
- Distribution	0.05		
- Heating	0.8	0.1	0.09,(0.6)
- Boiler	0.5	0.1	
- Air-handling unit	0.25		
- Distribution	0.05		

Table I—Continued

	$\dot{A}_c/\dot{A}_s$	$\dot{A}_\ell/\dot{A}_s \cdot (\dot{E}_\ell/\dot{E}_s)$	$\eta_{II}=\dot{A}_p/\dot{A}_s$ $(\eta_I=\dot{E}_p/\dot{E}_s)$
Coal Gasification, (Koppers-Totzek) (8)	0.3	0.06	0.65
- Coal preparation	0.06	0.005	
- Gasifier	0.15	0.02	
- O_2 production	0.04	0.003	
- Heat recovery	0.02		
- Gas cleanup	0.035	0.03	
Coal Gasification (Synthane) (27)	0.5	0.05, (0.45)	0.46, (0.55)
- Coal preparation	0.04	0, (0.01)	
- Gasification	0.15		
- Steam, Pwr, O_2 Prod'n	0.25	0.02, (0.40)	
- conversion, Metha- nation, Treatment	0.06	0.03, (0.04)	
Ammonia Production (28,29)	0.35	0.03	0.62
- Methane Reformation	0.28		0.68
- Ammonia Plant	0.07		0.91
Steel production, U.S. Average (30, 31)			0.21
- Coking		← 0.08 →	0.9
- Blast Furnace		← 0.11 →	0.8
- Steam & Pwr Genera'n		← 0.17 →	
- Steel-making		← 0.08 →	
- Steel-processing		← 0.27 →	
- By-product wastes		← 0.01 →	
Paper Produc'n (30, 32)			∿ 0
- Cogen, Steam & Power			0.33
- Pulping			0.08
- Paper-making			0.25
Cement Production (30, 33)			0.1
Petroleum Refining (30, 31)			0.1

* The entries are by-an-large for systems operating under optimal conditions - design loads, careful maintenance. For example, the effect of varying loads on the year-round performance of an air-conditioning system would reduce the overall effectiveness to below 1%. For heating, the reduction, not so drastic, is to 5% or less.

Table II. Exergy (And Energy) Outputs, Losses And Consumptions
As A Fraction Of Supply For Several Devices

	$\dot{A}_c/\dot{A}_s$	$\dot{A}_\ell/\dot{A}_s \cdot (\dot{E}_\ell/\dot{E}_s)$	$\eta_{II}=\dot{A}_p/\dot{A}_s$ $(\eta_I=\dot{E}_p/\dot{E}_s)$
Boiler (high-pressure)(18)	0.45	0.05,(0.1)	0.5,(0.9)
- Combustion	0.3		
- Heat transfer	0.15		
- Chemical effluent		0.04,(0.01)	
- Thermal effluent		0.01,(0.09)	
Furnace (comfort heating) (18)	0.65	0.25,(0.4)	0.1,(0.6)
- Combustion	0.3		
- Heat transfer	0.35		
- Chemical effluent		0.05,(0.05)	
- Thermal effluent		0.20,(0.35)	
Water Htr.(Gas-fired)(36)	0.6	0.38,(0.6)	0.24,(0.40)
- Combustion	0.3		
- Heat Transfer	0.3		
- Exhaust & Heat Losses		0.38,(0.6)	
Water Heater (Elec.) (36)	0.9	0.08,(0.85)	0.017,(0.30)
- Electric Generation	0.6	0.05,(0.70)	
- Heating	0.3		
- Heat Losses		0.03,(0.15)	
Domestic Cooking (19)			0.2,(0.7)
Clothes Drying (19)			0.1,(0.5)
Refrigerators (19)			0.1
Engine (lrg, recip.)(18,37)	0.35	0.25,(0.6)	0.4,(0.4)
- Combustion	0.20		
- Heat transfer	0.15	(0.3)	
- Exhaust		0.25,(0.3)	
Gas Turbine Engine (18)	0.3	0.4, (0.7)	0.3,(0.3)
Steam Turbine (lrg)(18)	0.1		0.9
Compressor (recip.) (18)	0.3		0.7
Heat Exchangers (23)			
- Cooling coil	0.75		0.25
- Steam heating coil	0.6		0.4
Blast Furnace (30)			0.8
Open Hearth Furnace (30)			0.78
Basic Oxygen Furnace (30)			0.92
Tobacco Dryer (36)	0.7	0.25,(0.6)	0.04,(0.4)
- Gas Combustion	0.3		
- Heat Transfer	0.4		
- Exhaust		0.25,(0.6)	
Paper Machine (32, 36)	0.65	0.1	0.25
- Drying	0.65		
- Effluent & Heat Losses		0.1	

The references referred to in Tables I and II usually con-
tain more or less elaborate exergy flow diagrams.

Conclusions -- Specific and General
This chapter is not intended to make proposals of specific means
for improving energy systems. The aim, here, is to show how to
ascertain how inefficiently today's systems perform, and how to
pinpoint and evaluate the true losses and consumptions. Then,
needs and opportunities for improvement can be accurately assessed;
the real prospective savings of "energy" that could accrue from
proposed improvements -- whatever they might be -- can be ration-
ally determined. In turn, decisions and investments (of time and/
or money) can be made judiciously. Whereas, analyses made from an
energy-efficiency viewpoint can be very misleading.

For example, is there hope for saving significant amounts of
energy by conservation? Some have scoffed at conservation, as a
significant means for relieving our energy problems -- at least
over the long term. Another group has recognized the importance
of conservation, but has placed most all of the emphasis on better
"end-use". Thus, in the Heating and Air-conditioning sector the
stress is on the prevention of losses (and gains) with better insulation, fenestration, exhaust air heat recovery, etc. Neither
group is aware of how great the potential savings really are, with
conservation, because they do not realize how inefficient the conversion processes are (as shown in Tables I and II). The bigger
prospective savings are in the conversion processes, not in end-
use. The important point is that energy analyses recognize losses
only, they do not correctly evaluate the relative importance of
different losses, and they fail to recognize consumptions -- which
are generally much more important than the losses. The 10 or so
per cent effectiveness with which exergy is utilized in this coun-
try, though improved greatly over the 1 or 2 per cent of a century ago, is very low; basically, this is encouraging inasmuch as
it shows that there is real opportunity for improvement remaining.

Another important point is that the production of electricity
is one of our most efficient energy conversions. As elaborated
upon earlier, the great losses commonly ascribed to the stack
gases and cooling water are hardly losses at all; the actual
"losses" (really consumptions) are elsewhere in the plant--primar-
ily in the boiler.

Furnaces, which are considered to be very efficient, are very
inefficient -- as shown by Table II, which also shows that the
primary cause is the large exergy consumption in heat transfer
(which is a consequence of the large temperature difference between combustion products and the heated medium).

For example, these comments have considerable negative impact
on the desirability of high-Btu coal gasification and of the "Hydrogen Economy" for the purpose of distributing these synthetic
fuels about for combustion in low-temperature furnaces and boilers.

The results shown in Table I for medium-BTU gasification are for a state-of-the-art medium-Btu process; conversion of low- or medium-Btu to high-Btu gas is relatively more inefficient. The medium-Btu process, alone, would be useful for consumers -- electrical utilities clearly are the most likely prospects -- who would be willing to install their own gasification equipment locally, in order to produce a clean fuel which could be burned with minimal environmental impact. The 65% effectiveness shows the added fuel cost which must be borne: 54% greater than without the gasification. And then there are the additional capital (and other) costs. Nevertheless, this technology may be competitive with other methods of contending with the emissions problem, especially when included as a part of combined cycle power plants.

It is evident from the foregoing information that the processes which are the great consumers of "energy" are combustion and heat transfer. Research and Development, motivated by the economic need, will develop new technology for overcoming these wastes -- directly or indirectly. Indirect improvements will come from the development of new systems combining processes -- old and new -- in novel ways. Direct improvements will be achieved with more efficient heat production, transport and transformation processes, and with more efficient chemical processes.

Detailed Second Law analysis of the combustion process shows that the dissipations result from an imbalance between the chemical potentials (μ) of the reactants and of the products, just as the heat transfer losses result from the great imbalance between the temperature of combustion products and heated medium. Fuel cells improve the balance of the chemical potentials, and yield an electric potential energy output upon reducing μ of the reactants. But, thus far, any savings of the losses due to chemical reactions have been counteracted with other losses; and the costs of capital, and of fuel and its preparation, are unattractive. However, it is predicted here that prospective schemes for reducing the chemical potential of oxygen before it reacts with the fuel will be forthcoming, to improve performance of power stations and other combustion engines.

The intent of MHD is to capitalize on potential energy now lost in the boiler heat transfer processes.

Solar energy, of course, has no combustion consumption associated with it. The concept of a solar furnace wherein conventional fuels and oxidants are preheated by solar energy, thereby reducing the imbalance of chemical potentials and hence reducing the exergy consumption of combustion, is an interesting one. The concept can be carried a step further by oxidizing the fuel using solid-electrolyte fuel cells, using the solar energy to achieve the high temperatures required by such cells. It is interesting to conjecture that concepts -- such as solar energy, combustion, fuel cells -- which alone may not be economical might combine to be competitive before any one of them would be, alone. Clearly, the complexities of such combinations are "overwhelming," but in the long run the gains could be worthwhile.

As mentioned earlier, energy analyses are misleading. The
exergy methods for analyzing "energy" systems are the key to pinpointing the losses and consumptions, for measuring their magnitudes and resultant per cent inefficiencies, in order to determine
where opportunities for improvement and conservation lie, for the
purposes of decision-making for allocation of resources -- capital, R&D effort, and so on. The exergy methods, which involve
exactly the same kinds of calculations as energy analyses, are a
valuable first step for ascertaining likely prospects (opportunities) for cost effective capital expenditures for "energy" conservation. Then, once improved systems and/or processes are con-
ceived for conservation, benefit-cost studies provide the ultimate
test of the proposed improvements -- of the usefulness of Second-
Law efficiency analyses for finding the opportunities.

III. Benefit-Cost of Energy Conservation

During the nineteenth century the efficiency with which energy resources were utilized was low and relatively constant. During the
late 1800's through about 1950, the efficiency increased dramati-
cally, five- to ten-fold depending upon the use, rising to a relatively flat plateau (38). What are the prospects for further
large "step" increases in efficiency? If we were to believe that
the First Law efficiency is truly representative, we would not
hold out hope for large increases inasmuch as the average First-
Law efficiency is 50 per cent (21). However, the true (Second-
Law) efficiency of energy resource utilization is about 10 percent
(19,20,22)   with very low efficiencies in all of the principal
end-use sectors. The prospects for energy conservation are very
significant by improving the efficiency of the conversions from
raw-fuel to end-use supplies.

Economics -- the ultimate motivation for efficiency improvements -- will determine the level of any efficiency increases and
the rate at which they are made. Economy was the driving force
for the developments leading to higher efficiencies in the first
half of the century. The need for larger quantities of fuel from
more remote sources made the fuel relatively expensive; therefore
it was more worthwhile to develop better technology and to invest
more capital in more efficient equipment. It can be safely predicted that fuel costs being experienced in this era will lead to
large increases in efficiency. Presuming that the appropriate
economic balance between capital and fuel costs had been achieved
when efficiencies reached their post-1950 plateau, it should be
expected, because of the rapid rise of the price of fuel, that now
there are many opportunities for achieving economies by investment
in more efficient state-of-the-art hardware. Furthermore, even
more efficient new technologies will be developed.

Of immediate interest are improvements in system performance
that may be accomplished with state-of-the-art technology. In

these cases allocation decisions can be based on objective economic cost-benefit calculations. References (32, 36, 39-45) outline
case studies that begin with an exergy efficiency analysis in order to pinpoint opportunities for economic "energy" conservation,
followed by the conception -- motivated by the results of the efficiency analysis -- of prospective technical schemes for conservation, and then by benefit-cost studies of various alternatives.
Following is a representative illustration of the results of other
such benefit-cost studies reported in more detail elsewhere (42-
45). The system improvements were motivated by Second-Law efficiency studies. The results corroborate the contention made above:
that there are currently many economic opportunities for "energy"
conservation.

Current Economics of Energy Conservation. Gyftopoulos and Widmer
presented thirteen examples of the cost effectiveness of energy
conservation measures (42-44) including an analysis of heat recovery in cement production.
     There are large amounts of thermal exergy in the "waste heat"
from cement kilns. A specific proposal would install a 4700 kW
power cycle driven by the otherwise wasted exergy at a cost of
$2.7 million. The resultant power would be used in-plant, re-
placing electricity that would need to be purchased at 2.6¢kWhr.
A life-cycle costing analysis showed that the savings on electricity would give a 22 per cent after-tax return on the $2.7 million
investment. This ROI exceeded the 15 per cent that top management
had established as a criterion for capital investments that increased production, and was far above the company's average return on assets of 10 per cent. Nevertheless the company rejected
the proposal, indicating that a 30 percent ROI was the policy for
projects that do not increase production. This decision, not un-
typical, was unfortunate for both the proposing firm and the
nation.
     It can be argued the decision is unfortunate for the cement
company, too. Evidently the justification for the two-tiered ROI
criterion is that "market penetration" has an intangible but real
value. That is, the per cent of the market which the company
holds now and in the near future has a significant impact on the
future sales. Undoubtedly this is true. However, such a drastic
difference in ROI criteria hardly seems appropriate. After all,
the company's future production capacity will be dependent upon
the availability of capital at that time, and the availability is
enhanced if the profit is greater now and in the near future. Also,
the savings from the proposed investment could be spent, in part,
on other methods of increasing sales. One wonders if the strong
preference (prejudice) on the part of management toward "production" is not motivated by the preoccupation with growth and, in
turn, with sales as "the" measure of the growth of a firm, supposedly gauging the effectiveness of management.
     To put it differently, the preference may be a matter of

inertia. In the era of the immediate past, capital investment
into increased production almost invariably paid off. However,
that was an era of expanding markets. Were the markets expanding
simply because production was being increased? Hardly. The most
important reason for expansion was that products were becoming
less expensive because they were being produced more efficiently.
This more efficient production was a result of capital being invested in "automation" in order to make more efficient use of a
critical resource that was suddenly becoming much more expensive
(in real terms), namely man-power. With the increased efficiency
of manpower utilization, products became less expensive, and concurrently the consumers could afford more -- because, being more
efficient labor, they could be paid more. Consequently, markets
expanded. During the era, a good "rule of thumb" for management
was "increase production." The mentality persists, even while
Japan is showing that investment in efficiency is key.

(1)   It seems that an "inertial preference" for production could
      have dire consequences since (a) the prospect is small that,
      in the near future, U.S. markets can continue to expand like
      they have in the recent past, and (b) this "philosophy" of
      expanded production a fortiori actually is counter to the key
      to success in the recent past, namely efficiency.

(2)   In the present era, it is our second-most valuable resource,
      energy-power that is being used very inefficiently while be-
      coming expensive.

(3)   Meanwhile, there are forces at play that would make both of
      these resources not only expensive but scarce -- at least in
      some places for a time.

      (Clearly, world markets could continue to expand greatly; the
      needs and hence demand are there. But the prospective con-
      sumers cannot afford their needs -- and won't be able to, un-
      til investments are made to utilize their prospective work.)

Conservation vs. New Supply. Gyftopoulos and Widmer pointed out
the impact on the nation of negative decisions like that of the
cement firm: "...the proposal was rejected; the plant continues
to purchase electricity from a utility. This electricity is generated by consuming the equivalent of about 188 barrels of petroleum per day." Clearly, this is a small amount; but when it is
multiplied by the number of instances in which such an opportunity
is passed over.... Furthermore, from a national viewpoint it is
noteworthy that the replacement cost -- the marginal cost -- of
the electricity that could have been saved is more than the 2.6¢/
kWhr paid by the company. Because the electricity is not saved,
the utility will need to spend over $5 million of capital to add
new generating capacity with greater capital costs (per kW) than
its existing facilities. Also, additional capacity will consume
new supplies of fuel, which are more expensive than the average
cost. Consequently, and ironically, Gyftopoulos and Widmer can

say, "If, however, the company had been paying for electricity the
4.1 cent replacement cost appropriate to new electricity supply,
the generator would have earned a return well above 30 per cent
and management would have undoubtedly approved the project."

Including the cement kiln case just reviewed, the thirteen
cases presented in (43) show the capital cost of various "energy"
conservation projects in comparison with the capital required to
provide an equivalent new supply.

Barriers to Energy Conservation. The importance of energy conservation to the future of our economy is discussed in References
(42-45); the authors list and explain "institutional barriers" to
energy conservation:

(1) Limited capital.

(2) Two-tier criteria for "discretionary" versus "mainstream"
     investments.

(3) Average costs for energy supplies are substantially lower
     than marginal costs.

(4) Regulatory constraints upon producers of by-product
     electricity.

(5) Different investment criteria for energy users versus
     regulated utilities.

An additional factor persists:

(6) The uncertainty regarding the impact of impending legisla-
     tion upon current decisions.

Regulatory constraints sometimes preclude cogeneration and
often undermine its benefits. In one notable case (24, 32) a
company was cogenerating substantial amounts of electricity,
serving its own needs plus those of the surrounding community.
The company was unsuccessful at getting a rate increase that was
needed to meet its costs. One factor was that the utilities com-
mission viewed the value of the plant outputs, steam and electricity, to be energy not exergy. Consequently, as will be shown in
Part IV, the allocation of costs to steam was much greater than
it should have been, and the price allowed for electricity was
less than its true cost. Subsequently, out of futility, the cogenerating firm submitted a petition requesting to be relieved of
the requirement to serve as the local utility. Ultimately the
petition was granted, rather than allow a rate increase -- and the
community was then served by a larger utility, at higher rates
than those requested by the cogenerating industry.

The matter of different investment criteria for energy users
versus regulated utilities is exemplified by comparison of the
feedwater heater project (41) to the cement kiln heat recovery
project referred to above. The utility chose to replace the de-

graded feedwater heater because the investment would earn more
than 8.5 per cent after taxes, whereas the cement company would
not even make a mainstream investment with an expected return less
than 15 per cent. Of course the reason utilities can make such
investments is that their profit is assured, perhaps at a low
level, but assured. (It is also interesting to see utilities
"forced" into the installation of inefficient peaking facilities
-- burning scarce hydrocarbon fuels with the assurance that the
fuel costs will be covered by "fuel cost adjustment clauses" in
their regulated prices.)

Conclusions

(1)   Energy resources are being used very inefficiently, and
      there is a wide margin for technical improvement.

(2)   The great amount of room for improvement signals opportuni-
      ties for improving systems by (a) the use of existing tech-
      nology, and (b) the development of new technology.

(3)   The motivation for acquiring and developing more efficient
      energy conversion equipment and systems will be economic --
      whether socialized or free enterprise -- responding to the
      rising value of fuels.

(4)   The key to acquiring and developing the better equipment and
      systems will be investment of money -- taxes or capital --
      in response to the rising value of fuels.

      If supply prices are held down by regulation, the investments
will need to be subsidized -- tax money. But at what subsidy
level?   In turn, the required use of more efficient systems will
have to be legislated. But at what efficiency levels?

      Sant (45) presents the results of an interesting econometric
analysis, which indicates that efficiency improvements are a
key to resolving the current energy shortage -- making it tempor-
ary, like previous manpower and energy shortages -- provided that
the artificial constraints and the "hangups" which cause the a-
forementioned barriers are removed.
      Part IV of this chapter will present the elementary theory of
Second-Law cost accounting and its application to several practical cases, showing its usefulness for (i) costing "energy" commodities, (ii) plant operation decisions, and (iii) plant designs
-- new and retro-fit. These methods can be used to prescribe efficiencies and investments, traded-off against each other, but
they will yield truly optimal results only to the extent that the
real value (cost) of the "energy" supply is known.

IV. Exergy Economics - Cost Accounting Applications

The objectives of cost accounting (46,47) are: (1) Determination
of the actual cost of products, (2) Provision of a rational basis
for pricing products, (3) To provide a means for controlling expenditures, and (4) To form a basis for operating decisions and
the evaluation thereof.
   Because exergy is the commodity of value, it provides the key
to the cost accounting of "power" in various forms (48-56, 24).
Costing with energy is impossible with many systems, leads to erroneous results with most, and works only when it can be manipulated so that it is tantamount to exergy costing.
   Exergy accounting is useful for several purposes:

(1) Costing. If a utility system and/or chemical process deliv-
     ers several products, then the proper apportionment of "en-
     ergy" costs to the various products requires use of exergy.
     This apportionment is critical inasmuch as (a) the "energy"
     costs are usually major ones, and (b) the costs are needed
     in order to establish prices and/or evaluate profits of each
     product. Or, if the utilities are used in-plant, the appor-
     tionment is needed to establish costs to be charged to the
     different manufactured goods, which consume the utilities in
     various proportions.

(2) Optimal Design. When an energy system or a processing plant
     is being designed, each system component or process unit can
     be selected optimally by proper balance of "fuel" costs and
     capital (plus other) costs. Exergy costing provides the
     means for establishing the unit cost of a "fuel" at any junc-
     ture within a system, needed in order to establish the opti-
     mal expenditure of capital for more efficient components.

(3) Feasibility Studies and Preliminary Design. During the feas-
     ibility study and the preliminary design phases of a design
     project (57), it is useful to calculate the unit costs of the
     major exergy flows at points within a complex. When alterna-
     tive technologies are being considered, comparison of the
     unit costs of exergies with the various alternatives can
     serve as a useful criterion for selection. Once the plant
     concept has been selected -- once the technologies have been
     chosen -- then the costs of major exergy flows can be com-
     pared to "standards," namely typical costs for such exergies
     in modern, economic plants.

(4) Operating Economics. For existing plants knowledge of the
     exergy costs allows economic analysis of each unit to be made;
     thus the tradeoffs between "power" costs and other operating/
     capital expenses can be evaluated properly. Thereby, operat-
     ing decisions can be made rationally -- regarding maintenance,
     replacement, revision, and operation.

Basic Principles

The first step in costing "energy" flow streams is the application of money balances to the system of interest. The <u>cost</u> (C) of the product of an energy-converter or process equals the <u>total</u> expenditure made to obtain it -- the fuel expense, plus the capital and other expenses:

$$C_{product} = C_{expended} = C_{fuel} + C_{capital}, etc.$$

For example, the products shown in Figure 1 cost

$$C_A + C_B = c_1 P_1 + c_2 P_2 + C_{capital}, etc.$$

where c_1 and c_2 are the unit prices paid for Feeds 1 and 2. The average unit costs $c_A = C_A/P_A$ and $c_B = C_B/\Delta P_B$ therefore must satisfy

$$c_A P_A + c_B \Delta P_B = c_1 P_1 + c_2 P_2 + C_{capital}.$$

When there is only one product from a plant a money balance like the foregoing yields a unique value for the unit cost of product. (It can be seen that it is then irrelevant whether energy or exergy or any other commodity is used to measure "power," P.)

When there is more than one product from a plant, a money balance cannot be solved for unique unit costs of the several products. An additional equation is needed for each unknown unit cost, besides one. The additional equations are determined by cost <u>accounting</u> not thermodynamic, considerations (52). Before discussing some alternative <u>methods</u> for determining the necessary additional equations, it will be illustrated by the following example that when there are multiple products, the use of energy to measure power leads to radical errors; exergy yields rational results.

Costing Applications

Example: Cogeneration. When steam and electricity are cogenerated (Figure 7), it is often critical to know how much of the costs should be attributed to each commodity. The first step in determining the unit cost of the turbine shaft work and that of the "back-pressure steam" (the exhaust steam from the turbine) is to obtain the unit cost of the high-pressure steam supplied to the turbine. The cost can be determined by applying the foregoing expression to the boiler, to get the unit cost of high-pressure steam:

$$c_{HP\ steam} = \frac{c_{coal} P_c + C_{boiler}}{P_{HP\ steam}}$$

which can be solved for c_{HP}.

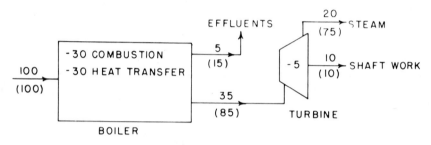

Figure 7.   Cogeneration power plant showing exergy and energy flows at
one operating condition.

In turn, the outputs of the turbine, driven by the high-pressure steam, are shaft-work and low-pressure steam. A money balance yields the total cost of these products as

$$c_{shaft}P_{shaft} + c_{LP}P_{LP} = c_{hp}F_{hp} + C_{turbine}$$

If the purpose of the steam turbine is to convert power supplied in steam to shaft power, then the shaft power should be debited for the cost of the power <u>extracted</u> from the steam, $[P_{HP} - P_{LP}]$, as well as for capital and other costs:

$$c_{shaft}P_{shaft} = c_{HP}[P_{HP} - P_{LP}] + C_{turbine}$$

which corresponds to the money balance, with

$$c_{LP} = c_{HP}.$$

The last two equations can be solved for c_{LP} and c_{shaft}.

For a simplified cogenerating plant, the resultant unit costs (51, 52) are presented in Table III for the two alternative cases, with power flows P measured (1) by energy, and (2) by exergy. Now,

Table III. Results of Energy Costing and Exergy Costing of Electricity and Steam from a Typical Cogenerating Plant.

	Energy Costing	Exergy Costing
High pressure steam	0.32¢/lbm	0.32¢/lbm
Shaft work	0.9¢/kWh	2.4¢/kWh
Low pressure (50 psig) steam	0.276¢/lbm	0.19¢/lbm

if energy-based costing is valid under the above circumstances, with back-pressure steam being delivered at 50 psig, then it should be appropriate when the back-pressure is 40 psig, or 30, 20, 10.... The upper curve in Figure 8 shows the results of just such costing. The curve indicates that steam at pressures like 0.5 psia (T ≐ 80°F) would be worth about 0.21¢/lbm -- almost as much as 50 psia steam -- even though it has virtually no usefulness for heating. Certainly this is wrong; there would be no buyers of 0.5 psia steam at 0.21¢/lbm. There is no logical way to decide at which back-pressures energy costing would be appropriate and at which pressures it would not be. Essentially, this shows that the use of <u>energy</u> as the measure for the power

flow is in error. On the other hand, the lower curve on Figure
8, derived from exergy costing, yields that the cost per pound of
low-pressure steam does go to zero as its usefulness goes to zero.
This is precisely the result that any rational costing scheme
should provide.

This example has clearly illustrated that energy-based costing can lead to errors and that unit costs should be obtained from
exergy based costing methods. (The "lost kilowatts" method commonly used by utilities to cost co-generated steam is not an energy method, but an exergy method in disguise.)

Costing Methods. The cogeneration example is the simplest type,
with only two products. In many instances there are several utilities (perhaps including chilled water, compressed air, hot water,
steam at different pressures) being delivered; and/or there may be
several products (say chemical) being produced. The money balance must then be complemented by additional equations, so that
the number of equations equals the number of unknown unit costs
to be determined.

As is usual in cost accounting practice, there ae a variety
of methods for establishing the complementary equations--for allo-
cating the costs--and which method is most equitable depends upon
the circumstances.

The method used in the previous example, called the extraction method, assumes that the sole purpose of the turbine is to
produce shaft power. Therefore, the shaft work is charged for the
capital cost of the turbine and for the exergy extracted from the
steam by the turbine to produce the work. With this rationale,
the additional equation is obtained by equating the unit costs of
high- and low-pressure steam exergy, $c_{LP} = c_{HP}$. The result is
that the shaft work bears the entire burden of the costs associated with the turbine process and the capital expense.

Some other methods are discussed in (24, 52); e.g., the equality method would charge the shaft exergy and the low-pressure
steam exergy equally for the cost of high-pressure steam and turbine capital: $c_{LP} = c_{shaft}$. When the equality method is used the
numerical results are different from those shown by Figure 8 for
the extraction method. However, the general features are the
same; the energy-based cost curve is relatively flat, with an in-
tercept (at $T = T_0$) equal to about 0.15¢/lbm. Whereas, the exergy cost curve drops to zero unit cost as turbine exhaust temperature is reduced to T_0.

Design Applications

Design Optimization. The goal in design optimization of energy
conversion systems is to select the equipment which strikes the
best balance between overall capital (and other) costs and the
cost of the exergy input--for the particular type of system opera-

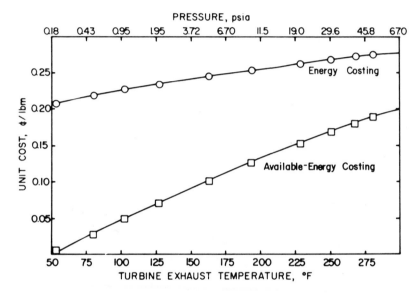

Figure 8.   Unit steam costs as a function of steam exhaust pressure on the basis of energy and exergy.

ting under the given constraints. That is, the aim is to minimize
the total expenditure for capital and fuels. The rationale for
the use of exergy in such optimization is that it allows a large
complex system to be optimized by parts, splitting the overall
system into much less complex subsystems (perhaps unit operations
and/or individual components thereof). Then, rather than optimizing the whole system at once, each subsystem can be optimized individually, minimizing the total expenditure for subsystem capital
and for fuel supplied to the subsystem--for exergy supplied to the
subsystem. This subsystem optimization is made possible because
the unit costs of exergy flows at the various junctures between
subsystems can be calculated (iteratively). When optimization can
be done subsystem by subsystem, much insight can be applied because of dealing with the much simpler subsystems, leading to
easier optimization. Typical parameters (decision variables) of a
component that can be adjusted in order to attain the optimum
system are efficiency, operating pressure and temperature, speed,
and so on and so forth.

Benedict applied exergy costing to the design of an air separation plant in 1949; see (58). Gaggioli used the methods for
the optimal selection of steam piping and its accompanying insulation (50, 59). Evans, Tribus, El-Sayed and co-workers (53, 54)
have developed the fundamentals of the subject, under the title of
Thermoeconomics, and present chapters in this volume.

Design Optimization Example: Optimal Piping and Insulation
Sizing. The problem to be treated here is the optimal sizing of a
steam line and its insulation as presented by Wepfer, Gaggioli,
and Obert (60). As a specific illustration, a bleeder line that
delivers steam from a turbine to a feedwater heater is designed
The flow process is accomplished at the expense of exergy in steam from the turbine, some of that exergy is consumed
to drive steam through the pipe. Furthermore, some is lost by
heat transfer through the pipe and its insulation. The optimal
combination of pipe size D_n (nominal pipe diameter) and insulation
thickness, θ, would be that which minimized the overall costs consisting of: (1) the cost of the exergy consumed and lost, plus
(2) the capital (and other) expenses for the piping and insulation.
The key is to determine the unit cost of steam exergy -- in a manner analogous to that used above to get c_{LP}.

The function to be minimized, the objective function, is the
yearly total cost of the piping system--the sum of the amortized
capital cost of the pipe and insulation plus the operating costs

$$\emptyset_T(D_n,\theta) = \emptyset_P(D_n) + \emptyset_I(D_n,\theta) + \emptyset_F(D_n) + \emptyset_Q(D_n,\theta)$$

where $\emptyset_P$ is the amortized pipe cost, $\emptyset_I$ is the amortized insulation
cost, $\emptyset_F$ is the operating cost resulting from viscous dissipation
of exergy, and $\emptyset_Q$ is the operating cost from heat losses of exergy.

By evaluating ϕ_p, ϕ_I, ϕ_F, and ϕ_Q as functions of D_n and θ, the pipe size and insulation thickness which minimize the annual total cost are found (59). The principal results for the particular steam line are embodied in Figure 9, which shows total cost as a function of nominal pipe diameter and insulation thicknesses. The optimal combination for this design problem is a nominal pipe diameter of 12 inches, and an insulation thickness of 3.5 inches.

Application to Preliminary Design: Coal Gasification. An extensive efficiency and exergy costing analysis has been carried out on a preliminary design (62) of the Synthane coal gasification process (63-65). The results provide useful information for design decision-making, to improve the economics.

Figure 10 shows the flows of exergy between the major groups of equipment in the Synthane process. Also shown is the consumption in each group and its efficiency. Furthermore, unit costs of exergy are shown in parentheses. A notable feature of the SYNTHANE process is that more electric power is produced from the gasifier by-product char than is consumed by the process itself; hence the excess is available for "export" from the plant.

It is evident from Figure 10 that the largest inefficiencies in the process are associated with the production of "utilities" -- especially important since the efficiency with which each utility is produced is substantially lower than the state-of-the-art efficiency whereby the utility can now be produced economically. The 26% overall efficiency should be compared with an expected efficiency of at least 35% and up to 40%.

Further details regarding the consumptions and losses are discussed in (63-65), along with possible means of improvement.

Consider the unit costs of the exergy flows between major component groups shown in parentheses on Figure 10. Notice the inordinately high costs of steam, shaft power, electricity and oxygen -- especially considering the inexpensive cost of fuel to the group, 0.35¢/kWhr = $1.03/10^6$ Btu. These high unit costs clearly point out that the design of the utility plant is far from optimal. It can be surmised that the low-efficiency utility production is a critical factor leading to an inordinately high unit cost of product gas (compared to other coal gasification processes (62)). In order to appraise the importance of the power plant inefficiency, assume that steam and electric power could be produced at any typical cost, such as steam at $6.30/10^6$ Btu of exergy and electricity at 2.35¢/kWhr. Then (even with several conservative approximations) the unit cost of product gas is reduced from $6.10/10^6$ Btu of exergy (2.1¢/kWhr) to $4.97/10^6$ Btu (1.7¢/kWhr).

Evidently, the design philosophy that was adopted for the conceptual design of the Synthane process was this: Inasmuch as "unlimited" amounts of char are available -- as an inherent by-product of the Synthane gasifier and therefore at "no cost" --the

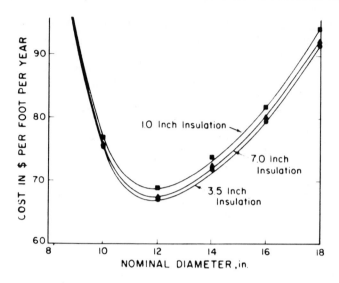

Figure 9.   Total costs as a function of pipe diameter and insulation thickness.

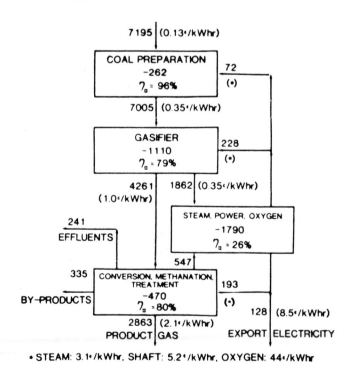

Figure 10.   Exergy flows and unit costs for SYNTHANE process.

efficiency of utility production is not important -- especially
with the (unrealistic) design constraint that had been imposed:
any excess electricity can supposedly be sold only at 0.45¢/kWhr.
Therefore, it seemed, capital expenditure for utility systems can
and should be kept to a minimum in order to minimize the cost of
the product gas. The Second-Law costing analysis, however, shows
that it was crucial to the viability of the SYNTHANE process (now
defunct because this point was not recognized) to build a better
utility system, and it is somewhat important that a reasonable
price be acquired for electricity.

Thus, it should be noted that, with exergy accounting, the
effect of prospective design changes upon product cost can be estimated accurately, without a complete system re-design and re-
costing.

Maintenance and Operation Decisions

The determination of an appropriate cost of exergy at various
junctures of a system in a manner similar to that described above
for Design Optimization is useful not only in design but also allows decisions regarding the repair or replacement of a specific
subsystem to be readily made. The amortized cost of such improvements can be easily compared with the cost of the additional exergy that will be dissipated if a component is left to operate in
the given condition. The proper decision then becomes very apparent; for example, see (41) for an application to a feedwater
heater. Similarly, such costs can be employed for the purpose of
making decisions regarding the control -- either manual or auto-
matic -- of a system. Consider the following case study.

Boiler Feedpump Drives. Second-Law techniques are applied to a
practical problem: determing when to drive feedpumps with electric motors and when to use steam turbines. This analysis was
performed by Fehring and Gaggioli (40) for the Wisconsin Electric
Power Company. The Second-Law costing procedures are straight-
forward, avoiding the extremely laborious calculations associated
with traditional methods, and/or avoiding the erroneous conclusions
often drawn when those methods are simplified.

Two of the units at the Oak Creek Power Plant have full-sized
steam turbine driven boiler feed pumps, along with half-sized motor driven boiler feed pumps for use at low loads and unit start-
up. The boiler feed pump turbines are supplied with steam from a
particular turbine "bleed-point".

Turbine bleeder pressure is roughly proportional to unit load;
at less than one half unit load, the bleeder pressure becomes too
low to power the boiler feed pump turbines. Therefore, either the
motor driven boiler feed must be employed, or a "booster" supply
must be used to provide steam to the boiler feed pump turbine
using steam which is throttled from a higher pressure.

The power plant operating staff, believing that it was always

more efficient to utilize the steam turbine driven boiler feed
pumps, have attempted to use an existing booster steam line to
drive these turbines when the units were below half load. However, a high pressure drop must be taken across the pressure reducing valve in the booster line. Although there is no energy decrease, this throttling does destroy exergy--and causes excessive
wear on the valve trim.

Because of the problems associated with the operation of
these original booster steam supplies, the plant had requested
that a replacement booster steam source be installed. As contem-
plated, the new source would be supplied from the throttle steam
of the main turbine, reduced in pressure to approximately 50 psia.
The installed cost for the new booster steam source was estimated
at $75,000 per unit (for mid-1976 installation).

On the basis of an energy analysis, the project was tentative-
ly approved. It was realized, however, that the results of a simple energy balance may be misleading, and that if a First Law analysis were to be used, it should encompass the whole power plant.
Rather than do such an exhaustive energy analysis, it was decided that a Second-Law, or exergy, analysis should be performed
on the process of interest, as a check against the initial results.

There are various conceivable sources of booster steam within
the plant. The amount which would be needed from each source was
determined from the turbine and pump performance characteristics,
at several different plant loads. Then, the corresponding exergy
requirements were calculated. Then, employing money balances, the
unit cost of exergy from the various sources were calculated, at
the various loads. In turn, the exergy costs per hour associated
with the use of the alternative sources were calculated. According to the results, presented in Figure 11, Bleeder 5 provides the
least expensive exergy source for driving the boiler feed pump
over most of the operating range of the unit. The use of auxiliary electrical power is the second lowest alternative, followed
in turn by Bleeders 6, 7 and then main throttle steam. It is obvious, then, that when the Bleeder 5 pressure is too low to power
the boiler feed pump turbine, the electric driven pump should be
employed.

The use of the Second Law has properly accounted for the substantial loss of exergy associated with the pressure and temperature reduction necessary in the booster steam station. It has
correctly predicted the relative operating costs associated with
the two different modes of operation, eliminating the need for the
installation of the proposed new booster steam supplies. This has
prevented the uneconomical operation of the units and resulted in
a $150,000 savings in first costs and an annual operating cost
savings of at least $65,000.

Closure
The foregoing results illustrate four simple, practical applications of Second-Law costing methods: (i) costing, (ii) optimal
design, (iii) preliminary design, and (iv) operation and mainte-

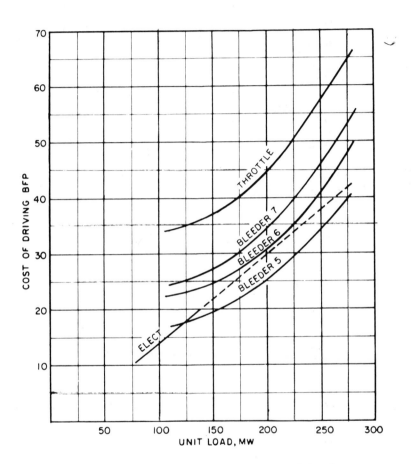

Figure 11.   Cost of driving the boiler feed pump with alternate energy sources.

nance. The key is the assignment of costs to the commodities
(such as steam, electric charge, etc.) which carry exergy at any
juncture within a complex system.

It is exergy that fuels the processes occurring in any device.
For any subsystem or component in a complex system, knowledge of
the cost of the exergy supplied to a device allows an economic
analysis of that device to be made, in isolation from other components of the system. Thus, for design, maintenance, and operation purposes, decisions can be made more simply, without contending with the whole system. Furthermore, when dealing with one
unit at a time, more insight into system operation is obtained
and, for example, creativity is enhanced for improving the system
via process modification.

These applications hardly scratch the surface of the potential of exergy costing. Thermoeconomics is only in its infancy.
It is first of all very important that the fallacy of energy
methods become well known, in order to avoid inevitable mistakes
that will otherwise arise. In turn, it is equally important that
the validity of exergy methods, and the methodology for applying
them -- no more difficult than the energy methodology -- become
well known. Indeed, it is by illustrating the exergy methods
that the errors of energy methods will become apparent. (Also,
the growth of exergy applications will be hand in hand with better, more comprehensible presentations of the Second Law.)

Thermoeconomics needs to flourish in order to avoid misconceptions and erroneous statistics which would lead to bad mistakes by engineers involved in design and operating decisions,
and by managers and politicians who are involved with "energy"
use and development in the private and government sectors. Thermoeconomics needs to flourish not only to avoid bad decisions, but
also in order to make good decisions -- optimally.

It is hoped that the illustrative examples presented here
will serve to crystallize the basic principles and will motivate
the readers to apply exergy efficiency analysis and exergy accounting -- taking advantage of these methods for their own practical
purposes, while also advancing the state of the art.

Literature Cited

1. Gibbs, J.W., The Collected Works Volumes I, II, New York:
    Longmans, Green and Co., 1931 (p. 33, p. 77).
2. Maxwell, J.C., Theory of Heat, 10th ed. London: Longmans,
    Green and Co. Ltd., 1891, p. 195.
3. Goodenough, G.A., Principles of Thermodynamics, 3rd ed.,
    Henry Holt & Co., 1920.
4. Keenan, J.H., Thermodynamics, 1st MIT Press Ed: Cambridge,
    MA, 1970. Originally published by Wiley, New York, 1941.
5. Dodge, B.F., Chemical Engineering Thermodynamics, McGraw-
    Hill, 1944.
6. Obert, E.F., Elements of Thermodynamics and Heat Transfer,
    McGraw-Hill, 1949.

7.   Hatsopoulos, G.N. and Keenan, J.H. Principles of General
     Thermodynamics, New York: Wiley, 1965.
8.   Gaggioli, R.A., "Principles of Thermodynamics," Thermody-
     namics: Second Law Analysis, A.C.S. Symposium Series, 122,
     3-13, 1980.
9.   Wepfer, W.J. and Gaggioli, R.A., "Reference Datums for
     Available Energy", Thermodynamics: Second Law Analysis,
     A.C.S. Symposium Series, 122, 77-92, 1980.
10. Rodriguez, Luis S.J., "Calculation of Available-Energy
     Quantities", Thermodynamics: Second Law Analysis, A.C.S.
     Symposium Series, 122, 39-60, 1980.
11. Wepfer, W.J., Gaggioli, R.A. and Obert, E.F., "Proper
     Evaluation of Available Energy for HVAC", Trans. ASHRAE, 85,
     1 (1979), 214-230.
12. Shieh, J.H. and Fan, L.T., "Estimation of Energy and Exergy
     Contents in Structurally Complicated Materials," Energy
     Sources, 6, 1-46, 1982. (Also, see the chapter in this book
     by the same authors.)
13. Petit, P.J. and Gaggioli, R.A., "Second Law Procedures for
     Evaluating Processes", Thermodynamics: Second Law Analysis,
     A.C.S. Symposium Series, 122, 15-38, 1980.
14. Obert, E.F., Thermodynamics, New York: McGraw-Hill, 1948.
15. Szargut, J. and Styrylska, T., "Angenaherte Bestimmung der
     Exergie von Brenstoffen", Brenstoff-Warme-Kraft, 16,
     589-596, 1964.
16. Ahrendts, J., "Reference State", Energy The International
     Journal, 1980.
17. Szargut, J. and Dziedziniewicz, C., "Energie Utilisable des
     Substances Chimiques Inorganiques", Entropie, 40, 14-23,
     1971.
18. Gaggioli, R.A., Yoon, J.J., Patulski, S.A., Latus, A.J., and
     Obert, E.F., "Pinpointing the Real Inefficiencies in Power
     Plants and Energy Systems", Proc. Amer. Power Conf., 37,
     1975, 671-679.
19. Reistad, G.M., "Available Energy Conversion and Utilization
     in the United States", Trans. A.S.M.E., J. Eng. Power, 97,
     (1975) 429-434.
20. Reistad, G.M., "Available-Energy Utilization in the U.S.",
     Thermodynamics: Second Law Analysis, A.C.S. Symposium
     Series, 122, 93-110, 1980.
21. Cook, E., "The Flow of Energy in an Industrial Society", Sci.
     Amer., 225, No. 3, 83-94, 1971.
22. ThermoElectron Corp. Annual Report, 101 First Ave., Waltham,
     MA, July, 1977.
23. Gaggioli, R.A., Wepfer, W.J., Elkouh, A.F., "Available Energy
     Analysis for HVAC", Energy Conservation in Building Heating
     and Air-Conditioning Systems, ASME Symposium Volume, H000116,
     (1978), 21-30..
24. Wepfer, W.J., "Applications of Available Energy Accounting,"
     Thermodynamics: Second Law Analysis, A.C.S. Symposium
     Series, 122, 162-186, 1980.

25. Reistad, G.M., Gaggioli, R.A., and Obert, E.F., "Available
     Energy and Economic Analysis of Total Energy Systems", Proc.
     Amer. Pwr. Conf., 32, (1970), 603-611.
26. Gaggioli, R. A. and Wepfer, W., "Second Law Analysis of
     Building Systems", Energy Conversion and Management, 21,
     65-76, 1981.
27. Rodriguez, S.J., L. and Gaggioli, R.A., "Second Law Effici-
     ency of a Coal Gasification Process", Can. Jour. Chem. Eng.,
     58, 376-381, 1980.
28. Riekert, L., "The Efficiency of Utilization in Chemical
     Processes", Chem. Eng. Sci., 29, (1974), 1613.
29. Cremer, H., "Thermodynamic Balance and Analysis of a
     Synthesis Gas and Ammonia Plant", Thermodynamics: Second Law
     Analysis, A.C.S. Symposium Series, 122, 111-120, 1980.
30. Gyftopoulos, E.P., Lazaridis, L. and Widmer, T., Potential
     Fuel Effectiveness in Industry, Cambridge, MA.: Ballinger
     Publ. Co., 1974.
31. Hall, E. and Hanna, W., "Evaluation of the Theoretical Poten-
     tial for Energy Conservation in Seven Basic Industries",
     Report from Battelle Columbus Laboratories to the Federal
     Energy Administration, Report No. FEA/D-75/CE1 (July 1975).
32. Gaggioli, R.A., Wepfer, W.J. and Chen, H.H., "A Heat Recovery
     System for Process Steam Industries", Trans. A.S.M.E. J. Eng.
     Power, 100 (1978) 511-519.
33. Appelbaum, B. and Lannus, A., "Available Energy Analysis of a
     Dry Process Cement Plant", Paper presented at the 85th
     National Meeting of the A.I.Ch.E., Philadelphia, June 8,
     1978.
34. Fitzmorris, R. and Mah, R.S.H., "Improving Distillation
     Column Design Using Thermodynamic Availability Analysis",
     A.I.Ch.E. J., 26. 265, 1980.
35. Funk, J.E. and Knoche, K.F., "Irreversibilities, Heat Penal-
     ties and Economics for the Methanol/Sulfuric Acid Processes",
     Proc. 12th Intersociety Energy Conversion Conference 1977, 1,
     Paper 779142, by ANS: LaGrange Park, Ill., pp. 933-938.
36. Gyftopoulos, E.P. and Widmer, T.F., "Availability Analysis:
     The Combined Energy and Entropy Balance", Thermodynamics:
     Second Law Analysis, A.C.S. Symposium Series, 122, 61-76,
     1980.
37. Patterson, D.J. and VanWylen, G.J., "A Digital Computer Simu-
     lation for SI Engine Cycles", SAE Progress in Technology
     Series, Vol. 7, p. 88, Soc. of Automotive Engrs., Inc., New
     York, 1964.
38. Starr, C., "Energy and Power", Sci.Amer ., 225, No. 3, 3-18,
     1971.
39. Gaggioli, R.A. and Wepfer, W.J., "Second Law Costing Applied
     to a Coal Gasification Process", Chemical Engineering
     Progress Technical Manual, 6, 140-145, 1980.
40. Gaggioli, R.A. and Fehring, T., "Economics of Boiler Feed
     Pump Drive Alternatives", Combustion, 49, 9 (1978), 35-39.

41. Fehring, T. and Gaggioli, R. A., "Economics of Feedwater Heater Replacement", Trans. A.S.M.E., J. of Eng. Power, 99, (1977) 482-489.

42. Gyftopoulos, E. and Widmer, T., "End-Use Energy Conservation", Proc. Int'l. Conf. on Energy Use Mgt., Vol. II, p. 44, Pergamon, 1978.

43. Gyftopoulos, E.P. and Widmer, T.F., "Benefit-Cost of Energy Conservation", Thermodynamics: Second Law Analysis, A.C.S. Symposium Series, 122, 131-142, 1980.

44. Hatsopoulos, G.N., Gyftopoulos, E.P.,Sant, R.W. and Widmer, T.F., "Capital Investment to Save Energy", Harvard Business Review, 56, 2 (1978), 111-122.

45. Sant, Roger W., "The Least Cost Energy Strategy", The Energy Productivity Center, Mellon Institute, Arlington, VA, 1979.

46. DeGarmo, P. and Canada, J., Engineering Economy, 5th ed., MacMillan, 1973.

47. Alford L. and Beatty, H., Principles of Industrial Management, Ronald Press, 1951.

48. Gaggioli, R.A. and Wepfer, W.J., "Exergy Economics", Energy - The International Journal, 5, 823-838, 1980.

49. Keenan, J.H., "Steam Chart for 2nd Law Analysis", Mech. Eng., 54 3, (March 1932), 195-204.

50. Obert, E.F. and Gaggioli, R.A., Thermodynamics, 2nd Edition, New York: McGraw-Hill, 1963.

51. Gaggioli, R.A., "Proper Evaluation and Pricing of Energy", Proc. Int. Conf. on Energy Use Management, II, (1977) Pergamon Press, 31-43.

52. Reistad, G.M. and Gaggioli, R.A., "Available-Energy Costing", Thermodynamics: Second Law Analysis, A.C.S. Symposium Series, 122, 143-160, 1980.

53. Evans, R.B. and Tribus, M., "A Contribution to the Theory of Thermoeconomics", UCLA, Dept. of Eng.: Report No. 62-36 (August 1962).

54. Evans, R.B. and El-Sayed, Y.M., "Thermoeconomics and the Design of Heat Systems", Trans. A.S.M.E. J. Eng. Power, 92 (1970) 27-34.

55. Szargut, J., "Zagadniene Racjonalnej Tareyty Oplat Za Pare I Goraca Wode", Gospodarka Paliwami Energia, 14, No. 4, 132-135, 1963.

56. Bergmann, E. and Schmidt, K.R., "Zur Kostenwirtschaftlichen Optimierung der Warmetauscher fue die regenerative Speisewasservorwarmung in Dampfkraftwerk - ein Storoungsrechenverfahren mit der Exergie", Energie und Exergie, VDI Verlag, Dusseldorf, pp. 63-89, 1965.

57. Asimow, M., Introduction to Design, Prentice Hall, 1962.

58. Benedict, M. and Gyftopoulos, E.P., "Economic Selection of the Components of an Air Separation Process", Thermodynamics Second Law Analysis, A.C.S. Symposium Series, 122, 195-203, 1980.

59. Gaggioli, R.A., "Thermodynamics and the Non-Equilibrium
     System", Ph.D. Dissertation, University of Wisconsin-
     Madison, 1961.
60. Wepfer, W.J., Gaggioli, R.A. and Obert, E.F., "Economic
     Sizing of Steam Piping and Insulation", Trans. A.S.M.E. J.
     Eng. Industry, 101, 427, 1979.
61. Fan, L.T. and Shieh, J.H., "Thermodynamically-based Analysis
     and Synthesis of Chemical Process Systems", Energy - The
     International Journal, 5, 955-966, 1980.
62. Braun, C.F., and Co., "Factored Estimates of Western Coal
     Commercial Gasification Concepts", ERDA Report No. FE-2240-5,
     October, 1976.
63. Gaggioli, R.A., Wepfer, W.J. and Rodriguez S.J., L. "A
     Thermodynamic - Economic Analysis of the Synthane Process",
     Report to Pittsburgh Energy Technology Center, November
64. Rodriguez (S.J.), L. and Gaggioli, R.A., "Second Law Effi-
     ciency Analysis of a Coal Gasification Process", Canadian J.
     of Chem. Eng., 58, 376-382 (1980).
65. Gaggioli, R.A., Rodriguez, L. and Wepfer, W., "A Thermo-
     dynamic-Economic Analysis of the Synthane Process", Report
     to Pittsburgh Energy Technology Center, U.S.D.O.E., Report
     No. COO-4589,A000-1.

RECEIVED August 29, 1983

Second Law Analysis to Improve Industrial Processes

WILLIAM F. KENNEY

Exxon Chemical Technology Department, Florham Park, NJ 07932

The methodology for Thermodynamic, or Second Law, Analysis of industrial processes has been discussed in the previous ACS symposium and in a number of literature articles. This chapter will attempt to provide some insight on using the data from such studies to achieve practical improvements in the process indus-tries. Examples will be drawn from two recent presentations by the author (1-2) dealing with related aspects of finding and implementing economic improvements to plant energy efficiencies. The interaction of energy efficiency and capital cost will be explored in some of these examples.

The terms used in this analysis are relatively conventional. A general thrust of the chapter will be SIMPLICITY. Differences in terminology have been eliminated wherever possible. In this analysis Availability, Available Energy, Exergy, and Work will be used as equivalent. This means that kinetic and potential energy effects and the potential work to be derived from the diffusion of chemical species into equilibrium with the environment have been ignored. This simplification may introduce significant inaccura-cies in some studies, but is not important here. The intent is to demonstrate that simplified - perhaps even approximate - analysis can have valuable practical applications.

Table I. Methodology of Thermodynamic Analysis

- Define Process Boundaries
- Obtain Consistent Heat and Material Balance
- Select Ambient Temperature, T_o
- Identify Consistent Source of Entropy Data
- Calculate Changes in Available Energy for the Overall Process
- Breakdown the Process into Subprocesses until Major Sources of Lost Work are Identified
- Prioritize Areas for Improvement

0097-6156/83/0235-0051$06.00/0
© 1983 American Chemical Society

The methodology used was also relatively standard, but simplified. The steps in the analysis were carried out in the order
above. On the assumption that the reader is familiar with the
general methods of thermodynamic analysis, these are presented
without explanation. For further background, one can consult
reference (3).

The calculation of process or equipment efficiencies via
second law analysis will not play a large part in this discussion.
In the authors' view, the role of efficiency calculations is to
help quantify the potential for improving a given segment of the
process. By combining the magnitude of losses with the efficiency
of a given segment, a feel for the priority of studying that segment can be developed. For example, in many cases there are large
losses associated with major compressors in industrial situations.
However, the efficiency of major machinery is generally very high.
Thus the potential for significant improvement in the machinery
section may be quite low.

In all calculations only as much detail was developed as was
needed to generate insight. The process was broken down into sub-
sections to identify the sources of loss. Subsequent breakdowns
continued only until a clear picture of the sources on inefficiency was developed. A check on accuracy was maintained by
insuring that the work balance, i.e., $W_{lost} = W_{actual} - W_{ideal}$,
was reasonably close. As indicated above, the various segments of
the process were listed in priority order based on the potential
for improvement. Idea generation was then carried out based on
the thermodynamic data.

Opportunities in Process Improvement

Consider the process shown in Figure 1 for the separation of
propylene and propane. Eight hundred moles/hr of a 65% propylene,
35% propane mixture are to be fractionated at a pressure of about
240 psig into 495 moles/hr of 99.9% propylene and 92.8% propane.
The tower is very large, containing about 275 actual trays with a
heat input of $72\frac{1}{2}$ million BTU's/hr provided by low pressure
(20 psig) steam. Essentially all of this energy is rejected to
cooling water in the condenser of the tower.

With the data given, little can be suggested for improving
the process. No enthalpy losses are identified. One might consider preheating the feed partially using the overhead stream.
Without other knowledge of the process, nothing more is apparent.

A simple thermodynamic analysis provides considerably more
data to work with. The required task is to separate propylene
from propane. On a theoretical basis the ideal work (the minimum
availability change) required for this separation is about 400 k
BTU's/hr, of which an appreciable fraction is needed to raise the
temperature of the products to the final values shown. The available energy (availability, exergy) supplied to this process from
the condensing low pressure (20 psig) steam is 18.6 M BTU's/hr.

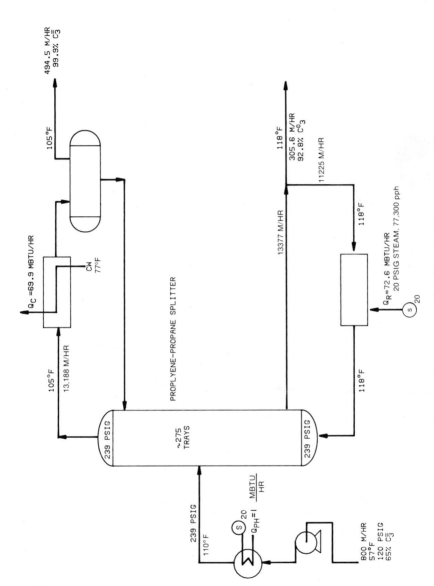

Figure 1. High Pressure C_3 Separation Option ($\underline{5}$).

The thermodynamic efficiency of the process, defined by the ideal work requirement over the actual is about 2%.

The thermodynamic analysis of the process and several alternatives is given in Table II, based on T_0 = 77°F (537°R). The sum of the ideal work and the lost work is close enough to the actual work (-4%) to proceed.

Can these data be used to suggest process improvements? The overall efficiency of the process is low, indicating good potential for improvement. The sources of lost work indicate priorities for brainstorming. Clearly the reboiler is the first segment on which to focus.

One idea would be to reduce the tower pressure to the point where ambient heat could be used in the reboiler. This also improves separation efficiency, so that less heat (65 M BTU/hr vs. 72 M) is required. The proposed flow diagram is shown in Figure 2. The price that has to be paid is the installation of a refrigerated condenser so that reflux can be produced at the top of the tower. In this case about 3000 hp (7.6 M BTU/hr) are required to drive the refrigeration compressor, and waste heat (at T_0 = 77°F) is used to drive the reboiler.

This proposal reduces available energy input (7.6 M BTU/hr work vs. 18.6 M BTU/hr steam), but does not reduce the total energy input (7.6 + 65.3 = 72.9 M BTU/hr). Thus the idea does not save energy from a first law viewpoint. From the second law viewpoint the heat supplied to the reboiler at 77°F (T_0) contains no useful energy.

The breakdown of lost work in the refrigerated process is compared to the high pressure option in Table II. Aside from the more uniform split of the losses, two other points are worth noting about these data. First, the ideal work is lower, primarily because the available energy of the products is lower. Second, there is still lost work associated with the reboiler. This is because of the temperature difference between T_0 and the process, even though there is no lost work between the heat source and T_0. This point brings up an additional opportunity to improve the high pressure option: use yet lower pressure steam. If steam saturated at 130°F (2.2 psia) were available, only 6.6 M BTU/hr of available energy (vs. 18.6 for 20 psig) would be consumed in the reboiler with a resultant reboiler lost work of roughly 1.4 M BTU/hr. Either the high pressure or the refrigerated process can be designed now for about the same fuel consumption.

More insight is yet available from the data in Table II. In the refrigerated process, the two condensers and the throttle valve involve more than 50% of the lost work remaining. One way to eliminate the inefficiencies of the condensers is to recycle the latent heat of the overhead vapor in a heat pump (vapor recompression) system, as shown in Figure 3. The distillation tower pressure, and hence its overhead temperature are kept the same, but the overhead, instead of being condensed, is compressed to a pressure at which it will condense at 77°F (about 180 psig).

Table II. Thermodynamic Analysis of All Options

| | Lost Work, MBtu/hr | | | |
| | High Pressure | | Refrigerated | Heat Pump |
	20 psig stm	2.2 psia stm		
Tower	1.8	1.8	1.1	1.1
Tower Condenser	2.2	2.2	1.7	-
Reboiler	13.2	1.4	1.4	2.1
Compressor	-	-	1.1	1.3
Refrigerant Condenser	-	-	1.3	-
Throttle Valve	-	-	0.9	0.2
Small Exchangers	0.3	0.1	-	0.1
Total Lost Work	17.5	5.5	7.5	4.8
Ideal Work	0.4	0.4	0.3	0.3
TOTAL	17.9	5.9	7.8	5.1
Actual Work	18.6	6.6	7.6	5.0

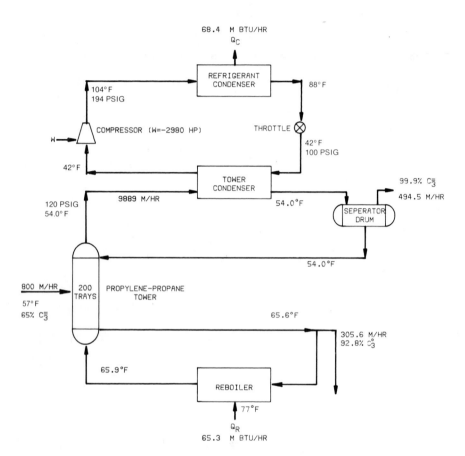

Figure 2. Refrigerated C₃ Separation Option (5).

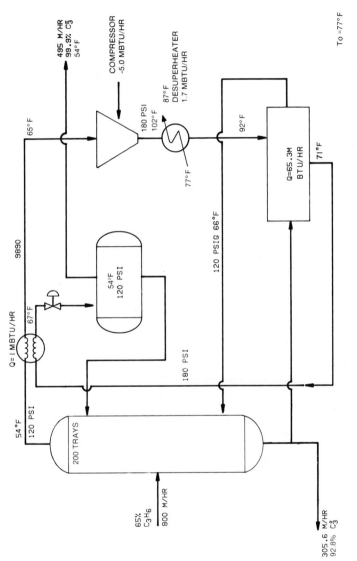

Figure 3. Heat Pumped C$_3$ Separation Option (5).

Sufficient vapor is compressed to provide the necessary 65 M BTU/
hr. This requires about 2000 hp, or 5.0 M BTU/hr, of work. The
lost work of the two condensers has been eliminated but two heat
exchangers are needed to maintain heat balance. Some throttling
must still be done.

Note in Table II that the heat pump alternative represents
about a 30% saving on available energy input compared to the
refrigerated process, and that both represent a very large reduction in the available energy requirement of the original process.
The heat pump process does indeed save the losses at both condensers, but has some compensating losses elsewhere.

The heat pump case also required significantly less capital
than the refrigerated case. The superheater is much smaller than
either of the condensers and requires no cooling water supply. In
addition, the compressor and pressure let-down valve are slightly
smaller, and no connection to a waste heat supply system is required. In fact, on a grass roots basis, which includes the
capital investment for steam supply and cooling tower, the heat
pump case has a lower total investment than the original high
pressure tower case, as well as the refrigerated case.

In one petrochemical plant a retrofit heat pump system has
been installed on an existing C_3 splitter and a grass roots heat
pump was included on a new C_4 tower. Both are highly economic.
However heat pumps are not always the choice when very low temperature heat is available to drive a reboiler directly.

Hardware Improvements Can Also Be Identified

An early attempt at second law analysis for design improvement was
to investigate the purification system in an ethylene plant. In
this process the cooled reactor effluent is compressed to about
500 psig and enters a chilling system leading to a cryogenic
product distillation train. In the plant in question the first
tower in the sequence was a demethanizer in which methane and
hydrogen were taken overhead, and ethylene, ethane, and heavier
products came out the bottom for further purification. The feed
gases are chilled against overhead product and refrigeration
streams producing four separate liquid feeds to different points
in the tower.

A simplified flow diagram for this segment of the process is
shown in Figure 4, and thermodynamic analysis in Table III. The
refrigerated exchangers and cold box represent about 30% of the
lost work of the process. However, the tower itself has a very
high percentage of the lost work in the system. Thus the details
of the tower heat and material balance were examined in search of
ways to improve its efficiency.

Inherent in the feed preparation system for the demethanizer
tower are a series of partial condensers. Several are shown on
the third (-30°F) vapor feed stream which condense liquid separated in the -144°F drum. Because the hydrogen and methane are

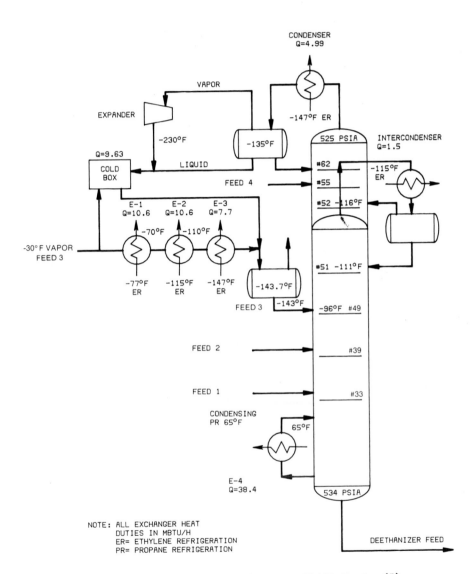

Figure 4. Original Demethanizer Chill Train ($\underline{5}$).

largely non-condensible at these pressures and temperatures, the
liquid is sub-cooled relative to its pure-component dew point.
Therefore, while the -143°F feed to tray No. 49 has the same composition as the liquid on that tray, it is some 50°F colder. Other
mismatches occur in the lower sections of the tower, but these
turn out to be the less important because the absolute temperature
level is higher.

Table III. Demethanizer Section Lost Work Analysis

Equipment	Lost Work, HP	Percent
Refrigerated Exchangers	586	9.6
Cold Boxes	1274	20.9
Tower	3244	53.2
Expander/Recompressor	677	11.1
Other	319	5.2
Total	6100	100%

    To evaluate the economic potential of eliminating this seem-
ingly theoretical inefficiency, a system to reheat the third feed
to the tower after the vapor was separated was explored. This
scheme is shown in Figure 5 where some 6 M BTU/hr of refrigeration are recovered by a simple feed-product heat exchange system
which produces a match between the feed temperature and the temperature on Tray No. 49. Detailed tower computer calculations
showed that to maintain the fractionation required in the tower
and the overall heat balance of the system, some of the 6 M BTU/
hr refrigeration savings have to be given back in the tower condensers (0.4 M BTU/hr @ -147°F and 1.4 M BTU/hr @ -115°F). In
addition, the reboiler duty, which is really a refrigerant recovery system, is somewhat reduced. However, because net savings
occur in -115°F and -147°F ethylene refrigeration loads, the
actual horsepower needed is 2200 hp. lower for the reheat scheme
than for the original design. The total lost work in the original
tower design was 3240 hp. While the basis is not the same,
clearly a big portion of the tower lost work must have been eliminated to achieve so large a saving in actual work. At the site
in question, a 2200 hp. saving puts $400,000 more into profits
and gives a one year payout at design conditions. In an action
typical of the industrial environment, a new thermodynamic analysis of the system with the reheat was never carried out.
    If the original design had incorporated the reheat scheme,
a lower overall plant investment would have been possible. The
savings in refrigeration compressors, steam turbine drivers, and
high pressure steam boilers would have more than compensated for
the increased investment in refrigeration heat exchangers. This
says that a second law analysis at the process design stage could
have made a significant contribution to project economics. Even

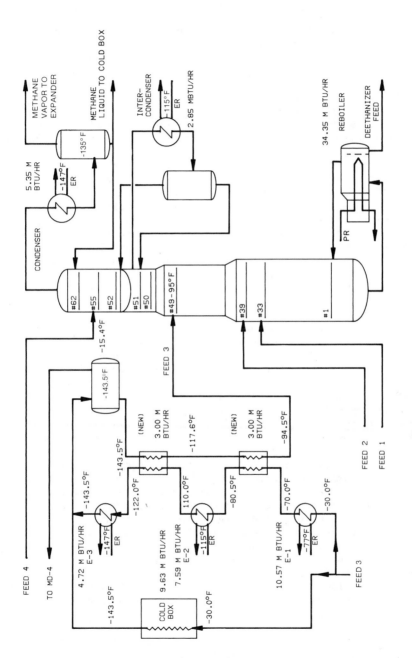

Figure 5. Third Feed Reheat System (5).

in a retrofit case, if a use exists (increased capacity for example) for the liberated utility and refrigeration system capacity, the net investment required for the heat exchanger installation can quite properly be reduced.

Consider The Entire Energy System In Analysis

In evaluating the economics of a process energy efficiency improvement, the investment impact in the utility system is often
ignored. Thus the fractionation tower designer views his optimization problem as trading the investment in more trays against
the increased operating cost for more reflux. In reality there
is also a utility system investment required to provide the increased reflux. The trade off must be evaluated against the net
investment for the plant system, i.e., the increase in tower cost
minus the decrease in exchanger, heat supply, piping, pump and
cooling system costs.
        Steam/power (utility) systems engineers have long recognized
this fact and tend to calculate both operating and investment cost
impacts back through the system to the primary source of available
energy, usually fuel. In effect, minimizing available energy
(fuel) consumption is the object of these studies, even though no
formal thermodynamic analysis is done. Consider the following
example:
        In a major ethylene plant the drivers for the process gas and
propylene refrigeration compressors were designed to be steam
turbines in the base case. The overall steam balance for that
system is shown in Figure 6. The power requirement for the two
major drivers was 52,640 kw. High pressure steam was to be
generated from the waste heat boilers on the cracking furnaces and
from auxiliary offsite boilers. When all was said and done, fuel
consumption was to be 1368 M BTU/hr, and approximately 155 t/hr of
steam would flow through the turbines directly to condensers. As
an alternative, a gas turbine driver was proposed for the process
gas compressor. Since the compressor was so large, an auxiliary
steam turbine driver was also required. The hot exhaust from the
gas turbine was used as preheated air for waste heat boilers.
Note also that the pressure level of the utility boilers was decreased from 1500 psig to 600 psig. The capacity and pressure
changes reduce fan and pump power requirements as well. Only
about 70 t/hr of steam need to be condensed from the turbines in
this scheme. The total fuel required was approximately 178 M
BTU/hr less than for the base case. This gave a savings of 13%
in total fuel fired and a large net second law efficiency improvement, even though generating 600 psig steam, as opposed to 1500
psig, was a step in the wrong direction.
        Table IV. gives the changes in investment required to switch
to the gas turbine driver. The additions and deletions from the
base case required to achieve the savings above are shown.
Because of the reduced fuel firing onsite, it was now necessary

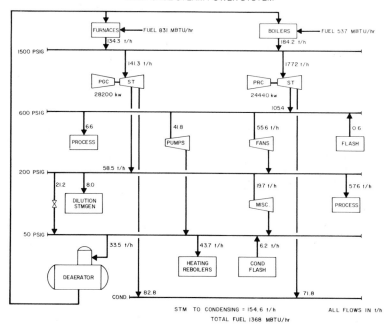

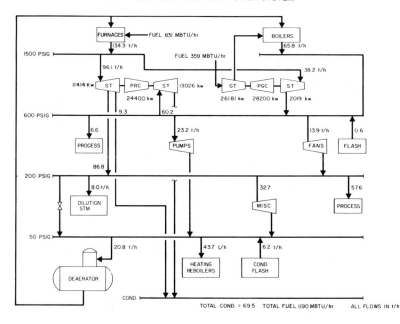

Figure 6.   Base Case and Improved Steam/Power Systems (5).

to spend appreciable capital for exporting liquid fuel and also
to compress the gaseous fuel system up to the pressure level required for the gas turbine. Many interrelated changes were required to reoptimize the steam/power system, but the net result
was a small saving in investment.

Table IV. Investment Summary-Change To GT Driver, $k (1978) (5)

	Add	Delete
Process Gas Compressor		
Steam Turbine	2300	4900
Gas Turbine	8350 (a)	
Refrigeration Compressors		
Steam Turbines	6350	3700
Steam System		
1. Waste Heat Boiler	11350 (a)	
2. 600 psig Boilers	7050 (a)	
3. 1500 psig Boilers		34960
4. Fuel System Modifications	6600	
	42000	43500

(a) Includes 5% extra contingency.

     This case lends support to the thermodynamic conviction that
if fuel can be saved, investment should be lower as well. Note,
however, that the thermodynamicist was disappointed by decreasing
boiler pressure to 600 psig. Clearly, there must be some limitation to the extent to which energy efficiency can be improved
at no increase in capital cost.

Relation to Process Efficiencies

In an effort to explore this aspect further, a paper written by
Gyftopoulos and Benedict concerning the maximum potential efficiency of an air separation plant provided some insight (4).
Compressed air is separated by cryogenic distillation into oxygen
and nitrogen. In a unique approach, the authors developed an
idealized process wherein all thermodynamic inefficiencies which
could be corrected by capital investment were eliminated. The
losses in the distillation tower were not much affected by this
approach. Their thermodynamic analysis for the practical and
idealized processes are compared in Figure 7.
     The point of this analysis was to characterize the source of
inefficiencies in the process as designed. The main heat exchanger was the key item. The authors' developed equations which
related the area of the heat exchanger and its irreversible
entropy change to two controllable design variables; namely, the
pressure drop, and the hot end temperature driving force between

feed air and exhaust N_2 in the exchanger. These equations were as follows:

$$\text{Exchanger Surface} = 1.886 \times 10^6 \triangle P^{-0.4} \triangle T^{-1.4} \tag{1}$$

$$\text{Compressor Power} = 0.27 \left[3165 \triangle P + 776 \triangle T \right] = 0.27 \left[\triangle S \right] \tag{2}$$

This author used these equations to explore the present day onsite and power plant investments associated with reducing the entropy change of the process, i.e., with improving its efficiency. Based on simplified investment curves, the average costs of equipment in 1981 dollars were calculated to be:

Heat Exchange Surface - 5\$/FT^2

Compressor and Driver - 1300\$/kw

Industrial Condensing Steam Turbo-Generation - 2500\$/kw

Figure 8 gives the results of these calculations. The lower set of curves shows that the investment for the heat exchanger and compressor generally decreases as the power requirement of the system increases. The three separate curves represent three different heat exchanger $\triangle T$'s. This is typical of the conventional "trade off" expectation, but a 4°F $\triangle T$ design has a 10% lower power requirement than either of the other choices at the same investment level.

The upper curves add the investment for onsite power generation to that for compressor and heat exchanger to show energy supply effects. For any exchanger $\triangle T$ there is a minimum investment. This point is between 1.8 and 2.3 times the $\triangle S$ of the idealized process. There is an optimum $\triangle T$ (4°F) for which both investment and energy efficiency are better than for the other two cases. Note that a plant can be designed to operate with a 3°F $\triangle T$ and use about 75% of the power required for a 5°F $\triangle T$ at the same capital cost.

The above results rest on the assumption that there are no significant changes in the investment of other components of the system as the configuration of the heat exchanger and compressor are changed within a relatively narrow range.

The unique contributions of Gyftopoulos and Benedict were to reduce a practical process to a simple energy model, and to define the energy consumption of the process when not restricted by capital. This is much different than the classic "Reversible Process" of thermodynamics and provides a basis for assessing how much energy efficiency can be improved before capital costs skyrocket. For this simple analysis it appears that energy consumption about twice that defined by the unrestrained capital case represents the range where overall plant investment must increase to improve efficiency further. Because the practical design operates at about seven times the energy consumption of the idealized process, it appears that there is much potential to be explored before reaching the break point.

Qualitatively these observations should also apply to other processes, but except for several unrelated additional examples, those studies remain to be done.

PROCESS	UNRESTRICTED CAPITAL		PRACTICAL	
	Work Input, 10^6 Btu/hr	$T_0 \Delta S_{irr}$, 10^6 Btu/hr	Work Input, 10^6 Btu/hr	$T_0 \Delta S_{irr}$, 10^6 Btu/hr
Air Blower		-	5.11	0.988
Air Cooler		-		0.268
N2 Compressor	7.72	-	23.88	5.109
N2 Coolers		-		2.967
Main Exchanger		0.927	-0.87	6.409
Expander		-		0.671
Tower		2.222		2.886
Reboiler		0.089		1.583
Reflux Cooler		0.161		1.448
Valve		0.101		0.427
Heat Leak		-		0.787
TOTAL	7.72	3.50	28.12	23.54
$T_0 \Delta S_{irr}$, 10^6 Btu/hr		3.50		23.54
Theoretical Work $= H - T_0 S$		4.21		4.61
Predicted Work, 10^6 Btu/hr		7.71		28.15

Figure 7.   Thermodynamic Analysis of Air Separation
Processes (4).

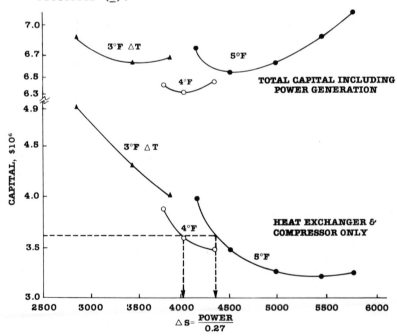

Figure 8.   Capital Cost-Efficiency Relations: Air Separation Process (5).

Conclusions

The author believes that a large number of practical applications of thermodynamic analysis are possible before a level of complexity which is intolerable to an industrial process engineer is reached. Process design and development, equipment design, energy system analysis, and research guidance are a few of the areas where potential uses exist.

The analysis can lead to identifying the sources of losses in a process, but provides no guarantee that useful ideas will be generated. This remains the challenge of the process engineer. It is hoped that the examples discussed here will provide some encouragement to the doubtful.

The relations between energy efficiency and capital cost must be evaluated from the analysis of the overall plant system. At some point, improved energy efficiency will require more investment. However, many of the practical processes of today may well be operating quite far from this point. Further process analysis along these lines may well be fruitful, particularly for grass roots design situations.

Legend of Symbols

A - Availability (Available Energy, Exergy) of a substance at T, P.

$\triangle$A - Change in Available Energy as state changes (work).

$$A = H - T_o \triangle S$$

H - Enthalpy.

HP - Horsepower.

S - Entropy.

T_o - Temperature of the environment (dead state).

W - Work, W_{ideal} = ideal work, W_{lost} = lost work, etc.

Literature Cited

1. Kenney, W. F., "Using Thermodynamic Analysis to Improve Industrial Processes," ASME Meeting, Phoenix, Arizonia, 11/15/82.
2. Kenney, W. F., "Some Observations on Energy Efficiency and Capital Cost," IECTC, April 1982, Houston, Texas.
3. Smith and Van Ness, "Introduction to Chemical Engineering Thermodynamics," Chapter 11, 2nd Ed. 1959, McGraw Hill, NY, for example.
4. Benedict & Gyftopoulos, ACS Symposium Series 122, "Thermodynamic Analysis," p. 195.
5. Kenney, W. F., "Energy Conservation in the Process Industries," Academic Press, NY.

RECEIVED July 12, 1983

EFFICIENCY ANALYSES
OF PROCESSES AND PLANTS

Reversibility of Combustion Processes

HORST J. RICHTER

Thayer School of Engineering, Dartmouth College, Hanover, NH 03755

KARL F. KNOCHE

Lehrstuhl für Technische Thermodynamik, RWTH Aachen, D-5100 Aachen, Federal Republic of Germany

Technical combustion processes are highly irreversible. Theoretical considerations show that the irreversible entropy production of combustion can be decreased if immediate contact of fuel with oxygen is prevented and intermediate chemical reactions are supported. Therefore, the potential for mechanical work output can be increased. Metal oxides can be used as reactants for these intermediate reactions.

In many of our energy conversion processes, like in conventional power plants, we rely on chemical energy being released in a combustion process. The combustion of fossil fuels, usually employed as an intermediate step, is in common practice highly irreversible. This is the main reason for the overall low efficiency of these energy conversion processes.

It is known that the less irreversible the chemical reactions, the closer they occur to the thermodynamic equilibrium. Unfortunately, the equilibrium of technical combustion processes is usually at such high temperatures that the materials which enclose the reaction volume cannot withstand these temperatures.

Therefore, the development of high temperature materials is a way to improve such efficiency. Technical combustion is usually performed by crude mixing of fuel and oxygen, thus the reaction is allowed to occur in a very disorderly way, inevitably resulting in irreversible entropy production.

The human body, as an example, produces mechanical energy from chemical energy by allowing many intermediate chemical reactions of the "fuel" and the "oxygen" separately before one molecule of "fuel" is united with the stoichiometric number of oxygen molecules in the muscle cells, where the mechanical energy is produced. Not all of these intermediate reactions are known, but physiological studies by Lehmann ([1]) indicate that the total efficiency of the human body is surprisingly high, for some labor it is in the range of 30 to 35%, thus exceeding the Carnot efficiency substantially. Therefore, it seems advantageous to

0097-6156/83/0235-0071$06.00/0

try to imitate nature and search for possibilities to improve the
technical combustion by preventing immediate contact of oxygen
with fuel and rather support intermediate reactions.

The intermediate reactions in the living organism are high
molecular organic reactions. But for technical purposes
anorganic substances like metal oxides are probably more
favorable as oxygen carriers. This was studied extensively by
Knoche and Richter (2).

Technical Background

The maximum work output of any thermodynamic system or process
can be obtained, if the material in the system or the working
fluid in the process is brought into equilibrium with the
environment reversibly. The actual work output of a technical
process with combustion is much smaller because the combustion
is highly irreversible. The work losses in a continuous combustion can be evaluated if the exergy (or available energy)
before and after the reaction is calculated. This exergy is
described by the equation:

$$e = (h_1 - h_o) - T_o(s_1 - s_o) \qquad (1)$$

In the above equation the exergy is written as an intensive
property, e.g. per mol of fuel. The subscript "1" represents
the original thermodynamic state, the subscript "o" describes
the state when thermodynamic equilibrium with the environment is
established. In these considerations it is arbitrarily assumed
that any reaction product is in equilibrium with the environment
when it has a temperature of T_o = 300K and a pressure of P_o = 1bar.

As was shown by Baehr (3) the available energy or exergy can
be represented graphically very easily in an enthalpy, entropy
diagram (see Figure 1).

Knoche (4) plotted the reactants and reaction products in one
enthalpy, entropy diagram by superposition, thus he was able to
evaluate immediately the irreversible entropy production, respectively the exergy losses of the combustion (see Figure 2).

For an adiabatic combustion process at constant pressure, the
enthalpy stays constant. Figure 2 shows the exergy of the
reactants, e_1, and the exergy of the reaction products, e_2, after
this reaction has taken place. The loss in exergy is:

$$w_L = e_1 - e_2 = T_o(s_1 - s_2) \qquad (2)$$

In technical combustion processes, this loss of exergy can be as
high as 50%. If the combustion process is adiabatic, but at constant volume, then the internal energy stays constant. Usually
the exergy losses are somewhat smaller than in the previous case
(see Figure 3).

A third way of combustion would be isothermal, isobaric as
indicated in Figure 4. This reaction requires a heat reservoir

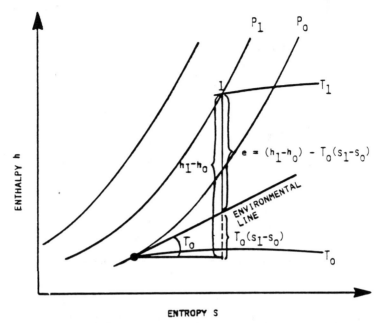

Figure 1.   Enthalpy, entropy diagram with graphical pre-sentation of exergy (3).

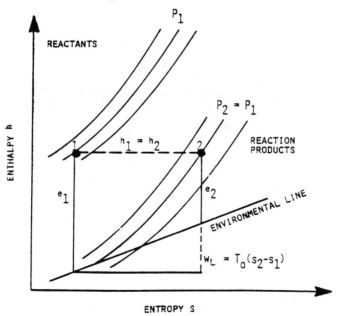

Figure 2.   Enthalpy, entropy diagram with exergy for reactants e_1 and reaction products e_2 after an adiabatic, isobaric combustion (4).

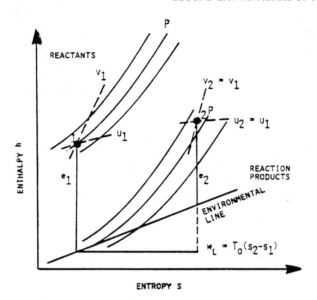

Figure 3. Enthalpy, entropy diagram with exergy for reactants e_1 and reaction products e_2 after an adiabatic combustion at constant volume (4).

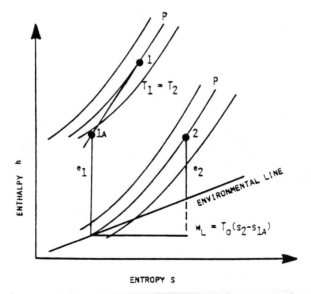

Figure 4. Enthalpy, entropy diagram with exergy for reactants e_1 and reaction products e_2 after an isothermal, isobaric combustion (4). The change of state from 1 to 1a is a release of heat at constant temperature, therefore the entropy decreases.

at the combustion temperature thus is not an adiabatic process. This case usually shows the smallest irreversible entropy production.

From these brief considerations an exergetic efficiency of the combustion process can be defined:

$$\zeta = 1 - \frac{\text{exergy loss during combustion}}{\text{exergy of reactants}} \qquad (3)$$

or

$$\zeta = 1 - \frac{T_o \Delta s}{e_{reactants}} = 1 - \frac{W_L}{e_{reactants}} \qquad (3a)$$

where W_L is the work loss. This exergetic efficiency will be the measure of performance for the following considerations.

Combustion with Intermediate Reactions

As was indicated above, the losses in exergy can eventually be diminished if intermediate reactions are employed rather than immediate contact between fuel and oxygen. A schematic of such a reaction scheme is shown in Figure 5.

In "reactor" A the fuel is oxidized, an appropriate carrier provides the oxygen. The carrier will be circulated between "reactor" A and reactor "B". In reactor B it will be recharged with oxygen. The mass balance performed at the surface of the control volume of the process is identical to a normal combustion. Only fuel and oxygen or air are transferred into the control volume and combustion products and nitrogen are exhausted from it; the latter one only if air is used for oxidation instead of pure oxygen.

The chemical energy released in a normal combustion process increases only the internal energy of the combustion products. According to the second law of thermodynamics only part of this energy can be harnessed. An improvement of the process should enable us to obtain a larger fraction of the available energy as useful work, thus the irreversible entropy production should become smaller.

In the following considerations, methane will be used as fuel. For intermediate reactions we could use the following possibilities for improvement:

a) The oxidation of fuels occurs by reduction of solid metal ("Me") oxides. This takes place such that the number of gaseous moles is increased substantially. As an example, the oxidation of methane:

$$CH_4 + 4"Me"O \rightarrow CO_2 + 2H_2O + 4"Me" \qquad (4)$$

In this reaction the number of gaseous moles is tripled. At higher pressures the reaction occurs closer to the equilibrium, thus the irreversible entropy production and consequently the work losses are smaller. But this is usually only a very small improvement.

b) Metal oxides are used as oxygen carriers which react with the fuel at relatively low temperatures endothermally. In this case heat has to be provided to the "reactor" A in Figure 5 to keep the temperature constant. Therefore, in the reoxidation of the oxygen carrier in "reactor" B more heat has to be released, since the overall energy balance has to be the same as for the usual combustion process. Work producing cyclic processes (Carnot heat engines) could be employed between a heat source, represented by the constant temperature reoxidation of the oxygen carrier, the heat sink of the endothermic fuel oxidation, and the heat sink of the environment (see Figure 6). In the ideal case, if the chemical reactions in both reactors are reversible, the work producing devices (heat engines) should theoretically deliver the maximum available energy or exergy of the fuel as mechanical work.

Both processes described above can be utilized at least partially.

Only the ideal cases of an isothermal-isobaric combustion process will be assumed. This combustion is superior to the usual isobaric-adiabatic process. Such an assumption can be verified more easily than in a normal combustion process, since in the cases studied here the chemical reactions take place at the surface of the oxygen carriers.

If, for the methane combustion, copper oxide is used as an oxygen carrier, the chemical reaction in the "reactor" A (see Figure 5) is:

$$CH_4 + 4CuO \rightarrow CO_2 + 2H_2O + 4Cu \tag{5}$$

and the reoxidation in "reactor" B is:

$$Cu + 0.5 \ O_2 + 0.5 \cdot \frac{0.79}{0.21} \ N_2 \rightarrow CuO + 0.5 \cdot \frac{0.79}{0.21} \ N_2 \tag{6}$$

The oxidation reaction of CH_4 with CuO is shown in Figure 7 and the reoxidation of Cu in Figure 8. In this case, both the reduction and reoxidation are irreversible; the only improvement is due to the increase in the number of gaseous moles, increasing the efficiency at higher pressures.

The evaluation of the irreversible entropy production is very easily performed in these enthalpy, entropy diagrams as was shown by Knoche (4). It is simply for one mole methane:

$$W_L = T_o(\Delta s_A + 4\Delta s_B) \ . \tag{7}$$

Δs_A is the irreversible entropy production during the copper reduction in reactor A and Δs_B is the irreversible entropy pro-

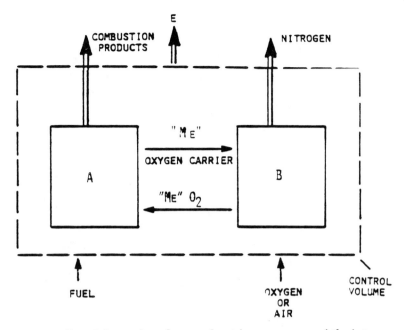

Figure 5. Schematic of a combustion process with intermediate reaction.

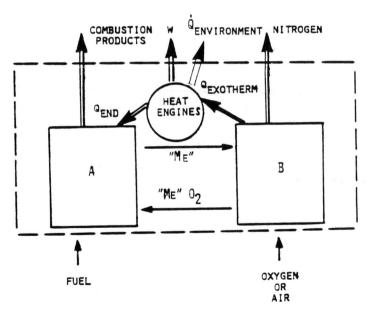

Figure 6. Schematic of a combustion process with intermediate reactions and heat engines.

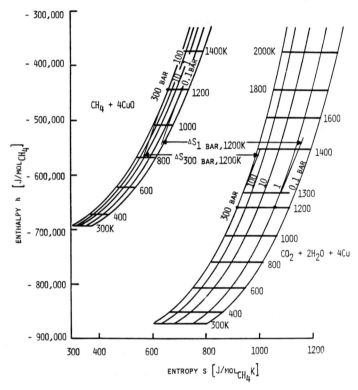

Figure 7. Enthalpy, entropy diagram of reduction of copper oxide with methane.

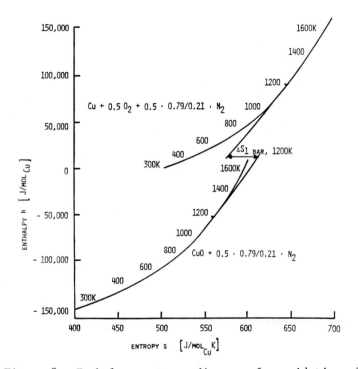

Figure 8.   Enthalpy, entropy diagram of reoxidation of copper.

duction in the subsequent reoxidation process in reactor B, both
at the desired process temperatures.

Another possible oxygen carrier is nickel oxide, the chemical
reaction in reactor A in Figure 6 would be:

$$CH_4 + 4NiO \rightarrow CO_2 + 2H_2O + 4Ni \tag{8}$$

and in reactor B

$$Ni + 0.5 \ O_2 + 0.5 \cdot \frac{0.79}{0.21} \ N_2 \rightarrow NiO + 0.5 \cdot \frac{0.79}{0.21} \ N_2 \tag{9}$$

The difference to the copper process is, that the reduction of
nickel oxide with methane is an endothermic process, thus a heat
engine could be employed. Figures 9 and 10 show the enthalpy,
entropy diagrams of the reactions outlined in equations (8) and
(9). (A metallurgist will not favor the reoxidation of nickel
since it is very difficult, but equilibrium thernodynamic considerations do allow it.)

Another very interesting oxygen carrier is cadmium. For the
methane combustion the following reactions are occurring in
reactors A and B (see Figure 6):

$$CH_4 + 4CdO \rightarrow CO_2 + 2H_2O + 4Cd \tag{10}$$

and

$$Cd + 0.5 \ O_2 + 0.5 \cdot \frac{0.79}{0.21} \ N_2 \rightarrow CdO + 0.5 \cdot \frac{0.79}{0.21} \ N_2 \tag{11}$$

At a temperature of about 1050K and a pressure of 1 bar, the
cadmium vaporizes, thus in the reduction process of CdO,
(equation 10), the number of gaseous moles increases from one
mole of CH_4 to 7 moles. In the ideal case the cadmium oxide
reduction is reversible at a temperature of about 800K at a
pressure of 300 bar (see Figure 11). The subsequent reoxidation
of Cd is reversible at a temperature of about 1600K, in which
solidification occurs (see Figure 12). Thus the total process
with cadmium oxide as an oxygen carrier is reversible if the
reoxidation could be handled at this relatively high temperature.
(Cadmium is probably not a potential oxygen carrier in a real
design due to its hazardous properties.)

The irreversible entropy production for isothermal, isobaric
chemical reactions is plotted in each diagram for an isothermal
irreversible process at 1200K, but can be evaluated as easily for
any other temperature.

The exergetic efficiency of the total process can then be
evaluated with equations (3a) and (7) as a function of the maximum process temperature.

Figure 13 shows the exergetic efficiency vs. the maximum
temperature for different combustion processes with and without
intermediate reactions. As a comparison, the adiabatic, isobaric

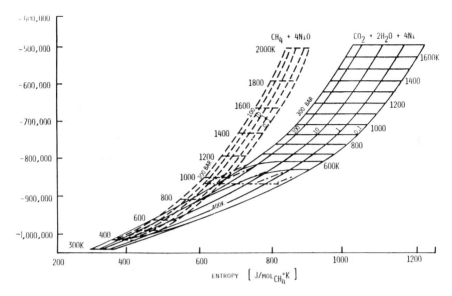

Figure 9. Enthalpy, entropy diagram of nickel oxide reduction with methane.

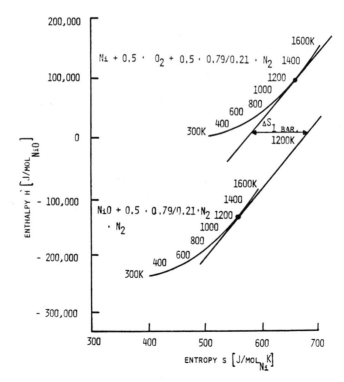

Figure 10. Enthalpy, entropy diagram of reoxidation of nickel.

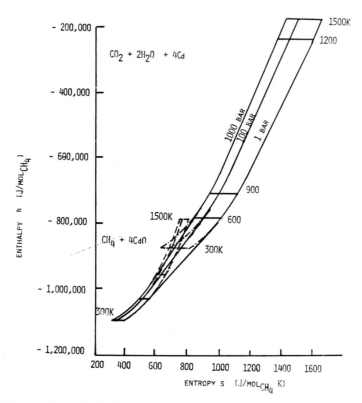

Figure 11.   Enthalpy, entropy diagram of cadmium oxide
reduction with methane.

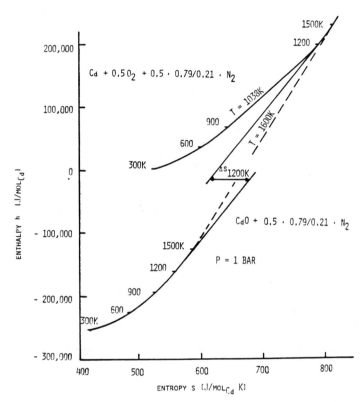

Figure 12. Enthalpy, entropy diagram of reoxidation of cadmium.

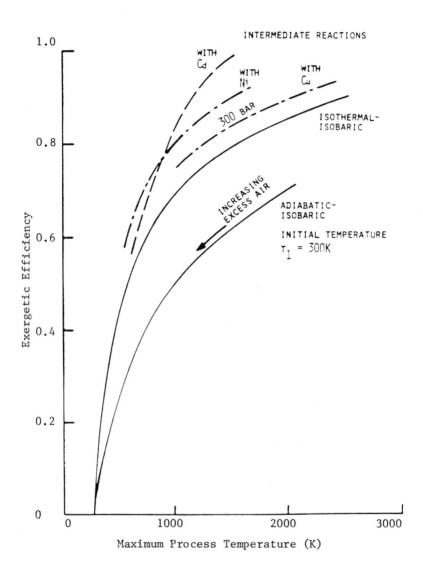

Figure 13. Exergetic efficiency of methane combustion for different processes as a function of maximum process temperature.

combustion of methane is plotted for an initial temperature of T_1 = 300K and different excess air ratios.

The exergetic efficiency of combustion processes with intermediate reactions is theoretically higher than combustion in usual practice.

For a maximum process temperature of 1600K a reversible combustion can be obtained theoretically with cadmium as the oxygen carrier.

Conclusions

These basic thermodynamic considerations show that intermediate reactions in combustion processes can be very advantageous and that in some cases most or all of the chemical energy could be harnessed as mechanical energy at least theoretically. Important questions of reaction kinetics, actual design and applicability of such a device of the selected oxygen carriers have not been included in these fundamental thermodynamic equilibrium studies.

Yet there is indication from the production of synthetic fuels that the reduction reactions are reasonably fast, see Terisawa and Sakikawa (5), and Lewis and Gilliland (6).

Literature Cited

1. Lehmann, G. "Praktische Arbeitsphysiologie"; Thieme Verlag: Stuttgart, 1953.
2. Knoche, K.F.; Richter, H.J. Brennstoff-Wärme-Kraft. 1968, 5, 205-210.
3. Baehr, H.D. "Thermodynamik"; 2nd Ed., Springer-Verlag. Berlin, 1966.
4. Knoche, K.F. Brennstoff-Wärme-Kraft. 1967, 1, 9-14.
5. Terasawa, S.; Sakikawa, N.; Shiba, T. Bull. Jap. Petrol. Inst. 1963, 3, 20-26.
6. Lewis, W.K.; Gilliland, E.R.; Reed, W.A. Ind. Engng. Chem. (1949), 6, 1227-1237.

RECEIVED July 7, 1983

Thermodynamic Analysis of Chemical Energy Transport

H. B. VAKIL

Corporate Research and Development, General Electric Company, Schenectady, NY 12301

This paper deals with several aspects of thermodynamic analyses of chemical energy transport systems using reversible endothermic/exothermic chemical reactions. The first application of thermodynamic availability (or exergy) analysis deals with the relative merits of properly selected chemical reactions as secondary energy carriers for process steam applications. Results show that a conversion of primary thermal energy to secondary energy forms of a high thermodynamic quality is always associated with efficiency penalty. The chemical energy systems are able to avoid large changes in exergy ratios by providing a secondary source with a quality lower than that of the primary source but sufficiently high to deliver the process steam. The second application of availability analysis is used to evaluate the nature and magnitude of thermodynamic irreversibilities in a methane reformer plant coupled to a high-temperature nuclear reactor. It is shown that a combination of thermal histograms and availability concepts are helpful not only in evaluating the net impact of irreversibilities in various chemical process steps on the steam power plant, but, more importantly, in suggesting process modifications that could improve the overall efficiency by avoiding unnecessary entropy production.

The use of reversible endothermic/exothermic chemical reactions to transport and/or store thermal energy has been the subject of many recent investigations. Such systems have been referred to as Chemical Heat Pipes (CHP) (1), Fernenergie, (2), or Thermochemical Pipelines (TCP). Their ability to transport high-grade thermal energy over distances exceeding 50 mi (80 km) without excessive energy losses makes these systems particularly attractive as

0097-6156/83/0235-0087$06.00/0
© 1983 American Chemical Society

thermal utility grids for such end-uses as industrial process steam and low-grade thermal energy.

The major components of a chemical energy transport system are shown in Figure 1. At the site of the primary thermal energy source, an endothermic chemical reaction is carried out in a chemical reactor where thermal energy is absorbed. The sensible heat required to heat the reactants from the ambient temperature to the elevated reaction temperature is provided mostly by a counter-current heat exchange with the hot product stream leaving the reactor. In the absence of the reaction catalyst, the high-enthalpy reactor products do not undergo a spontaneous reverse reaction upon cooling and may be transported or stored at the ambient temperature without any measurable degradation. This ability to exploit the large activation energy barrier for a spontaneous chemical reaction in order to freeze-in the high-energy state at the ambient temperature is a feature unique to the chemical energy transport systems and distinguishes them from other thermal processes (e.g., phase change or sensible heat).

At the delivery location, the high-enthalpy products are heated to the exothermic reaction temperature and passed through a catalytic reactor where they undergo the reverse reaction, thereby releasing the stored reaction heat. Once again a counter-current heat exchanger is used to exchange the sensible heats, and the low-energy reactants are transported back to the primary site to repeat the energy absorption step. The overall result is a closed-cycle process, without any net chemical production or consumption, that absorbs thermal energy at the primary source location and delivers it to the end-use location.

While there have been many investigations concerned with exploring various chemical reactions for chemical energy systems, for the purpose of this paper we will restrict our consideration almost exclusively to the methane-based high-temperature system (HTCHP), with only a brief mention of the lower-temperature hydrogenation/dehydrogenation reactions (LTCHP).

The methane reforming reactions have been among the first reactions to be proposed and analyzed for chemical energy transport. The main energy carrying step is the steam/methane reaction

$$CH_4(g) + H_2O(g) \rightarrow CO(g) + 3H_2(g)$$

$$\Delta H^o_{298} = 206.2 \text{ kJ/mole}$$

This reaction is generally carried out in the forward, endothermic direction at temperatures up to 1100K and pressures of around 40 bars. The reverse, exothermic reaction (generally referred to as the methanation reaction) is capable of yielding thermal energy at peak temperatures in excess of 800K. The water-gas shift reaction

$$CO(g) + H_2O(g) \quad CO_2(g) + H_2(g)$$

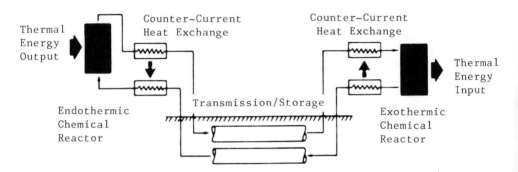

Figure 1. Chemical heat pipe schematic.

$$\Delta H^o_{298} = -41.2 \text{ kJ/mole}$$

also takes place simultaneously, but is not a major contributor to the overall energy carrying capability of the system, especially if the latent heat of water is taken into account.

In view of the fact that the most important goal of the CHP systems described here is to achieve the interconversions of chemical, thermal, and mechanical energies with the highest efficiency and the lowest losses, these systems are ideal subjects of thermodynamic analyses. Since there are no chemical raw materials consumed and since the only delivered products are heat and work, thermodynamic efficiency of the overall process plays a vital role in the design and economics of such systems. It is the purpose of this report to present the results of two separate applications of the second law analysis to these chemical energy systems.

The first application deals with the broader question of the relative merits of properly selected chemical reactions as intermediate or secondary energy carriers, specifically for industrial process steam applications. Major alternatives considered in the comparison consist of electrical generation and the generation of secondary fuels (e.g., substitute natural gas (SNG) or hydrogen).

The second application deals with a typical process design for the endothermic end of the HTCHP coupled to a very-high-temperature, helium-cooled nuclear reactor (VHTR). Exergetic considerations are used to investigate major sources of irreversibilities, as well as the potential effects of significant energy exchanges between the chemical process and the steam power plant. A potential improvement of the process suggested by the thermodynamic analysis is described and evaluated.

A COMPARISON OF ENERGY TRANSPORT ALTERNATIVES

Thermodynamic Framework

In order to have a meaningful comparison of various forms of energy conversion processes in a thermodynamic framework, it is necessary to define a quantitative measure not only for the energy or enthalpy content of an intermediate energy carrier, but also for its thermodynamic quality. This is especially important in the present case since we are comparing such diverse alternatives as the generation of electricity, the synthesis of a secondary fuel (e.g., SNG or H_2) for subsequent combustion, and the upgrading of low-enthalpy reactants to high-enthalpy products in order to transport the enthalpy of chemical reactions. The concepts and definitions of thermodynamic availability in general and steady-flow exergy of extraction and delivery (3) in particular are very useful in providing a quantitative measure for the quality. The basic definitions used in this report are summarized below:

The exergy of a process flow stream is defined as

$$Ex \equiv H - T_0 S - \Sigma n_j \, \mu_j 00$$

The change in exergy resulting from a change from State 1 to State 2 without any exchange of matter with the dead-state environment is given by

$$(\Delta Ex)_{1 \to 2} = (H_2 - H_1) - T_0(S_2 - S_1)$$

$$= (\Delta H)_{1 \to 2} - T_0(\Delta S)_{1 \to 2}$$

In addition to these standard definitions, it is useful to define a quantitative measure of the quality of energy that can be delivered as a result of a change from State 1 to 2; this is represented by the exergy ratio (α_E) defined as

$$\alpha_E \equiv \frac{(\Delta Ex)_{1 \to 2}}{(\Delta H)_{1 \to 2}}$$

According to this definition, the quality of mechanical or electrical energy is equal to unity and that of thermal energy at a temperature, T, is equal to the Carnot factor, $1 - T_0/T$. For chemical reactions, the exergy ratio (α_E) represents that fraction of the delivered energy that could be converted to thermodynamic work by a reversible process and has a value most often (but not always) between zero and unity.

Exergy Ratios of Primary and Secondary Energy Sources

In order to assign a reasonable value of the exergy ratio to such primary sources as nuclear reactors, one needs to make some realistic assumptions. Even though the primary nuclear reaction has a high value of α_E, the degradation inherent in the conversion of the nuclear reaction energy to the useful thermal energy in the reactor coolant appears to be unavoidable. The calculations presented here have been made based on the thermal energy delivered by the coolant stream. For fossil fuel sources, the thermodynamic calculations have been simplified greatly by assuming that coal is represented by carbon in the graphite state, natural gas and SNG by $CH_4(g)$, and oil by liquid toluene.
The calculated values of exergy ratios for various primary and secondary sources are shown in Table 1. It is to be noted that all the fossil fuel sources are inherently of very high quality. The value of α_E greater than unity for coal stems from the fact that a reversible oxidation of carbon at ambient temperature has a negative $T_0 S$ and results in production of more work than the enthalpy of combustion, with an associated absorption of heat from the environment. In spite of these high values of α_E, when fossil fuels are combusted, say for electrical generation, the maximum temperature is limited not by thermodynamic considerations but by materials constraints. The effective α_E under this limitation is restricted to values below 0.65.

Table I. Exergy Ratios for Various Types of Energy Sources

Primary Sources			Exergy Ratios (α_E)
VHTR	- Reformer heat	(1225 K to 875 K)	0.71
	Boiler heat	(875 K to 625 K)	0.60
	Average	(1225 K to 625 K)	0.67
HTGR		(1050 K to 680 K)	0.65
LMFBR		(835 K to 670 K)	0.60
LWR		($\sim$555 K)	0.46
Coal		$C(S) + O_2(g) \rightarrow CO_2(g)$	1.002
Natural Gas		$CH_4(g) + 2O_2(g) \rightarrow CO_2(g) + 2H_2O(l)$	0.92
Oil		$C_7H_8(l) + 9O_2(g) \rightarrow 7CO_2 + 4H_2O(l)$	0.98
Secondary Sources			
Electricity			1.00
SNG		(same as natural gas)	0.92
Hydrogen		$H_2(g) + 1/2\text{-}O_2 \rightarrow H_2O(l)$	0.83
HTCHP:	EVA-ADAM	$CO(g) + 3H_2(g) \rightarrow CH_4(g) + H_2O(l)$	0.60 (0.70)*
	HYCO	$2CO(g) + 2H_2(g) \rightarrow CH_4(g) + CO_2(g)$	0.69 (0.75)*
LTCHP		$C_6H_6(l) + 3H_2(g) \rightarrow C_6H_{12}(l)$	0.48 (0.61)*
Steam			
Powerplant	- Superheat	(165 b; 625 K to 813 K)	0.57
	- Boiler	(165 b; 523 K feed)	0.50
Process steam condensation at	- (70 b; 559 K)		0.47
	- (35 b; 515 K)		0.42
	- (15 b; 471 K)		0.37

*Numbers in parentheses are for 40.5 bars and include entropy of mixing.

Another point to be noted from Table 1 is that the secondary sources, with the exception of CHP candidates, also show extremely high values of exergy ratios. Since the industrial process steam to be made available to the user (with condensing pressures in the range of 15 b to 70 b) has values of α_E from 0.37 to 0.47, it is evident that these secondary energy sources are of a much higher quality than that required to do the job. A conversion to an intermediate form of energy with an unnecessarily high α_E is detrimental to the overall efficiency, as will be shown by the following comparisons.

Comparisons of Alternatives for Process Steam Delivery
One of the major advantages of CHP systems is the ability to utilize the thermal energy from large, remotely located energy sources (e.g., VHTR or coal-based power plants) and deliver it in the form of process steam to several small-scale users of process steam. The overall energy efficiency for these systems is projected to be as high as 80 to 85% (1).

For the HTCHP case, the primary source is assumed to be a VHTR capable of supplying the required high-temperature heat to a methane reformer. The overall efficiency for this combination is compared with that for the alternative -- VHTR → electricity → process steam. These two options are shown as (a) and (b), respectively, in Figure 2. Also shown in the figure are two more possibilities, using coal as the primary source. The lower-temperature CHP, represented by the reaction of benzene hydrogenation as the energy delivering step, has a lower value of α_E than the HTCHP and is more suited to coal and lower-temperature nuclear sources because of the reduced temperature nuclear requirements for its endothermic step. This is shown as case (c) in the figure and is compared with the alternative of coal → SNG → process steam, shown as case (d).

The overall energy conversion efficiencies show clearly the significant advantages offered by the two CHP systems. While options (b) and (d) are only two of the many alternatives for energy transport, similar results would be obtained for options such as coal → hydrogen → steam.

In order to understand the underlying reasons for the higher overall efficiencies afforded by the CHP systems, it is necessary to re-examine the four options in the framework of the thermodynamic quality. Figure 3 shows semi-quantitative plots of individual steps for each of these options, using exergy ratios as the ordinate.

Several important points can be noted from the figure:

● Both non-CHP options include a step during the primary conversion in which a large increase in α_E takes place. The existence of this step, with an uphill thermodynamic climb, is due mainly to the high quality of the secondary energy form. As a consequence of the second law of thermodynamics, a significant fraction of the primary energy has to be

a)   HTCHP CASE

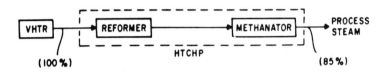

b)   ELECTRICAL TRANSMISSION

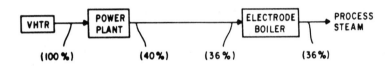

c)   LTCHP CASE

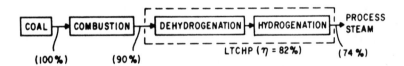

d)   SNG OPTION

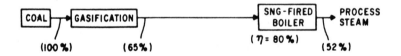

Figure 2.   Overall energy delivery efficiencies.

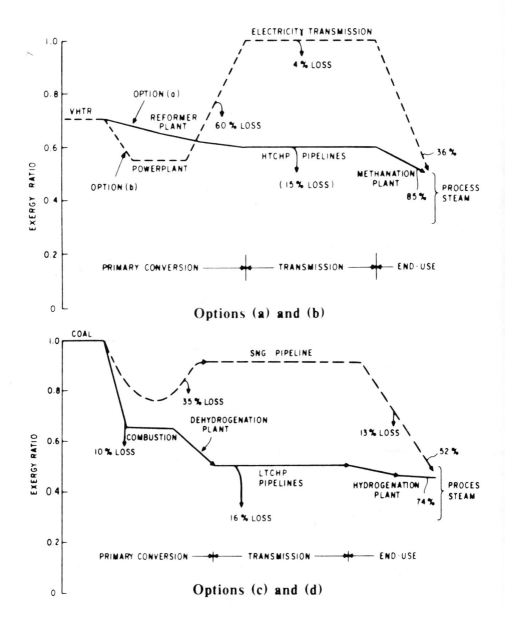

Figure 3. Exergy ratios of energy conversion steps.

rejected to a lower (ambient) temperature during such an
upgrading process. This leads inevitably to a low energy
efficiency for such steps.

● The two CHP options totally avoid such an uphill climb
thereby allowing a large fraction of the original thermal
energy to be delivered. This results from the fact that the
transported energy has a lower α_E than that of the primary
source; the drop in α_E is sufficient to provide the thermodynamic driving force necessary to accomplish the conversion.

● If the energy expended in creating a high-quality secondary
energy in options (b) and (d) were fully recovered during the
final conversion to the process steam with a significantly
lower α_E, the overall system efficiency would be comparable
to that for CHP options. In practice, however, the conversion from a very high α_E to a low one is thermodynamically
very inefficient; it is simply not possible to make process
steam from electricity (or SNG) with an efficiency comparable
to that of a reversible process.

There are several general conclusions to be drawn from these comparisons:

1. If primary thermal energy has to be converted, transported,
   and re-converted to deliver low- or medium-quality steam, it
   is inadvisable and inefficient to generate intermediate
   energy forms with a very high value of α_E.

2. The overall first-law efficiencies of CHP systems are high
   because they provide a path with only moderate drops in α_E
   and avoid excessive upgrading or downgrading of thermodynamic
   quality.

3. While we have been fortunate to be able to extract clean
   fossil fuels with a high α_E from the ground, it is not always
   desirable to snythesize such high-quality intermediate energy
   carriers, especially from thermal sources. This is especially true for those applications where the synthetic fuels
   are to be used for generating thermal energy.

THERMODYNAMIC ANALYSIS OF HTCHP METHANE-REFORMER PLANT

In the design of a VHTR-based HTCHP, the thermal energy from the
reactor helium in the high-temperature range (1225K to 875K) is
used for carrying out the methane reforming reaction. The remaining lower temperature heat (875K to 625K) is used to generate
steam for electrical generation. Typical process designs for the
chemical plant and the power plant have been presented elsewhere
(1,4,5). The purpose of the second half of this report is to
demonstrate the usefulness of thermodynamic analyses, using one
example of the reformer process design. The major features of

this design are shown in Figure 4. The flow rates for this design
are scaled to give a total net enthalpy transport rate of 1000
MJ/s at the pipeline conditions.

The water flow from the return pipeline (20 tonne-moles/hr)
and the condensate from the condenser (48 tonne-moles/hr) are
mixed and evaporated in the feedwater reheater and the boiler
(Units 3 and 4*). The steam thus produced is then mixed with a
preheated mixture of methane-rich gases from the pipeline and
heated to the reactor inlet temperature of 725K in (6). The
reforming reaction is carried out in the helium-heated reformer,
and the reaction products are sent to a gas cooler (1) and a partial condenser (2) for the removal of the excess, unreacted water.
The cooled CO-rich gases are sent to the pipeline for delivery to
the methanation plants. As shown in the figure, the total cooling
requirements in Units 1 and 2, and the total heating requirements
in Units 3 and 6 are substantial in comparison to the total heat
absorbed in the reformer. Thermal histograms, which show a
detailed breakdown of the cooling and heating duties for each temperature increment, have been shown (5) to be of particular value
in investigating the extent to which an internal heat exchange
between these two duties can be accomplished.

A breakdown of the heating and cooling duties for the example
case, using a temperature increment of 20K, is shown in Figure 5.
The inherent mismatches in the temperature levels of the heating
and cooling duties are clearly visible in the figure. The
reformer plant shows a surplus of thermal energy in the temperature range of 875K-725K (as a result of excess sensible heat in
the reactor effluent) and at temperatures below 475K (as a result
of partial condensation). The plant also shows a huge deficit of
thermal energy at 525K due to the latent heat requirements in the
boiler (4). In order to improve the overall thermodynamic efficiency, it becomes necessary to trade thermal energy between the
reformer plant and the steam cycle of the power plant. The thermal deficit at 525K is provided by steam extraction from the power
plant in return for the thermal surplus, which is used by the
power plant for steam reheat and for feedwater preheat.

It is important to evaluate the net effect of these exchanges
on the total production of electricity by the power plant in order
to determine the true energy efficiency of the reformer plant. It
is also important to estimate the net effect of any changes in the
chemical plant design on the electricity production so that the
overall system efficiency can be improved. Exergy analyses of
typical steam power cycles have yielded a simple method for
estimating these effects without requiring a detailed power cycle
design for each and every design case for the chemical plant.
This simple procedure is based on the observation that steam power
plants generally produce an amount of electricity that is roughly
85% of the exergy input into the steam, regardless of whether this

---

* Numbers in parentheses refer to units of the reformer shown
  in Figure 4.

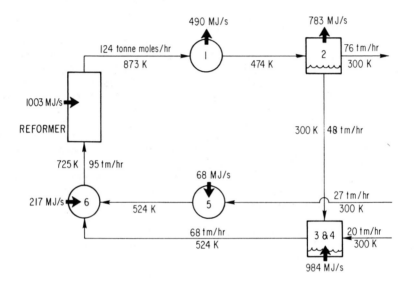

Figure 4. Reformer plant design. Key: 1, output cooler;
2, partial condenser; 3, feedwater heater; 4, boiler;
5, gas heater; 6, final gas heater.

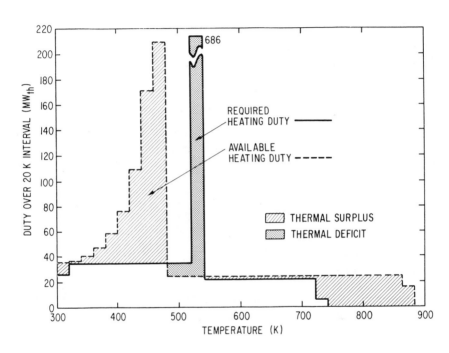

Figure 5.   Thermal histograms of reformer heat exchanges.

input is in the boiler, superheater, or feedwater heater (6); this is despite the fact that the first-law energy conversion efficiency varies significantly among these designs.

Following this procedure for the reference case design with a 20K temperature difference in heat exchangers, the net flow of exergy from the power plant to the reformer plant amounts to 65 MW. If the heat exchanger temperature difference is increased to 40K as a result of a reduction in heat transfer area, this net exergy flow increases to 112 MW. This simple calculation demonstrates the use of exergy exchange concepts to determine the quantitative effect on the power plant generation resulting from increased thermodynamic losses in the reformer plant.

Thermodynamic Losses in the Reformer Plant

A continuation of this application of the second-law analyses is an examination of the various irreversibilities in the reformer process for potential improvements. The chief sources of thermodynamic irreversibilities (with the associated exergy destruction) are (1) frictional losses, (2) heat transfer with a finite temperature difference, (3) chemical reaction far from equilibrium, and (4) diffusion.

The first two effects are calculated directly from the overall process design. The frictional losses are obtained from the compressor power requirements. They amount to a total exergy destruction of 10 MW. Heat exchanger losses due to heat transfer in Units 1-6 (excluding the reformer) can be evaluated from the exergy exchange calculations; they amount to 49 MW for a 20K temperature difference.

The various individual losses inside the reformer reactor are particularly difficult to evaluate. However, the total, integrated exergy destruction in the reactor can be calculated as a difference between the exergy extracted from the reactor helium and the exergy gained by the process stream. This overall exergy destruction in the reformer amounts to 43 MW. It should be noted that this is a particularly low amount, in comparison to other chemical reactors, due primarily to the efficient reformer tube design with counter-current flows and to the internal pigtail (1,4).

In addition to these three obvious sources of thermodynamic losses, the analysis also shows that a simple, innocuous process step leads to an exergy destruction comparable to that in the reactor or the heat exchangers; this step being the irreversible mixing of steam and methane-rich gases in preparing the reactor feed. Using the assumption of gas mixture ideality, the exergy destruction resulting from this mixing is calculated to be 40 MW. It is interesting to note that the efficiency penalty resulting from this irreversibility is not evident until it is necessary to recover the unreacted water molecules from the product stream. It then becomes clear that the high-temperature thermal energy invested in the boiler is recovered at considerably lower temperatures in the partial condenser. This is the primary cause of the net exergy flow from the power plant to the reformer plant that was calculated earlier.

Potential Improvements in the Reformer Process

From the foregoing analysis it is clear that a process modification that helps eliminate or minimize the irreversibility associated with the reactor feed preparation will lead to a major reduction in the thermal mismatch, reduce the exergy dependence on the power plant, and increase the overall energy efficiency. In the author's opinion, this conclusion would not be evident as readily without the thermodynamic analysis of process irreversibilities, which attests to the value of such exergy analyses.

One method for eliminating the irreversible mixing is to evaporate the water gradually, in intimate contact with the methane-rich gas stream. The net result would be a continuous humidifcation of the gas stream, with a minimum of irreversibilities due to diffusion or mixing. In such a mixed-feed-evaporator (MFE) the gas stream would be mixed with the liquid water and passed through heat exchanger tubes that are heated. Such two-phase heat exchanger designs are well within the current state-of-the-art (7).

The thermal histograms, once again, provide a convenient method to evaluate the modified exergy flows from the power plant. The changes in the heating duty resulting from the use of MFE are evident in Figure 6. The net impact of this design modification can be evaluated by re-calculating the exergy contents of the revised thermal exchanges between the reformer plant and the power plant. The results show that the net flow of exergy from the power plant is reduced (from the previously calculated value of 65 MW) to 24 MW for the same 20K temperature difference in the heat exchangers. This decrease results from a reduction in the amount and the temperature of extraction steam from the power plant; a large fraction of the condensation heat is now utilized within the reformer plant for evaporation. The net effect of the improvement is an increase in the electricity produced by the power plant of roughly 34 MW -- an amount sufficient to provide the total compressor power requirements for the reformer and methanation plants, as well as for the gas pipelines over a distance of 160 km.

SUMMARY

Second-law thermodynamic analyses have been shown to be of considerable value when applied to systems where an efficient energy interconversion is important. Using the chemical energy transport systems as examples, the use of exergy ratios as a measure of thermodynamic quality has been shown to give important insights into the efficiency or inefficiency inherent in any conversion of one energy form to another.

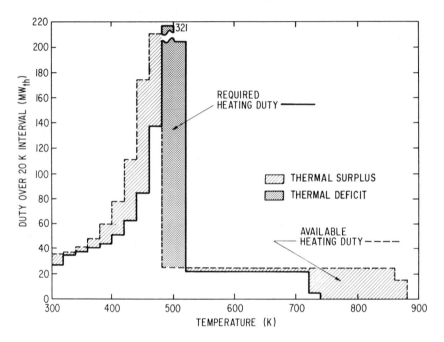

Figure 6.   Modified histograms with MFE.

In applications to chemical process design, the concepts of entropy production and exergy exchanges have been used to gain a better understanding of the overall energy efficiency and to suggest potential modifications to process design that would improve system efficiency.

It is hoped that these thermodynamic tools will prove useful in other applications as well. For example, the use of exergy ratios to evaluate the merits of an energy system could be applied to fuel interconversions for use in fuel cells or gas turbines. Similarly, the use of thermal histograms combined with exergy exchange analyses could prove useful in understanding and improving the design of an integrated chemical plant where energy exchanges among various processes play an important role.

ACKNOWLEDGMENT

The analyses presented in this report were performed as a part of a study of closed-loop chemical systems for energy storage and transmission that was funded by the U.S. Department of Energy.

NOMENCLATURE

Ex   Steady-flow exergy of extraction and delivery

H   Enthalpy

ΔH^o_{298} Standard enthalpy of reaction at 298K

n_j Moles of component j in a mixture

S   Entropy

T_o Ambient or dead-state temperature

α_E Exergy ratio

Energy efficiency (first-law efficiency)

$\mu_{J_{00}}$ Chemical potential of component J in the dead-state

Literature Cited

1. H. B. Vakil and J. W. Flock, "Closed Loop Chemical Systems for Energy Storage and Transmission," Final Report on Contract EY-76-C-02-2676 for U.S. Department of Energy, Report No. COO-2676-1, August 1978.
2. T. Bonn et al., "Nuklear Fernwarme und Nukleare Fernenergie," Kernforschungsanlage (KFA) Report Jul-1077, 1974.

3. R. W. Haywood, "A Critical Review of the Theorems of Thermodynamic Availability with Concise Formulations," J. of Mech. Eng. Sci., vol. 6, Nos. 3 and 4, 1974.
4. "High Temperature Reactor for Process Heat Applications," Nuclear Engineering and Design, vol. 34, No. 1, 1975.
5. H. B. Vakil and P. G. Kosky, "Design Analyses of a Methane-Based Chemical Heat Pipe," 11th Intersociety Energy Conversion Engineering Conference, Lake Tahoe, Calif., September 1976.
6. H. B. Vakil and J. W. Flock, "Closed Loop Chemical Systems for Energy Storage and Transmission," Final Report on Contract EY-76-C-02-2676 for U.S. Department of Energy, Report No. COO-2676-1, (Appendix II), August 1978.
7. H. B. Vakil, P. G. Kosky, U.S. Patent 4,345,915 (1982).

RECEIVED July 7, 1983

Thermodynamic Analysis of Gas Turbine Cycles with Chemical Reactions

H. B. VAKIL

Corporate Research and Development, General Electric Company, Schenectady, NY 12301

Thermodynamic exergy analyses of gas-turbine cycles show that the major losses occur neither during the compression of air nor during the expansion of hot combustion products, but rather during the combustion reactions. Main reasons for these losses stem from the peak temperature limitations imposed by the materials of construction, coupled with the very high thermodynamic quality of the fuel source.

This paper investigates the potential for reducing exergy losses during combustion by lowering the thermodynamic quality of the fuel through endothermic chemical reactions, utilizing the low-grade heat from the turbine exhaust gases. Using methanol as an example, it is shown that steam reforming or methanol cracking reactions could yield higher energy products of lower thermodynamic quality, with a subsequent reduction of entropy production during combustion.

Using the "second law" analyses, it is shown that the use of low-grade exhaust heat for the chemical conversion of high-quality fuel into medium quality gaseous products offers a higher overall efficiency than even a reversible conversion of the exhaust thermal energy to mechanical work. This apparent violation of thermodynamic laws can only be explained by taking into consideration the entropy production during the combustion step. These results suggest the possibility that lowering the exergy ratio of fuels through chemical reactions using low-grade heat could provide an easier technical route to higher efficiencies than the search for higher temperature materials.

0097–6156/83/0235–0105$06.00/0
© 1983 American Chemical Society

Clean fossil fuels in either liquid or gaseous form have played, and will continue to play, a major role in supplying our energy needs. The fact that the primary energy in the fossil fuels can be converted easily to either mechanical or electrical energy has led to their widespread use in the Industrial and Transportation sectors. While there are many different routes for converting the chemical energy of fossil fuels into useful work, almost all of them involve an initial conversion of the fuel energy into thermal energy through combustion; the thermal energy is then used in combination with a thermodynamic cycle to produce useful work. The gas turbine in an excellent example of such systems for power generation, in that it is simple, efficient, and relatively inexpensive. The basic gas-turbine cycle Figure 1 consists of three steps:

1. Ambient air is compressed (usually, adiabatically) to a pressure of roughly 10-12 bars in a compressor.

2. The compressed air is used to carry out combustion of the fossil fuel in a high temperature combustor to generate pressurized hot gases.

3. The hot combustion products are expanded in a turbine to generate power, a part of which is made available to the compressor.

The efficiency of a gas turbine is usually defined as the fraction of fuel combustion enthalpy change that is delivered as net mechanical (or electrical) energy after subtracting the energy requirements of the compressor. The continuing push for higher efficiencies has resulted in several modifications of the basic cycle described earlier; most of these involve the use of recuperative heat-exchange with the turbine exhaust. For example, the sensible heat of the hot exhaust gases can be used either to preheat the compressed air prior to combustion, or to generate steam in a boiler for additional power generation. More recently, a different use of the recuperated heat has been proposed -- one where the heat is supplied to carry out an endothermic chemical reaction with the fuel (1,2). The net result is the generation of a different, secondary fuel for combustion. An example of such chemical recuperation with methanol as the primary fossil fuel is the use of exhaust heat to carry out either the steam-methanol reforming reaction or the methanol cracking reaction.

It is the aim of this paper to present a comparison of thermal and chemical recuperation options in a thermodynamic framework. The paper will begin by identifying the major irreversibilities in a simple gas-turbine cycle with liquid methanol fuel; continue with a comparison of thermodynamic losses and overall efficiencies among various options utilizing thermal and/or chemical recuperation; and conclude with an examination of the thermodynamic "quality" of the primary and secondary fuels to understand the implications of fuel modification. It is not the purpose of this paper to evaluate the suitability of methanol as a fuel for gas turbines. Consequently, no attention will be given to such factors as the cost of methanol fuel, safety considerations of exchanging heat between hot exhaust gases and fuel, and the dynamics of the complex cycle with recuperative chemical reactions. The purpose of this paper is to outline the thermodynamic implications of chemical recuperation using methanol fuel as an example.

CYCLE CALCULATIONS

In order to simplify and standardize the cycle calculations, the following conditions were assumed:

Inlet air to compressor	1 atm; 77F
Compressor outlet air	12 atm; 650F
Combustor outlet/turbine inlet	12 atm; 2000F
Turbine exhaust	1 atm; 1000F
Regenerator outlet/stack gases	1 atm; 200F

The calculations of the energy flows are relatively straight forward once these conditions are specified; the same is true for the calculations of entropy production and the associated exergy destruction. The only area of complication is the need for a "dead-state" definition in order to calculate the enthalpy and exergy losses associated with the stack exhaust. The difficulties associated with the choice of an appropriate "dead-state" have been discussed extensively in the literature ($\underline{3}$). In the present context, however, the exact definition of the "dead-state" in general, and the chemical potential of CO_2 and H_2O in particular is important only for determining the exergy content of the stack exhaust. As will become apparent later, this is not a major factor in the overall efficiency. Consequently, it is not necessary to have a detailed and sophisticated description of the "dead-state"; any reasonable choice of reference state is adequate for the purpose of this paper. The resultant ambiguity in the exergy losses in the stack exhaust is a consequence of the effect of "dead-state" composition on the expansion work of CO_2 and H_2O -- a phenomenon of little interest here.

Thermodynamic properties were calculated using ideal gas mixture assumption and the standard reference states of 77F (298.15K) and 1 atm for air (molar composition - 80% N_2, 20% O_2*), for CO_2 (pure gas at 1 atm), and for H_2O (pure liquid at 1 atm). It

---

* Molar ratio of 4:1 was assumed in order to simplify the stoichiometry and thermodynamic calculations.

should be noted that in view of the assumed combustor outlet temperature of 2000F, the air/fuel ratio is determined directly from air and fuel conditions at the inlet of the combustor assuming complete combustion under adiabatic conditions.

IRREVERSIBILITIES IN A SIMPLE GAS-TURBINE CYCLE

The simplest example of a gas-turbine cycle is one in which no heat regeneration takes place; the compressor outlet air is sent directly to the combustor at a temperature of 650F and liquid methanol at ambient temperature is combusted to produce hot gases at a temperature of 2000F.

Thermodynamic properties of the flow streams at various locations as well as the thermal and mechanical energy flows for the cycle are shown in Figure 2 using 1 g-mole of liquid methanol as the basis. Also shown in the figure are the magnitudes of entropy production and exergy destruction in each of the process steps.

The overall cycle produces net turbine work (after accounting for the compressor work) of 253.5 Kj per g-mole of liquid methanol. Without any heat recovery from the turbine exhaust, the major energy loss (based on the first-law of thermodynamics) appears to be due to the hot stack gases. However, calculations of entropy production clearly show that the irreversibilities in the combustor give rise to the largest exergy loss in the entire cycle--an amount exceeding either the total net turbine work or the exergy loss resulting from throwing away the hot turbine exhaust gases. By comparison, the losses associated with compression/expansion inefficiencies are almost insignificant.

The combustion exergy loss is even more prominent if a thermal regenerator is used for extracting heat from the turbine exhaust in order to generate steam to form a combined gas turbine/steam cycle. The regenerator is capable of extracting 336 Kj of thermal energy from the exhaust gases with an exergy content of 154 Kj. Previous studies of steam cycle thermodynamic efficiency (4) have shown that roughly 80% of input exergy to the steam cycle can usually be made available as steam turbine work output. Thus, the net exergy loss associated with the exhaust gases is reduced to 30.8 Kj in the steam cycle and 9.5 Kj as unrecovered exergy* in the stack gases at 200F (366.5K) from the regenerator.

The main reason for the substantial irreversibility in the combustion process is the peak temperature limitation imposed by the materials of construction in the first stage of the gas-turbine, coupled with the high thermodynamic "quality" of the fuel combustion energy. One way to reduce this irreversibility is to develop turbines that can withstand higher inlet temperatures--a

---

* This amount will vary depending on the particular definition
  of the "dead-state" as discussed earlier.

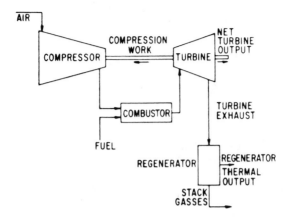

Figure 1.   Schematic of gas turbine cycle.

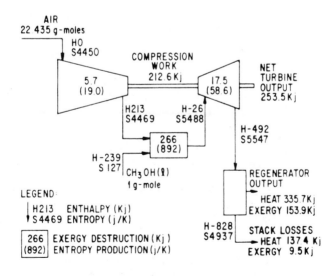

Figure 2.   Thermodynamics of the Base Case.

subject of considerable research interest to materials scientists and turbine designers. A totally different approach would be to manipulate the fuel itself in order to lower its "quality" and thereby make it more suitable thermodynamically to existing materials. The use of thermal and/or chemical recuperation are examples of the latter approach of reducing the "quality" of the fuel to be burned. It should be emphasized, however, that a lowering of the thermodynamic "quality" of the fuel is desirable only if the total amount of irreversibilities--those in the combustor as well as those inherent in the fuel processing step-- is reduced.

In order to clarify these ideas, we need to compare the irreversible entropy productions (or the exergy destruction) in cycles that utilize regenerative heating of compressed air, thermal recuperation in the form of evaporation and superheating of the methanol fuel, and chemical recuperation through either reforming or cracking reaction with methanol. The next section presents such a comparison in a simplified form to illustrate the utility of thermodynamic analyses.

ANALYSES AND COMPARISON OF CASES

In addition to the simple cycle without any form of recuperation (Base Case) that was analyzed in the previous section, we will consider five separate cases with thermal and/or chemical recuperation. In the interest of maintaining uniformity, we will assume a constant temperature of 800F as the exit temperature of any stream undergoing such recuperation. A description of the five cases is as follows:

Case 1   Regenerative heating of compressed air from 650F to 800F with liquid methanol at 77F as the fuel.

Case 2   Preheating of compressed air as in Case 1, combined with the evaporation and preheating of methanol to give methanol vapor at 800F as the fuel.

Case 3   Preheating of compressed air, combined with evaporation and preheating of an equimolar mixture of methanol and water to give $CH_3OH + H_2O$ vapor mixture at 800F as the fuel.

Case 4   Thermal recuperation as in Case 3 with the addition of catalytic reforming of methanol/steam mixture to give $CO_2 + 3H_2$ at 800F as the fuel.

Case 5   Thermal recuperation as in Case 2, with the addition of catalytic cracking of methanol to give $CO + 2H_2$ at 800F as the fuel.

The calculation procedure consists of a determination of the air/fuel molar ratio that will result in a combustor outlet temperature of 2000F, followed by the calculation of enthalpy and entropy of flow streams at the various state points in the cycle. The net work output from the turbine is calculated by subtracting the compressor work requirements from the total turbine expansion work determined from the enthalpy balance. Irreversible entropy productions are calculated by a routine entropy balance around each process step.

The total amount of regenerated heat for each of the cases is broken down into three separate thermal energy flows; the amount needed to preheat the compressed air (Q_a), the amount absorbed by the fuel during thermal or chemical recuperation (Q_f), and the surplus made available to the steam cycle (Q_{st}). The work output from the steam cycle is calculated from the net exergy flow to the steam cycle, after supplying the required exergy flows associated with Q_a and Q_f with an exergy-delivery efficiency of 80%. Earlier studies (5) have shown such 'second law' methods for estimating steam-cycle outputs to be more valid and useful than an assumption of a constant 'fist-law' conversion efficiency*. This is especially so when the steam cycle is coupled to different processes requiring heat at different temperature levels, as in this analysis.

The magnitudes of the mechanical and thermal energy flows for each of the cases are given in Table 1. Also shown in Table 1 are the values of the overall conversion efficiency, which is defined as the ratio of total work output ($w_T + w_{st}$) to the molar heat of combustion of liquid methanol (727 Kj).

The results indicate very clearly the progressive increases in efficiency with increasing degree of recuperation exemplified by the sequence: Base Case → Case 1 → Case 2 → Case 5. The most important point to note is that even with a reversible conversion of Q_{st}, the total work is lower with only thermal recuperation than that with chemical recuperation. For example, a comparison of Case 5 with Case 2 (the only difference being the presence or absence of the decomposition reaction) shows an increase in the net turbine output by 42.2 Kj—an amount nearly twice as large as the corresponding decrease in the steam cycle output. Even in the reversible limit, the decrease in steam cycle output would equal only 24 Kj. An analogous result is obtained by comparing Cases 3 and 4, thereby showing a similar behavior with the steam-reforming reaction. The fact that one can improve the overall efficiency by withdrawing some exergy from a reversible steam-cycle and carrying out irreversible fuel transformation with it appears, at a first glance, to be in violation of the second law. However, a more detailed look at the combustion irreversibilities shows that this

---

* This efficiency is generally referred to as the 'first law' efficiency.

Table 1. Mechanical and Thermal Energy Flows

	Moles Air per Mole CH_3OH	Thermal Flows (Kj)				Work Flows (Kj)			Efficiency
		Q_a	Q_f	Q_{st}	Q_{ex}	W_C	W_T	W_{st}	%
Base Case	22.44	-	-	335.9	137.4	212.8	253.7	123.2	51.83
Case 1	25.04	64.5	-	306.8	142.7	237.5	277.5	101.5	52.13
Case 2	27.83	71.7	61.8	275.7	148.3	263.9	303.0	91.7	54.29
Case 3	26.58	68.4	120.0	219.8	192.1	252.0	315.1	80.3	54.38
Case 4	29.47	75.9	184.1	187.5	197.9	279.5	341.6	71.4	56.80
Case 5	32.44	83.5	163.8	224.3	157.5	307.6	345.2	72.5	57.45

is not so, and that the efficiency gains result from large reduc-
tions in the entropy production during combustion of the derived
fuels.
    The magnitudes of entropy production during combustion and
the associated exergy losses for the various cases are shown in
Table 2 with two different normalizations: one based on 1 g-mole
of methanol, and the other based on 100 Kj of total net work (w_T +
w_{st}) produced. The progressive reduction in combustor irreversibilities with increasing amount of recuperation is evident from
the results. The decrease is even more prominent on the basis of
equal total work produced, which is the more realistic basis of
the two. The chemical recuperation using methanol cracking reaction appears to reduce combustor irreversibilities by nearly one-
third of that with no recuperation. In order to understand why
this should be so, we need a quantitative measure of the thermodynamic "quality" of fuels and of the required task of heating
compressed air from 800F to 2000F. In second law analyses of systems with thermal, mechanical, and chemical changes, we have found
the concept of exergy ratio - a ratio of exergy change to enthalpy
change of a given process - to be a particularly useful measure of
thermodynamic quality (4, 5). For a process with an enthalpy
change of H and an entropy change S,

$$\text{Exergy Ratio } (\alpha_E) \equiv 1 - T_0(\Delta S / \Delta H)$$

where T_0 is the ambient temperature (298.16K). It may be noted
that the exergy ratio of mechanical work is equal to unity, and
that of thermal energy at a temperature T is equal to the Carnot
factor $(1-T_0/T)$. For chemical reactions such as the combustion of
methanol, the exergy ratio depends on the relative magnitudes of
enthalpy and entropy changes; for exothermic reactions it is usually - but not always - in the range from zero to unity.
    The key energy transfer steps for which we need the exergy
ratios are:

● Compressed air heating from 650F to 800F

● Boosting the temperature of air from 800F to 2000F*

● Recovery of heat from turbine exhaust (1000F to 200F)

● Thermal and/or chemical recuperation for the fuel

● Idealized combustion of methanol and derived fuels.

---

* This is an approximation representing the idealized task of
  the combustor.

Table 2.   Entropy Production During Combustion
and Associated Exergy Losses

	Base Case	Example Cases				
		1	2	3	4	5
Basis: 1 g-mole Methanol						
Entropy Production (j/k)	892.5	864.1	788.4	768.3	686.2	672.6
Exergy Loss (kj)	266.1	257.6	235.1	229.1	204.6	200.6
Basis: 100 Kj Total Network						
Entropy Production (j/k)	236.9	228.0	199.8	194.4	166.2	161.1
Exergy Loss (Kj)	70.6	68.0	59.6	58.0	49.6	48.0

Exergy ratios for these steps are shown in Table 3. The average "quality" of the thermal exergy needed in the combustor is indicated by an exergy ratio of $\sim$0.7 (corresponding to a temperature level of roughly 1350F). By contrast, the "quality" of energy released by methanol combustion is extremely high as shown by an exergy ratio of 0.96 (i.e., equivalent to a temperature <14,000F). The implication of the limiting temperature of 2000F due to materials constraint is a significant degradation of the energy released by combustion; this explains the substantial entropy production in the combustor.

Thermal recuperation to a small extent, and chemical recuperation to a greater extent, can now be seen as a thermodynamically efficient way of reducing the "quality" of the fuel. This lowering of the exergy ratio is accomplished by utilizing regenerator heat at α_E = 0.3. In other words, chemical recuperation utilizes the excess thermodynamic potential of the fuel, which would otherwise be wasted, to upgrade the lower quality exhaust heat (with α_E = 0.3) into combustion quality heat (with α_E = 0.7). It does so by generating a secondary fuel with a theoretical delivery temperature of roughly 3000F. This thermodynamic "pumping" of exhaust heat explains why chemical recuperation leads to a greater overall efficiency than that for any other use of the regenerator heat.

One last point to be noted pertains to a comparison between the steam-reforming reaction (Case 4) and the methanol cracking reaction (Case 5). From the exergy ratio calculations, the reforming reaction appears to be superior in its ability to produce lower quality fuel. However, the overall efficiency calculations show a lower value for Case 4 than that for Case 5. The main reason for this reversal is due to the fact that nearly 25% of the recuperated energy for Case 4 is in the form of the heat of evaporation of H_2O and is not recovered from the exhaust gases. The result is an increase in the stack losses.

SUMMARY

A thermodynamic methodology for evaluating energy efficiency and entropy productions for gas-turbine cycles was presented. Using this approach, it was shown that the largest contribution to the overall irreversibilities of the cycle is from the fuel combustion step and is a direct consequence of the very high "quality" of the combustion energy and the relatively low limiting temperature of the materials of construction.

The use of regenerator heat from the exhaust gases to carryout endothermic reactions with the methanol fuel could lead to substantially lower entropy production in the combustor. The resultant increase in the overall efficiency was shown to be greater than that for any other use of the exhaust heat.

The option of lowering the exergy ratio of fuels through

Table 3. Exergy Ratios for Key Energy Transfer Steps

	Exergy Ratio (α_E)
Heating of compressed air (650-800F)	0.541
Heating of compressed air (800-2000F)	0.703
Heat Receovery from turbine exhaust (Base Case - 1000F to 200F)	0.458
Recuperation steps for the fuel	
Case 2	0.354
Case 3	0.237
Case 4	0.26
Case 5	0.342
Fuel combustion with air (Products at 298K, 1 atm)	
Case 1 - CH_3OH liquid at 298K	0.965
Case 2 - CH_3CH vapor at 800F, 12 atm	0.915
Case 3 - CH_3OH + H_2O vapor at 800F, 12 atm	0.862
Case 4 - CO_2 + $3H_2$ vapor at 800F, 12 atm	0.823
Case 5 - CO + $2H_2$ vapor at 800F, 12 atm	0.850

chemical reactions using low-grade heat could provide an easier technical route to higher efficiencies than the search for higher temperature materials.

Literature Cited

1. J. H. Omstead and P. S. Grimes, "Heat Efficiency Improvements Through Chemical Recovery of Waste Heat", IECEC #729046 (1972).
2. C. W. Janes, "Increasing Gas Turbine Efficiency Through the Use of a Waste Heat Methanol Reactor", IECEC #799423 (1979).
3. "Second Law Analysis of Energy Devices and Processes", Energy, vol. 5, #8-9 (1980).
4. H. B. Vakil and J. W. Flock, "Closed Loop Chemical Systems for Energy Storage and Transmission", Final Report on contract EY-76-C-02-2676 for U.S. Department of Energy, Report No. COO-2676-1.
5. H. B. Vakil, "Thermodynamic Analysis of Chemical Energy Transport", Presented at the AIChE Annual Meeting (Chicago, 1980); in ACS Symposium Volume on "Thermodynamics: 2nd Law Analysis" (to be published in 1983).

RECEIVED July 7, 1983

Available Energy Analysis of a Sulfuric Acid Plant

K. RAVINDRANATH and S. THIYAGARAJAN

Larsen & Toubro Limited, Bombay, India

Available energy concept is applied to
analyse a sulphuric acid plant. First
law and second law analyses are compared.
Second law analysis pin points available
energy consumptions and losses. Possible
improvements by reducing availability
consumptions and losses are presented.

Sulphuric acid plant is well known to produce along
with sulphuric acid an equivalent amount of steam.
The potential energy available with the basic raw
material sulphur calls for a high degree evaluation of
energy recovery in sulphuric acid plants.
   The efficiency of energy conversion and utilisation
in this process cannot be evaluated based on first law
of thermodynamics alone and true energy dissipations
can be brought out by using available energy analysis.
   Second law analysis is applied to a 100 tonnes per
day double-contact double-absorption (DC-DA) sulphuric
acid plant in order to bring out true energy conversion
efficiencies and consumptions based on work
availability of various streams. Second law efficiencies are compared with those of first law to pinpoint
true losses and inevitable consumptions in energy
conversion processes.
   Based on second law analysis alternatives are
worked out for improving overall energy conversion
efficiency by recovering thermal energy in acid coolers
for power generation and for preheating boiler feed
water.

Sulphuric Acid System

A block diagram of a typical DC-DA sulphuric acid

0097–6156/83/0235–0119$06.00/0

plant consisting of seven sections is shown in Figure
1. For carrying out total energy analysis of this plant
these seven sections are grouped into five major
blocks.

1. Sulphur preparation and combustion: Melting of
    solid sulphur to liquid, conversion of molten
    sulphur to SO_2 gas by using dry air and partial
    recovery of the combustion heat in a waste heat
    boiler.
2. Conversion I: First pass conversion of SO_2 to SO_3
    in three beds of V_2O_5 catalyst and intermediate
    heat removal in waste heat boiler, economiser and
    heat exchangers.
3. Drying & Absorption I: Moisture removal from
    process air by drying with sulphuric acid and SO_3
    removal in intermediate absorption tower after
    first pass conversion.
4. Conversion II: Second pass conversion of SO_2 to
    SO_3 after reheating the gases from intermediate
    absorber and subsequent heat removal in a heat
    exchanger.
5. Absorption II: SO_3 removal in final absorption
    tower before venting out inert gas through stack and
    acid cooling in various cascade coolers.

First Law Analysis

Input and output enthalpies of various streams across
each section for a 100 TPD DC-DA sulphuric acid plant
with 10% sulphur dioxide feed to converter, 99.8%
conversion efficiency and 99.9% absorption efficiency
are shown in Table I.
    If the efficiency is calculated based on thermal
energy entering and leaving each system, it works out
to be 94 to 98% for all sections, accounting for 2-6%
heat losses. On the other hand, if only net useful
energy from each system is considered it works out to
be 89 to 96% for sulphur preparation, combustion and
conversion sections (1, 2 and 4) and 5.9 to 0.1% for
drying and absorption sections (3 and 5). This denotes
that the energy efficiencies are at alarmingly low
levels for sections 3 & 5. Overall efficiency based on
net useful output is only 38.7%. Out of total thermal
losses of 61.3%, losses in warm water are as high as
52% based on first law analysis.

Second Law Analysis

The degradation in the quality of energy as it moves

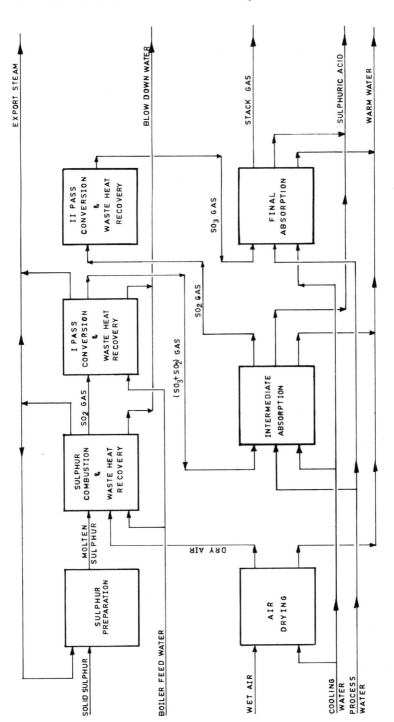

Figure 1. Block diagram of DC-DA Sulphuric acid plant.

Table I. Enthalpy Balance of A 100 TPD Sulphuric
         Acid Plant

Enthalpy values in KJ/hr ('000)   Datum Temp.:25°C

Input stream	Enthalpy	Output stream	Enthalpy
1. Sulphur preparation & Combustion:			
DM water	201.0	Steam export	7957.0
Dry air	268.5	Hot SO_2 gas	5234.0
Boiler I Water	3728.8	Deaerated water	1742.0
Reaction heat	12624.0	Heat losses	1889.3
Total	16822.3		16822.3
		Net useful	14933.0
		Efficiency % -	88.8
2. Conversion I			
Hot SO_2 gas	5234.0	SO_3 gas	2205.0
Dilution air	26.6	Boiler I water	3728.8
Deaerated water	1742.0	Steam export	3456.0
Cold SO_2 gas	454.0	Hot SO_2 gas	1632.6
Reaction heat	4055.0	Heat losses	489.2
Total	11511.6		11511.6
		Net useful	11022.4
		Efficiency % -	95.8
3. Drying & Absorption I			
SO_3 gas	2205.0	Cold SO_2 gas	454.0
Wet air	116.5	Dry air	292.0
Cooling water	5182.5	Sulphuric acid	89.3
Reaction heat	6638.0	Warm water	13306.7
Total	14142.0		14142.0
		Net useful	835.3
		Efficiency % -	5.9
4. Conversion II			
SO_2 gas	1632.6	SO_3 gas	1688.0
Reaction heat	118.0	Heat losses	62.6
Total	1750.6		1750.6
		Net useful	1688.0
		Efficiency % -	96.4
5. Absorption II			
SO_3 gas	1688.0	Sulphuric acid	2.6
Cooling water	751.6	Stack gas	451.3
Reaction heat	155.0	Warm water	2140.7
Total	2594.6		2594.6
		Net useful	2.6
		Efficiency % -	0.1
Net useful output	11505.0	(38.7%)	
Warm water loss	15447.4	(52.0%)	
Stack gas loss	451.3	(1.5%)	
Heat losses	2321.3	(7.8%)	

through various process units is assessed by calculating total available energy of input and output streams
of each system. Total availability constitutes basically chemical, thermal, pressure availabilities and
electric energy. The basis for detailed available
energy calculations of a sulphuric acid plant is given
in Table II.(1,2).
The available energy flow through five major sections of sulphuric acid plant is given in Figure 2. The
major inputs to this system are sulphur and power,
with demineralised (DM) water, wet air, process water
and cooling water from environment. The useful outputs
from the system are sulphuric acid and steam. Losses
to environment include heat losses from various
equipments,blowdown water, steam from deaerator vent,
warm water and stack gas.
Process information of various streams (streams 1
to 16) and their chemical, thermal and pressure
availabilities are given in Table III. Power inputs to
the system (streams 17 to 20) and availability losses
(streams 21 to 29) are given in Table IV. The streams
marked as inputs from environment in Figure 2 are
considered to have zero availability.
For each input and output stream of all sections,
availability is calculated. The difference between the
output availability of the total product (including
the losses) and the input availability is considered
as availability consumed in the process in order to
effect the conversion process. The ratio of availability of useful product to total input availability is
considered as effectiveness of the system. For each of
the five sections considered in available energy flow
diagram, the availability of input and output streams,
losses, consumption and effectiveness are calculated.
The five major sections are broken down further into a
number of components to pinpoint the areas of significant availability consumptions. These results are given
in Table V.

Results And Discussion

From the availability analysis, overall effectiveness of
a sulphuric acid plant works out to be 49% compared to
overall efficiency of 39% based on first law. Also
effectiveness in sections 2 and 4 (first and second
pass conversion) is as high as 86-92% thus leaving less
prospects for improvements in these sections. In
section 1 (sulphur preparation and combustion), the
effectiveness is 75% and the availability consumption

Table II.Basis for Availability Calculations

---

1. Chemical availability (a_c)

Atmospheric constituents	Equilibrium mole fraction	Stable condensed phases at standard conditions of $298^{\circ}K$ & 1 atm.
N_2	0.7567	H_2O
O_2	0.2035	$CaSO_4 \cdot 2H_2O$
H_2O	0.0303	$CaCO_3$
CO_2	0.0003	

Formulas for a_c in kcals/g.mole

a_S (solid)   = 191.09

$a_{SO_2}(g)$ = $122.04 + RT_o \ln X_{SO_2}$

$a_{SO_3}(g)$ = $105.605 + RT_o \ln X_{SO_3}$

$a_{H_2SO_4}(l)$ = 74.40

a_{H_2O} (g)   = $2.0717 + RT_o \ln X_{H_2O}$

a_{N_2} (g)   = $0.16518 + RT_o \ln X_{N_2}$

a_{O_2} (g)   = $0.94328 + RT_o \ln X_{O_2}$

2. Thermal availability (a_T kcal/g.mole)

$$a_T = (A - BT_o)(T-T_o) + (B/2 - \frac{CT_o}{2})(T^2-T_o^2) + C/3(T^3-T_o^3)$$

$$- AT_o \ln \frac{T}{T_o} + \frac{DT_o}{2}(\frac{1}{T^2} - \frac{1}{T_o^2}) - D(\frac{1}{T} - \frac{1}{T_o})$$

	A	B	$C \times 10^6$	$D \times 10^{-6}$
SO_2	7.70	0.0053	-0.83	0
SO_3	13.70	0.00642	0	-0.312
O_2	8.27	0.00258	0	-0.1877
N_2	6.5	0.001	0	0

3. Pressure availability (a_p)

$$a_p = RT_o \ln \left\{ \frac{P_{gas}}{P_o} \right\} \text{ kcals/g.mole}$$

Table III. Process Conditions and Availabilities of Various Streams   Basis: 1 hr

Stream no.	2	3	4	5	6	7	8	9
Description	SO2 from Boiler I	SO3 to Int.abs.	Gas from Int.abs.	Gas from final ht.ex.	Gas to IV bed	SO3 to final abs.	Combustion Air	Dilution Air
Flow kg	12329	13424	10108	10108	10108	10108	10963	1095
Temp. °C	470	200	70	315	425	190	50	50
Pressure Pascals	124577	113510	111381	107226	106104	1027767	128406	128408
Gas Comp. Mole %								
SO_2	11.0	0.32	0.36	0.36	0.36	0.02	-	-
SO_3	-	10.41	-	-	-	0.34	-	-
O_2	9.9	6.26	6.96	6.98	6.98	6.82	21.1	21.1
N_2	79.1	83.01	92.66	92.66	92.6	92.82	78.9	78.9
Availabilities '000 KJ								
A Pressure	194.5	112.00	83	49	40	13.2	223.3	23.5
A Thermal	2066.5	479	31	916	1550	348.5	10.7	1.1
A Chemical	21524	18698	736	736	736	653.5	38.5	3.9
A Total	23785	19289	840	1701	2326	1015	272.5	28.5

Stream no.	10	11	12	13	14	15	16	1
Description	Deaerated water	Feed water to Boiler I	Export steam Boiler I	Export steam Boiler II	Steam consumption	Acid from Int.abs.	Acid from final abs.	Raw sulphur
Flow kg	5413.3	4056	2963	1285.6	1007	4052.9	113.75	1365
Temp. °C	102	214	214	214	214	40	40	25
Pressure Pascals	101348	2026960	2026960	2026960	2026960	101348	101348	-
Availability '000 KJ	194	958	2720	1180	924	12837	360	34122

Table IV. Power Inputs and Availability Losses

Basis : 1 hr

Description	Stream no.	Power input KJ('000)	Availability losses KJ ('000)
1. Sulphur prepara-tion & combustion			
Power input	17	108	–
Heat losses	21	–	361.5
Boiler blow down	22	–	21
Deaerator	23	–	21
2. Conversion I			
Power input	18	7.2	–
Heat losses	24	–	83.5
Boiler blow down	25	–	7
3. Drying & Absorption I			
Power input	19	108	–
Warm water @ 45°C	26	–	473
4. Conversion II			
Heat losses	27	–	10
5. Absorption II			
Power input	20	36	–
Stack gas @ 70°C	28	–	172
Warm water @ 45°C	29	–	78
6. Air Blower	–	540	–

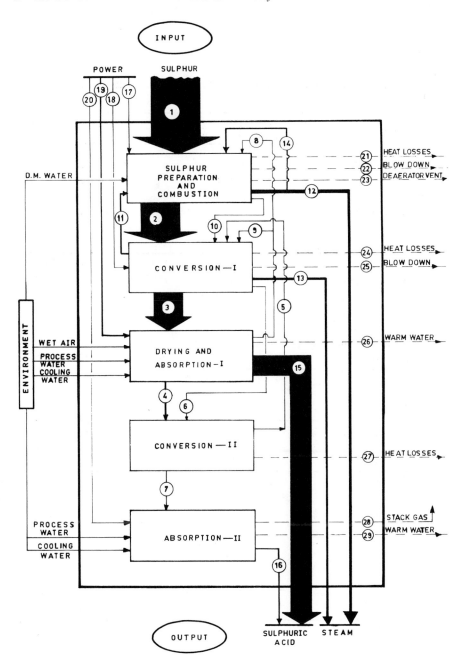

Figure 2. Available energy flow diagram of
Sulphuric acid plant.

Table V. Summary of Results from Availability Analysis

Basis: 1 hr

Section	Availability KJ ('000)				Effect-ivene-ss %	A_{loss} %	A_{cons} %
	A_{in}	A_{out}	A_{loss}	A_{Cons}			
1. Sulphur preparation and combustion	35461	26699	404	8358	75.29	1.14	23.57
Sulphur pits	34399	34147	252	–	99.27	0.73	–
Combustion furnace	34420	28768	–	5652	83.58	–	16.42
Waste heat boiler	29725	27430	131	2164	92.28	0.44	7.28
Deaerator	–	–	21	157	–	–	–
Pressure reducers	–	–	–	277	–	–	–
2. Conversion I	25739	23753	91	1896	92.28	0.23	7.1
First catalyst bed	23814	23342	–	472	98.02	–	1.98
Waste heat boiler	23577	23080	43	455	97.9	0.17	1.93
Second catalyst bed	21899	21792	–	107	99.5	–	0.5
Third catalyst bed	21048	20985	–	63	99.7	–	0.3
Intermediate heat ex.	23494	23373	15	107	99.5	0.05	0.45
Economiser	21179	20453	34	693	96	0.16	3.24
3. Drying & Absorption I	19709	13978	407	5324	70.92	2.07	27.01
Towers	–	–	–	4445	–	–	–
Coolers	–	–	407	761	–	–	–
4. Conversion II	3166	2717	53	396	85.8	1.7	12.5
Fourth catalyst bed	2326	2293	–	33	98.6	–	1.4
Final heat ex.	3143	2727	53	364	86.75	1.69	11.56
5. Absorption II	1051	360	247	444	34.2	23.5	42.3
Final absorber	–	–	172	274	–	–	–
Acid cooler	–	–	75	134	–	–	–
Sulphuric Acid Plant Overall System	34922	17098	1201	16623	48.96	3.44	47.60
Overall losses							
– Warm water	–	–	482	–	–	1.38	–
– Stack	–	–	172	–	–	0.5	–
– Heat losses	–	–	547	–	–	1.57	–

is about 23%. This high consumption can be related to
inherent dissipation during the conversion processes of
combustion of sulphur to SO_2 and heat transfer from
high temperature combustion products to steam. In
section 3 (drying and intermediate absorption) the
effectiveness is 71% and the consumption is about 27%.
This dissipation is in the conversion process of
gaseous sulphur trioxide to sulphuric acid. Similarly,
in section 5 (final absorption), the effectiveness is
34%, consumption is 42% and losses in acid cooler is
about 23%.
  First law analysis projects losses in sections 3
and 5 (drying and absorption) as 94.1% and 99.9%
whereas second law analysis yields 2% and 23.5%
respectively. Losses in warm water and stack gas are
insignificant in the range of 1.4% and 0.5% based on
second law analysis compared to 52% and 1.5% as
projected by first law analysis. Overall availability
analysis of a sulphuric acid plant clearly brings out
the fact that availability losses are only 4% and
what are hitherto considered as heavy losses in acid
coolers (52% as given by first law analysis) are quite
insignificant.

Areas of Improvement

Areas for improvement in effectiveness of availability
exist where large consumptions and losses occur.
Availability consumptions could be altered only by
changing basic process route whereas availability
losses could be reduced by suitable recovery processes.

Availability Consumptions: From Table V, it can be seen
that out of total availability consumption of 16.6
million KJ per hour two major areas of consumptions
are 50% in section 1 and 32% in section 3.

1. In section 1, 34% of total availability is
    consumed in combustion furnace alone where high
    level chemical availability is degraded to thermal
    availability. One way of reducing consumption in
    furnace is to increase SO_2 concentration (if
    necessary with oxygen enriched air). By increasing
    SO_2 concentration to 14.5%, availability consump-
    tion can be reduced by 0.7 million KJ/hr.
2. Another area of consumption is waste heat boiler
    (13%). This could be reduced by increasing steam
    pressure and superheat. By raising steam pressure
    to 60 kg/cm^2 and temperature to 600°C, availabil-
    ity consumption can be reduced by 0.8 million KJ/hr.

3. By incorporating a gas turbine before waste heat
    boiler, availability consumption can further be
    reduced.(6).
4. Availability consumptions in section 3 (absorption)
    are inherent in the formation of sulphuric acid
    from SO_3 gases. These cannot be eliminated complet-
    ely by any process route but part of consumption
    can be transferred to cooling water by hot
    cocurrent absorption system.

    In this system acid at 80^0C and gases at 200^0C
    enter cocurrently and leave the intermediate
    absorption tower at 110^0C. This helps in increasing
    temperature of warm water from acid coolers to 75^0C
    thus reducing availability consumption in acid
    coolers by 0.315 million KJ per hour.

Availability Losses: Major availability losses are heat
losses, stack losses and availability lost in warm
water. While it is not possible to reduce heat and
stack losses below the existing level, it is possible
to recover part of the availability from warm water by
either power generation using Rankine cycle system or
alternatively for preheating boiler feed water.

Power Generation: Rankine cycle approach can be used
here by vaporisation of a low boiling organic liquid
fluorocarbon or hydrocarbon using heat in acid coolers
and vapor is made to drive a turbine after which it is
condensed and recirculated to receive more heat. This
scheme is depicted in Figure 3 using propane as the
low boiling liquid.

Availability inputs and outputs of each unit and
the corresponding availability consumptions are given
in Figure 3. Net output from Rankine cycle system is
144 kw   (187-43). Availability analysis of this system
indicates an overall effectiveness of 62% with 30%
consumption and 8% losses.

By incorporating Rankine cycle system in a
sulphuric acid plant, the effectiveness of section 3
will improve to 73.1% from 70.9% and section 5 to 42%
from 34.2%. Overall effectiveness could be improved to
50.4% from 48.96%.

Boiler Feed Water Preheating: By preheating 5413 kg/hr
of boiler feed water, required from 100 TPD sulphuric
acid plant, to 68^0C using part of thermal energy
available in acid coolers, 312 kg/hr of additional
steam can be generated. This results in an additional
availability output of 0.287×10^6 KJ/hr. By this,

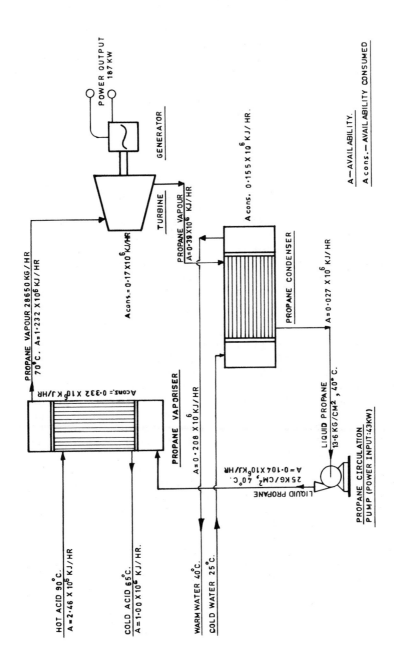

Figure 3. Available energy recovery by Rankine cycle system.

overall effectiveness could be improved to 49.80% from 48.96%.

By completely recovering the thermal energy available in acid coolers, 64,400 kg/hr of boiler feed water can be heated to 68°C. Out of this, only 5413 kg/hr is required in sulphuric acid plant and the rest can be sent to central boiler house in the complex. Availability output from sulphuric acid plant will increase by $(0.287 + 0.69)$ 10^6 KJ/hr, resulting in an overall effectiveness of 51.5%.

Summary

1. From second law analysis, overall effectiveness of a sulphuric acid plant is found to be 49% compared to overall efficiency of 38.7% based on first law.
2. Energy analysis based on first law of thermodynamics indicates that there are enormous heat losses (about 52%) in absorption and drying sections. But analysis based on second law shows that these losses (about 1.4%) are relatively less.
3. Energy consumptions and losses in each section of the process are calculated by second law analysis to pinpoint areas of significant availability consumptions and losses.
4. The major availability consumptions are in combustion furnace, waste heat boiler and absorption sections. It is not possible to eliminate these consumptions in sulphuric acid plant for the process system being followed presently. However marginal improvements can be achieved by high temperature combustion system, use of gas turbines, generation of high pressure superheated steam and cocurrent absorption system.
5. Even though availability losses are relatively insignificant compared to consumptions, recovery of this availability from acid coolers in the form of power generation and boiler feed water preheating is quite economical.

The potential energy of basic raw material sulphur makes sulphuric acid plants attractive even from energy point of view. Second law analysis of this plant gives an insight into energy transformation, quality degradation and pinpoints true efficiencies, inherent consumptions of conversion process and losses and helps in realising potential areas for improvements.

Literature Cited

1. Gaggioli, R.A.; Petit, P.J. Chemtech, 1977,
   496-505.
2. Sussman, M.V. 'Availability (Exergy) Analysis',
   Tufts University, 1980.
3. Reistad, G.M.; Gaggioli, R.A.; Obert, E.F.
   Proc. American Power Conference, 1970, 32, 603-11.
4. Gaggioli, R.A.; Fehring, T.H. Combustion, 1978,
   35-9.
5. Gaggioli, R.A.; Jae-Joon Yoon; Patulski, S.A.;
   Latus, A.J.; Obert, E.F. Proc. American Power
   Conference, 1975, 37, 656-69.
6. Harman, R.T.S.; Williamson, A.G. Applied Energy,
   1977, 3, 23-40.

RECEIVED July 7, 1983

Exergetic Analysis of Energy Conversion Processes
Coal Hydrogasification

KARL F. KNOCHE and G. TSATSARONIS

Lehrstuhl für Technische Thermodynamik, RWTH Aachen, D-5100 Aachen, Federal Republic of Germany

Energy conversion processes become increasingly important as oil and natural gas production decrease. Coal conversion processes are most important as future alternatives for liquid and gaseous fuels. These processes are rather complicated chemical plants with a great number of different reactors and separation units. Even for experts it is very difficult to estimate the influence of the existing irreversibilities on the overall energy conversion efficiency. Second law analysis is a very powerful tool in order to localize such irreversibilities and to improve the overall flow chart.

Second Law Analysis for Steady State Flow Processes

Under steady state conditions mass and energy flow continuously through any unit of a plant. The first law of thermodynamics yields for unit i, Figure 1

$$\dot{Q}_i + P_{ti} = - \sum_j \dot{n}_{ij} H_{mij} \qquad (1)$$

where $\dot{Q}_i$ is the heat flow exchanged with the unit, P_{ti} the necessary electrical or mechanical energy, $\dot{n}_{ij}$ the molar flow (stream j) and H_{mij} its molar enthalpy. The heat $\dot{Q}_i$ is considered to be exchanged with the surroundings at constant temperature T_o only. This is

0097-6156/83/0235-0135$06.00/0

not a restriction, because all internal heat exchange is linked to an appropriate mass flow included in the $\dot{n}_{ij}$. From the second law we have

$$\frac{\dot{Q}_i}{T_o} + \dot{S}_{pr,i} = -\sum_j \dot{n}_{ji} \, S_{mij} \tag{2}$$

and with equation (1)

$$P_{ti} = T_o \dot{S}_{pr,i} - \sum_j \dot{n}_{ij} \, (H_{mij} - T_o \, S_{mij}) \tag{3}$$

Considering the total plant, the summation over all units results in

$$\sum_i P_{ti} = T_o \sum_i \dot{S}_{pr,i} - \sum_i \sum_j \dot{n}_{ij} \, (H_{mij} - T_o \, S_{mij}) \tag{4}$$

In this equation

$$P_t = \sum_i P_{ti} \tag{5}$$

represents the total electrical energy consumed by (or delivered from) the plant and

$$\dot{S}_{pr} = \sum_i \dot{S}_{pr,i} \tag{6}$$

the total entropy production.

From all the internal flows between the different units only a few - the raw material flows $\dot{n}_{R1}, \ldots \dot{n}_{Rm}$ and the product streams $\dot{n}_{P1}, \ldots \dot{n}_{Pn}$ - enter respectively leave the total plant.

$$\sum_{k=1}^{m} \dot{n}_{Rk} - \sum_{l=1}^{n} \dot{n}_{Pl} = \sum_{i} \sum_{j} \dot{n}_{ij} \qquad (7)$$

The product streams include valuable materials as well as useless flows of exhaust gases and other wastes, which may have to be penalized for environmental degradation.
Also the right hand side of Equation 4 includes various flows leaving a certain unit and entering the next. Therefore according to Equations 4-7 the second law analysis of the total plant leads to

$$P_t + \sum_{k=1}^{m} \dot{n}_{R,k} (H_{mk} - T_o S_{mk})$$

$$= \sum_{l=1}^{n} \dot{n}_{p,l} (H_{ml} - T_o S_{ml}) + T_o \cdot \sum_{i} \dot{S}_{pr,i} \qquad (8)$$

Exergetic analysis of a coal hydrogasification process as an example

The coal hydrogasification process discussed below was developed within the scope of a cooperation agreement on the development of processes for the conversion of solid fossil fuels by means of heat from high-temperature nuclear reactors, agreed among five german companies (Bergbau-Forschung GmbH, Gesellschaft für Hochtemperaturreaktor-Technik mbH, Hochtemperatur-Reaktorbau GmbH, Kernforschungsanlage Jülich GmbH and Rheinische Braunkohlenwerke AG) (1).
The total plant of this process can be divided into three plants: The nuclear heat production plant, Figure 2 , the brown coal hydrogasification plant, Figure 3, and the steam power plant, Figure 2.

The nuclear heat production plant consists of the high-temperature gas cooled nuclear reactor as heat source as well as the steam reformer and the steam ge-

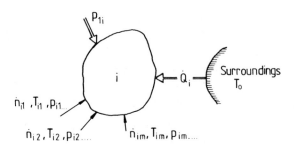

Figure 1. Mass and energy balance of the i-th unit.

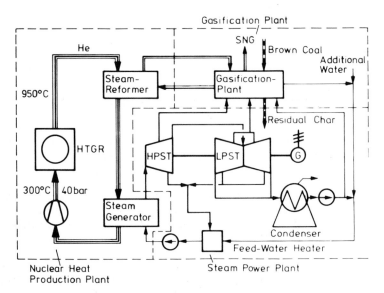

Figure 2. Simplified flow diagram of the nuclear heat production plant and the steam power plant. For the hydrogasification plant see Figure 3.

nerator in which the helium heat can be transferred into the gas and the steam process, respectively. The capacity of the total plant is fixed by the thermal power of the nuclear reactor of 3ooo MW. In the steam power plant, electrical energy and steam are produced in order to cover the demands of the hydrogasification plant for electrical energy, high-pressure steam (steam reforming) and low-pressure steam (coal drying).

The feed coal, Figure 3, is milled and dried to a residual moisture content of about 1o %, before it is introduced into the hydrogasification unit (unit 1). In the fluidized bed gasification reactor, the pressure amounts to about 8o bar and the temperature to 92o $^\circ$C. After cooling, the crude gas of the hydrogasification process is cleaned first with water (unit 4) and subsequently in an amisol unit (unit 12) where the components H_2S and CO_2 are removed. In the stretford unit (unit 13) elementary sulphur is obtained from the sulphur containing components. In the low-temperature separation unit (units 14-15), the pure gas is separated into four different fractions; the two methane fractions, the carbon monoxide, the nitrogen and the hydrogen fractions. The SNG product is compressed to 7o bar and given to the user, while about 4o % of the methane produced in the gasifier is mixed with high-pressure steam coming from the steam power plant. The steam/methane mole fraction at the inlet of the steam reformer (units 2o-21) must be equal to about 3-4 in order to avoid soot formation on the catalysts and in the following units. However, from the exergetic point of view a smaller mole fraction is desirable. In the shift conversion units (33-54 and 24), carbon monoxide reacts with steam to hydrogen and carbon dioxide. The product gas is cleaned in the gas cleaning units 55 and 29. Thus, the hydrogen fed into the gasifier comes from the low-temperature separation unit (this is more than 5o%) and from the pure gas coming from the two gas cleaning units 29 and 55. A significant part of the sensible heat of the crude gas (unit 3) and of the reformed gas (units 23 and 25) is used in order to generate low-pressure steam for the coal drying. The remaining low-pressure steam required for the coal drying comes from the steam power plant.

The detailed mass, energy and exergy balance for this complex thermodynamic system was performed with the aid of a set of computer programs named THESIS (Thermodynamic and Economic Simulation System), which was developed at the Lehrstuhl für Technische Thermodynamik of the Technical University Aachen for the ther-

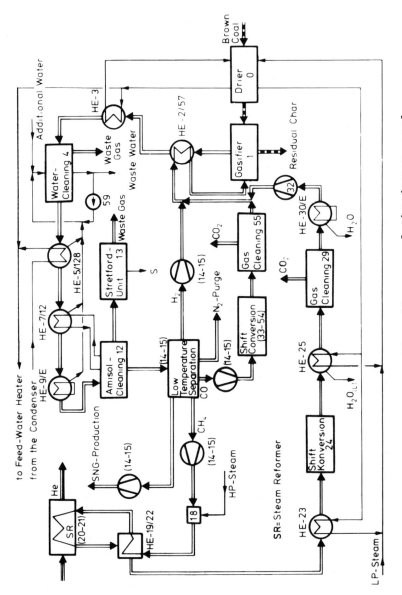

Figure 3. Process flow diagram of the brown coal hydrogasification plant.

moeconomic analysis of chemical plants and energy conversion systems (2-8). The most important results of
the exergy analysis of this process are given in Figures
4 and 5. For the calculation, a raw brown coal throuput
of 2315 t/h and a carbon gasification ratio in the
gasifier (unit 1) of 65 % were assumed. The steam/me-
thane mole fraction at the steam reformer inlet was
equal to 4. These assumptions yield a thermal power
of the nuclear reactor of about 6oo MW transferred
from the helium stream to the steam reformer gas, whereas the rest of the nuclear energy is used for steam
generation. The residual char production amounts to
261.8 t/h. The total exergy losses of the hydrogasification plant and of the steam power plant amount to
182o MW, Figure 4. About 73 % of them result from the
gasification plant and the remaining exergy losses
result from the steam power plant.

The highest exergy losses of the hydrogasification
plant appear in the gasifier (416.8 MW and in the low-
temperature separation unit (198.4 MW). The considerable
exergy losses in the gasifier are due to the irreversibility of the chemical reaction, to the heating of the
reactants and to the vaporization of the water contained
in the coal.

In order to judge the quality of the heat exchanger
i from exergetic point of view, we calculate the ratio
ε_i

$$\varepsilon_i = \frac{\dot{E}_{L,i}}{\dot{Q}_i}$$

where $\dot{E}_{L,i} = T_o \dot{S}_{pr,i}$ and $\dot{Q}_i$ denote the exergy losses
and interchanged heat in the i-th exchanger. The heat
exchangers 23, 3 and 2/57 as well as the steam generator (SG) are characterized by a disadvantageous ratio
ε_i : ε_{23} = o.296, ε_3 = o.226, $\varepsilon_{2/57}$ = o.185, ε_{SG}=o.135.
This is due to the high temperature differences between
the flows in the heat exchangers.

The exergy analysis of the considered process leads
to the following conclusions concerning a process improvement:
1. A part of the sensible heat of the crude gas and the
   reformed gas can be used in order to produce high-
   pressure steam for a steam power plant instead of
   low-pressure steam for coal drying.
2. More helium heat should be coupled in the steam reformer. In that case we were able to gasify more
   brown coal, because more hydrogen would be produced
   in the steam reformer, and in parallel we would reduce the irreversibilities in the steam generator

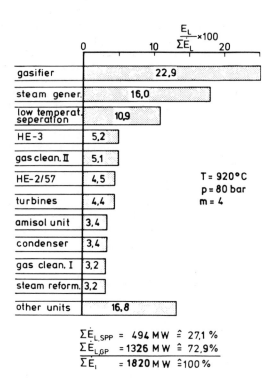

$$\Sigma\dot{E}_{L,SPP} = 494\ MW \cong 27{,}1\ \%$$
$$\Sigma\dot{E}_{L,GP} = 1326\ MW \cong 72{,}9\%$$
$$\overline{\Sigma\dot{E}_I \quad\ \ = 1820\ MW \cong 100\ \%}$$

Figure 4. Relative exergy losses of some units of the coal hydrogasification process (m = n_{H_2O}/n_{CH_4} at the inlet of the steam reformer).

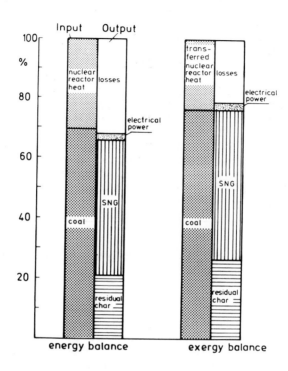

Figure 5. Energy and Exergy balance of the hydro-gasification plant.

because the inlet helium temperature into the steam
generator would be lower. The ratio coal throughput/
nuclear reactor thermal heat should be optimized.
3. Investigations from the exergetic and economic point
   of view should be made in order to find out the extent to which
   - the brown coal should be dried
   - the gas separation and gas cleaning units could be
     replaced by other units.

The thermal efficiency and the exergetic efficiency
of the considered process are 68.3 % and 77.7 %, respectively, Figure 5. This deviation between the thermal and exergetic efficiency can be explained by the
fact that the difference between energy and exergy for
coal, residual char and SNG is relatively small, while
the exergy of the helium heat between steam reformer
inlet and steam generator outlet amounts to about 67%
of the corresponding heat quantity.

Thermoeconomic analysis for stationary flow processes

Most effectively second law analysis of stationary flows
can be combined with cost analysis. The product cost

$$\sum_{l=1}^{n} \dot{n}_{P,l}\, c_l$$

($\dot{n}_{P,l}$ is the molar product flow and c_l the specific
cost) from a steady state flow process include
- the sum of annualized capital cost for all units
  $I_i a_i$ (investment cost I_i times annuity a_i for unit i)
- the total cost

$$P_t c_{El} + \sum_k \dot{n}_{R,k} (H_{m,k} - T_o S_{m,k}) c_k$$

for all energy and raw material input (P_t represents
the electrical energy and c_{EL} the specific cost of
electricity; $\dot{n}_{R,k}$ is the molar flow rate for raw
material stream k, $H_{m,k} - T_o S_{m,k}$ its molar exergy
and c_k the specific exergetic cost)
- the fixed cost F.

Therefore the total product cost is

$$\sum_l \dot{n}_{P,l} c_l = \sum_i I_i\, a_i + P_t\, c_{EL}$$

$$+ \sum_R \dot{n}_{R,k} (H_{m,k} - T_o S_{m,k}) c_k + F$$

(9)

Introducing the mean specific exergetic cost of the input energy

$$c_E = \frac{P_t\, c_{EL} + \sum_k \dot{n}_{R,k}\,(H_{m,k} - T_o\, S_{m,k})\, c_k}{P_t + \sum_k \dot{n}_{R,k}\,(H_{m,k} - T_o\, S_{m,k})} \tag{1o}$$

we obtain from Equation 8 - 1o

$$\sum_l \dot{n}_{p,l}\, c_l = \sum_l \dot{n}_{p,l}\,(H_{m,l} - T_o\, S_{m,l})\, c_E$$

$$+ \sum_i (I_i\, a_i + T_o\, \dot{S}_{pr,i}\, c_E) + F \tag{11}$$

The first term on the right hand side represents the value of the product exergy, $\sum_i I_i\, a_i$ and F, the contribution of capital and fixed cost whereas $T_o \sum_i \dot{S}_{pr,i}\, c_E$ indicates the product cost penalty according to the irreversivilities of the process.

The mean specific exergetic cost depends on the ratio of different energy input and therefore it is subject to changes in the energy input pattern. This value needs to be altered with any substantial variation in the flow sheet. By such a procedure it is not necessary (and also not meaningful) to distinguish the exergy losses in the different units with respect to the origin of the energy sources except the mean specific exergetic value of c_E.

Equation 11 can be considered as a fundamental equation for minimizing product cost by variations of process parameters or by changing the flow sheet during the design procedure.

Literature Cited

1. PNP-Projekt: Statusbericht zum Ende der Konzeptphase vom 1.8.1975-3o.11.1976. Ergebnisbericht der Planungs-, Forschungs- und Entwicklungsarbeiten, Januar 1977.

2. K.F. Knoche, H. Cremer, W. Eisermann: "Balance and Optimization Procedure for Thermochemical Cycles for Hydrogen Production", First World Hydrogen Energy Conference, Miami 1975

3. W. Eisermann: "Bilanzierung von Eisen-Chlor-Prozessen zur Thermischen Wasserzersetzung, Ph.D. Thesis, RWTH Aachen, 1977.

4. P. Schuster: "Experimentelle Untersuchung und Bilanzierung von Vanadium/Chlor- und Vanadium/Brom-Mehr-
   stufenprozessen zur thermochemischen Erzeugung von
   Wasserstoff",Ph.D. Thesis, RWTH Aachen 1980.
5. G. Tsatsaronis, P. Schuster, H. Rörtgen: "Bilanzierung des Verfahrens zur hydrierenden Vergasung von
   Braunkohle", Brennst.-Wärme-Kraft 32 (1980) No. 3,
   pp. 1o5-111.
6. Tsatsaronis, P. Schuster, H. Rörtgen: "Exergy Analysis
   of the Nuclear Coal Hydrogasification Process", Thermodynamic Availability Symposium, AICHE Annual Meeting,
   Detroit, Michigan, August 16-19, 1981.

7. G. Tsatsaronis, P. Schuster, H. Rörtgen: "Thermodynamic Analysis of a Coal Hydrogasification Process
   for SNG Production by using Heat from a High-Temperature Nuclear Reactor, 2nd World Congress of Chemical Engineering, Montreal, Canada, October 4-9, 1981,
   Vol. II, pp. 4o1-4o4.
8. K.F. Knoche, G. Tsatsaronis, F. Löhrer: "Thermody-
   namische Analyse eines ADAM/EVA-Prozesses zur Strom-
   erzeugung", Brennst.-Wärme-Kraft 34 (1982) Nr. 1,
   pp. 22-26.

RECEIVED August 29, 1983

Application of Exergy Analysis to the Design of a Waste Heat Recovery System for Coal Gasification

EBERHARD NITSCHKE

UHDE GmbH, Dortmund, Federal Republic of Germany

In the gasification of coal, oxygen and steam are added to convert the chemical energy of the coal to the chemical energy and sensible heat of the synthesis gas generated. The hot raw gas is either cooled in a steam generator (external steam generation) or in a quench (internal steam generation). The exergetic efficiency of the two alternatives is shown in relation to the gasification parameters (pressure, temperature). A downstream waste heat boiler has great exergetic advantages particularly at low and medium gasification pressures.

Due to the increase in energy costs in recent years, it has become imperative to minimize energy losses in chemical plants. The first step is to identify the causes of energy losses via elaborate energy balances for individual plant sections as well as for the plant as a whole. Such an analysis permits a ranking of process concepts in terms of energy utilization.

However, an analysis based on the first law of thermodynamics has considerable weaknesses in that the energy losses are only evaluated according to their absolute amount. The second law of thermodynamics has proved to be a very useful additional instrument because it permits the ranking of energy losses not only according to their quantity, but also according to their quality, i.e., an amount of heat at a higher temperature is of more value than the same amount of heat at a lower temperature. In the case of an energy balance, it is irrelevant whether a source of heat is used for generating high-grade steam which may serve for driving machinery, for example, or whether the heat is used for preheating water. Exergetic analyses can be used for ranking various

process concepts and have in fact already been employed for this
purpose.

Exergetic analyses seem to be particularly important for
processes which are still in the development stage, i.e. coal
gasification. They permit certain important factors to be recognized at an early stage, which means that certain aims can be established with regard to research and development. The extent to
which these aims can be achieved will, of course, depend on limiting conditions of a technical as well as an economic nature.
However, the knowledge of a theoretically achievable aim will be
sufficient incentive to come as close as possible to this goal.

An exergetic analysis of coal gasification has already been
performed for individual processes (1, 2). This showed that high
exergetic losses may occur in the waste heat section of such a
plant. In this article, an attempt is made to determine the influence of the chosen concept and the gasification parameters on
the exergy balance by systematic investigation.

Process Engineering Model

The coal gasification process is represented by a reactor, to
which coal, including ash, water and oxygen are fed. Since the
gasification reactor is not part of the balance chosen for this
analysis, the question of whether water is fed to the reactor in
the form of a liquid or in the form of steam can be disregarded.

The reactants, coal, water and oxygen are converted to hydrogen, carbon monoxide, carbon dioxide, methane, water vapour
and hydrogen sulphide at a given temperature and pressure according to thermodynamic equilibria and the kinetics of gasification.
A particular coal composition which is characteristic of a German
hard coal was taken as a basis (Table I).

Table I.   Composition and Calorific Value of the Coal

	maf	mf	
C	89.81	80.83	% by weight
H	4.33	3.90	
O	3.53	3.18	
N	1.47	1.32	
S	0.86	0.77	
ash	-	10.00	
	100.00	100.00	
GCV	35.0		MJ/kg

Apart from the temperature, the ratio of water to coal is an
essential parameter affecting the composition of the gas. These

parameters were modified within the limits set by industrial processes. The pressure has a relatively minor effect on the composition of the gas. At high temperatures the methane content in the raw gas is very low (<1 %), even when the pressure is high. It is only at temperatures at or below 1000 °C that a higher methane content is obtained at higher pressures.

Via gasification, the chemical energy contained in the coal is converted to the chemical energy of the synthesis gas, together with a considerable amount of sensible heat as well as the latent heat of vaporization of the water. Heat losses in the reactor, which are due to radiation and, depending on the gasification temperature, the melting of the slag are relatively low. As already mentioned, these losses are not taken into consideration. There are two basically different methods of utilizing the waste heat in order to obtain a high overall efficiency. These are:

Waste Heat Utilization by means of External Steam Generation in a Heat Exchanger (Figure 1). Raw gas is cooled in a heat exchanger, the heat being transported through a wall to an external system, where it is used for generating steam. In principle, the state of this steam can be arbitrarily selected. However, for practical reasons it is necessary that the pressure of the steam is higher than that of the gas. In the present case, a steam pressure of 100 bar was assumed. The boiler feedwater should be preheated to the evaporation temperature, while the steam leaves the waste heat boiler in a saturated state. It was further assumed that there is no temperature difference at the outlet of the heat exchanger, i.e. the gas leaves the waste heat system at the same temperature at which the boiler feedwater was supplied. It was assumed that no other heat losses occur in this system. The term "external" steam generation was selected for this because the steam generation takes place outside the actual gasification system.

Waste Heat Utilization by means of Internal Steam Generation in a Quench. The hot gas leaving the reactor is contacted directly with water in the quench vessel. Depending on the enthalpy of the gas, a certain amount of water evaporates and the gas leaves the vessel saturated with water vapour. In this case, it was assumed that the quench water is supplied to the quench at the saturation temperature. While a quench water system is necessary in commercial practice, heat losses from this system were neglected. Because this steam is generated within the quench vessel and leaves the vessel together with the gas, this alternative was designated "internal" steam generation.

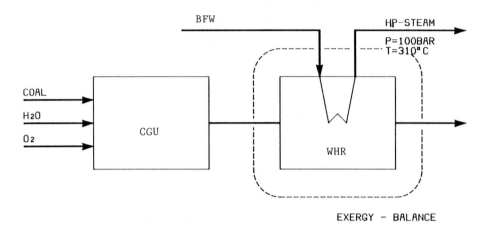

Figure 1. Alternative I : coal gasification unit (CGU) with waste heat recovery (WHR).

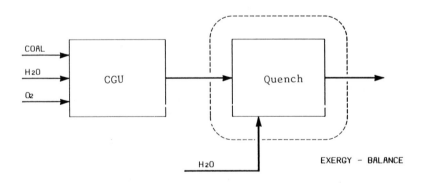

Figure 2. Alternative II : coal gasification unit (CGU) with quench.

Limits of the Model

As described above, the model only covers the gasification reactor, the waste heat system and the quench. The effect of the downstream sections on the design will be discussed in item 6.

The effect of different coal compositions and of the ash content was not investigated.

It was also assumed that gas generated is to be used as synthesis gas, e.g. for ammonia or methanol synthesis. It is therefore advisable to standardize on the generation of a particular amount of synthesis gas, which is characterized by the sum of the carbon monoxide and the hydrogen. Accordingly, it was decided that the plant should produce 1 kmol (H2 + CO)/s. 1 kmol/s is equivalent to the stoichiometric requirement of an ammonia/methanol plant with a capacity of approx. 1000 tpd or, more precisely, 1 kmol/s = 921.6 tpd methanol or 979.4 tpd ammonia.

The model does not refer to any particular coal gasification process. However, in view of the fact that the coal/water ratio is an essential parameter in the following calculations, it is obvious that in these cases it is the Texaco coal gasification process that was used as a basis.

Mathematical Model

The composition of the gas in an autothermal gasification process using oxygen was calculated by means of an equilibrium model like the one used in previous investigations (3, 4). The model determines the amount of oxygen required for a given gasification temperature by iteration. The gas composition itself is calculated on the basis of the simultaneous equilibria. Such a model can basically only be applied to coal gasification processes in which the coal and the gasification agent flow in co-current. Any deviations from the equilibria due to the nature of the coal or for process reasons were disregarded.

The molar exergy of the gas mixture at temperature T and pressure p is determined according to the following formula:

$$E(T,p) = E(T_u, p_u) + H(T,p) - H(T_u,p_u) - T_u [S(T,p) - S(T_u,p_u)]$$

Index "u" refers to the state of the environment. It is assumed that the mixture can be treated as an ideal gas, in which case the enthalpy and entropy can be calculated as the sum of the molar enthalpies and entropies of the individual components.

In view of the high water vapour concentration and the relatively high pressure, this assumption is by no means true throughout the entire range of the parameters considered. However, it can be assumed that any corrections will have no effect on the basic conclusion.

To estimate the exergy of the gas mixture at ambient temperature and ambient pressure, we took Schmidt/Baehr's model (5, 6)

·as a basis and adopted the methods described by Baehr for calculating the exergy. This model, which assumes the reference environment to be saturated moist air, does not allow for the presence of sulphur compounds. While it is possible, in principle,
to include such compounds in the exergy calculation, this was not
done here.

The exergy of the gas mixture is therefore calculated according to the following formula:

$$E(T,p,x_i) = \Sigma \; x_i \; E_i^O(T_u) + \Sigma \; x_i \; [H_i^O(T) - H_i^O(T_u)]$$

$$- T_u \; \Sigma \; x_i \; [S_i^O(T) - S_i^O(T_u)] + R_m \; T_u \; \ln(p/p_o) + R_m \; T_u \; \Sigma \; x_i \; \ln x_i$$

Index "i" refers to the individual components, the amounts
of which are determined by the mole fraction "x". The exponent
"o" characterizes the standard state (p = 101.325 kPa; T =
298,3 K).

The essential term in the formula is the first term, which
contains the sum of the standard exergies of the components. It
basically corresponds to the calorific value of the gas.

The values used in Tables II and III were taken into consideration for the standard exergies. All the other data were obtained from the Uhde thermophysical properties program package.

Table II.   Standard exergy of some gaseous compounds related to
the reference atmosphere (T_u = 298.15 k, p = 1.01325 bar)

i	N_2	O_2	H_2O	CO_2	Ar
E^O	0.692	12.214	8.592	20.06	11.68

Table III.   Standard exergy and calorific value of product gas
(T = 298.15 k, p = 1.01324 bar)

i	H_2	CO	CH_4	C_2H_4	C_2H_6
E^O	235.2	275.3	830.45	1356.6	1493.7
GCV	285.8	283.0	890.4	1410.6	1559.8
NCV	241.84	283.0	802.35	1322.6	1427.9

Results

Exergy as a function of the gasification conditions. The exergy
of the product gas as a function of the gasification pressure at
various temperatures is shown in Figure 3. It can be seen that
the exergy increases with the pressure. It should be noted that,
for the calculation selected here, the calorific value of the
synthesis gas accounts for about 70 - 80 % of the total exergy.
Whereas the total energy only increases by approx. 7 %, the
exergy contained in the gas in the form of sensible heat, latent
heat of vaporization and the energy due to compression increases
by more than 40 %. The energy feed to the coal gasifier (i.e., in
the coal) is constant, or rather, is only dependent on the gasification temperature. This means that the exergetic efficiency of
the coal gasification process improves markedly as pressure
increases.

The exergy of the streams obtained after heat recovery is
shown in Figure 4. For alternative I, the exergy content of the
cooled gas as well as exergy of the generated steam are included.
It can be seen that in the case of alternative II, the exergy
content of the gas increases relatively steeply with pressure. As
was noted earlier, the heat of combustion of the gas is a constant term in the exergy. If one subtracts this calorific value,
it can be seen that in the case of gasification at atmospheric
pressure, the remaining portion of exergy is only slight. The increase in exergy for alternative II is due to the fact that the
water vapour concentration in the gas increases greatly.

Figure 5 shows the boiler feedwater requirement for the
waste heat boiler according to alternative I, and for alternative
II, the quench water requirement, i.e. the amount of water which
evaporates in the quench. The partial pressure of the water,
which increases as a function of the total pressure, results in a
corresponding increase in the steam concentration, which has a
favourable effect on the exergy balance.

Exergetic Efficiency. The exergetic efficiency is defined as
follows:

$$\eta (I) = \frac{Ex \ (product \ gas) \ out + Ex \ (steam)}{Ex \ (product \ gas) \ in + Ex \ (boiler \ feedwater)}$$

$$\eta (II) = \frac{Ex \ (product \ gas) \ out}{Ex \ (product \ gas) \ in + Ex \ (quench \ water)}$$

Figure 6 shows the exergetic efficiency as a function of
pressure at various gasification temperatures. In the case of alternative I (waste heat boiler), the efficiency is independent of

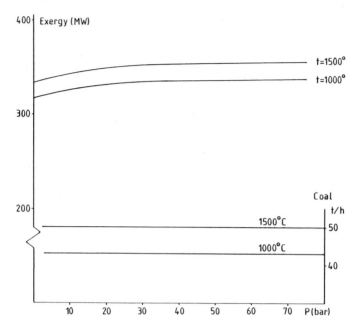

Figure 3. Exergy of product gas (H_2O/coal = 50/50).

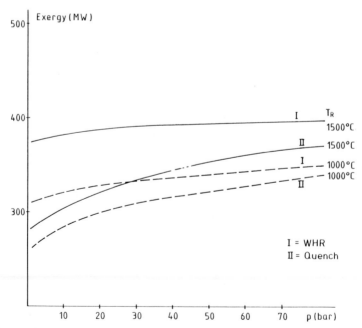

Figure 4. Exergy after waste heat recovery (I) and quench (II) for different gasification temperature (H_2O/coal = 50/50).

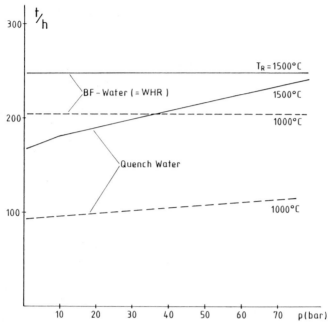

Figure 5. Boiler feed water and quench water consumption
(H_2O/Coal = 50/50).

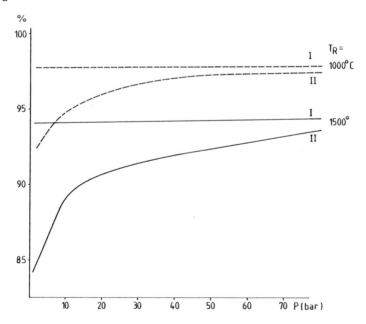

Figure 6. Exergetic efficiency for WHR (I) and quench (II)
(coal/H_2O = 50/50).

the pressure and only depends slightly on the temperature. As can
be expected, it is higher at low gasification temperatures. The
trend of the efficiency for alternative II is interesting. It is
relatively low at low pressures, then increases and approaches
the efficiency of alternative I at higher pressures. It should
also be borne in mind that the above formula also contains a
large constant term. If, for the sake of simplicity, the exergetic efficiency only referred to the exergy of the sensible heat
and the pressure, the exergetic efficiency at atmospheric pressure would only be a few percent and the increase with rising
pressure would be correspondingly steeper. This diagram confirms
that a waste heat recovery system incorporating the generation of
high-pressure steam is more favourable than a quench.

Technical Aspects

Influence of the downstream plants. Up to now, we have regarded
the coal gasification reactor with the waste heat recovery system
as an isolated unit. In the event that the gas generated is intended to be used as fuel gas, for example in a combined power
station, this approach is justified. If, however, the gas is to
be used as synthesis gas, the effect of the downstream units must
be taken into consideration. In such cases it is necessary to
feed the gas to a CO shift conversion unit in order to obtain the
CO/H_2 ratio required for the synthesis process. Apart from gasification at atmospheric pressure, which requires an intermediate
compression step, it has proved advisable to locate the CO shift
conversion directly downstream of the gasification section. A
stage in which dust particles are removed from the gas is situ-
ated between these two units. It is assumed that exergy losses do
not occur in this unit.

The two alternatives described above differ greatly from
each other with regard to the water vapour content in the gas.
The ratio of water vapour to dry gas as a function of the gasification pressure is shown in Figure 7. In alternative I, this
ratio is only dependent on the temperature and on the ratio of
H_2O to coal at the inlet to the gasification reactor, whereas in
the case of alternative II, it also depends to a large extent on
the pressure. The CO shift conversion section requires that its
feed gas has a certain water vapour/gas ratio in order to achieve
the required CO content in the tail gas. If, for example, an ammonia synthesis unit is located downstream, a water vapour/dry
gas ratio (S/G) of >1.1 is required. For a methanol synthesis
unit, this ratio is approx. 0.5 - 0.9, depending on the design of
the shift conversion section. The diagram shows that this ratio
is not achieved in alternative I and, in alternative II it is exceeded to a considerable extent. The question therefore arises as
to how the necessary ratio can be achieved. Figure 8 shows
various possibilities.

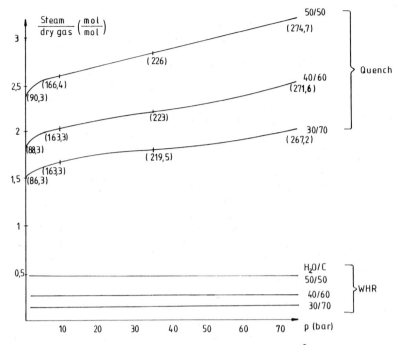

Figure 7. Steam to drygas ratio (TR = 1500 °C. Dew point of gas shown in parentheses.

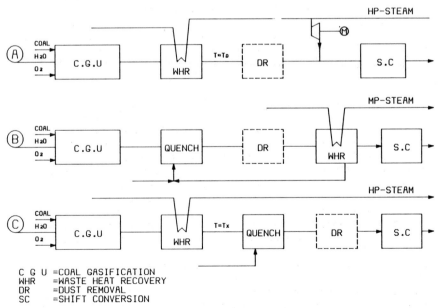

C G U = COAL GASIFICATION
WHR = WASTE HEAT RECOVERY
DR = DUST REMOVAL
SC = SHIFT CONVERSION

Figure 8. Different ways for the adjustment of the steam-to-gas-ratio.

It goes without saying that it is possible to mix the generated water vapour with the gas. However, such a step is only advisable if the exergy of the steam can be exploited in a back-
pressure turbine prior to mixing. This assumes that there is an
adequate pressure gradient, i.e. that the gasification pressure
is relatively low. It must be noted that, apart from efficiency
losses within the turbine and a pressure drop caused by the control system, an exergy loss cannot be avoided in principle because the steam can only be depressurized to the total pressure
of the gas and not to the pressure of the water vapour in the
gas. This means that mixing losses will occur.

    In the second alternative, i.e. with the quench, the task is
to reduce an excessive steam/gas ratio. Method B shows one way of
doing this. A waste heat boiler is located downstream of the
quench. It operates in the saturation range of the gas. The excess water is removed by means of partial condensation of the
water vapour, thus producing the desired S/G ratio. The heat of
condensation obtained can be utilized for generating steam. The
pressure of the steam generated in this way can, at the most,
only correspond to the partial pressure of the steam within the
gas. This method will therefore only be advisable when the total
pressure is at an appropriately high level.

    A combination of the two alternatives is shown in method C.
A waste heat recovery system is located downstream of the coal
gasifier. However, heat recovery stops at a certain temperature.
This temperature is determined by the enthalpy of the gas downstream of the waste heat boiler, which must be sufficient to produce the required S/G ratio in the downstream quench.

    Taking into consideration the results obtained with regard
to the exergetic efficiency of alternative I and alternative II,
method C should have a high exergetic efficiency because it has
the advantages of alternative I but avoids the exergy loss which
this entails due to the admixing of steam.

Technical and economic aspects. The above considerations show
that, particularly in the case of gasification at atmospheric
pressure and also at moderate pressure, it is necessary to use a
waste heat boiler in order to ensure a high exergetic efficiency.
It is therefore not surprising that processes which operate in
this pressure range provide for a waste heat boiler, i.e., the
Koppers process (7) and the Winkler process, which operates at a
pressure of 10 bar in the HTW version developed by Rheinbraun
(8). The design of such a system depends on whether the process
operates at temperatures above the ash melting point, like the
Koppers process, in which case a radiant boiler is required, or
below the ash melting point, like the Winkler process, in which
case convection boilers can be used. In the case of processes
which operate at a higher pressure such as Texaco (9), waste heat
systems have also been developed in recent years and have produced excellent results in practice. It is thus possible to fully

utilize the higher exergetic efficiency of gasification under
pressure. However, it must be pointed out that the specific cost
per ton of steam generated in such a waste heat system is considerably higher when compared with the specific cost of steam generated in a coal-fired boiler and which is to be used for driving
purposes in the plant. This relationship is slightly more favourable if there is a waste heat system downstream of the quench.
The gas is then already free of solids and the temperature is
considerably lower. This means that design principles can be used
which make fabrication costs more reasonable. However, the temperature is below the dew point and this fact must be taken into
consideration when selecting the materials of construction. Since
the exergetic efficiencies of alternatives I and II are roughly
the same at higher gasification pressures, the preferred method
must be selected carefully.

Summary and Conclusions

This investigation has shown that the exergetic efficiency of a
coal gasification reactor is greatly influenced by the type of
heat recovery system which is selected. The pressure of the gasifier, which is not included in a conventional energy balance, has
a considerable effect on the efficiency. The use of a waste heat
system for external steam generation is clearly more favourable
throughout the whole range of the parameters examined here. However, the extent to which the exergetic advantage can also be
turned into an economic advantage depends on a variety of technical, practical and economic constraints.

Literature Cited

1.   L. Rodriguez R.A. Gaggioli: Canad. J. of Chem. Eng. 58
     (1980) 3.328/38
2.   G. Tsatsaronis, P. Schuster, H. Rörtgen: Brennst. Wärme
     Kraft 32 (1980) 3. 105/111
3.   K.K. Neumann, E. Nitschke: Erdöl und Kohle, Compendium
     76/77, 328/347
4.   K.K. Neumann, F. Keil, E. Nitschke: Chem. Ing. Techn., 52
     (1980) 11, 908/909
5.   H.D. Baehr, E.F. Schmidt: Brennst. Wärme Kraft 15 (1063)
     375/381
6.   H.D. Baehr: Brennst. Wärme Kraft 31 (1979), 7, 292/302
7.   B. Beck: TVA Ammonia from coal symposium 1979, 72/78
8.   F.H. Franke, E. Pattas, E. Nitschke, J. Keller: TVA Ammonia
     from coal symposium 1979, 86/96
9.   B. Cornils, J. Hibbel, P. Ruprecht, R. Dürrfeld, J.
     Langhoff: Hydrocarb. proc. (1981), 1, 149/156

RECEIVED July 7, 1983

Thermodynamic Availability Analysis in the Synthesis of Optimum-Energy and Minimum-Cost Heat Exchanger Networks

F. A. PEHLER
Department of Chemical Engineering, Auburn University, Auburn, AL 36849

Y. A. LIU
Department of Chemical Engineering, Virginia Polytechnic Institute and State University, Blacksburg, VA 24061

This paper presents a thermodynamic availability analysis of an important process design problem, namely, the synthesis of networks of exchangers, heaters and/or coolers to transfer the excess energy from a set of hot streams to streams which require heating (cold streams). Emphasis is placed on the discussion of thermodynamic and economic (i.e., thermoeconomic) aspects of two recent methods for the evolutionary synthesis of energy-optimum and minimum-cost networks. These methods include the thermoeconomic approach of Pehler and Liu (1) and the evolutionary development method of Linnhoff and Flower (2).

Multiobjective Synthesis of Heat Exchanger Networks

A typical heat exchanger network has N_h hot streams S_{hi} ($i=1,2,...,N_h$) to be cooled and N_c cold streams S_{cj} ($j=1,2,...,N_c$) to be heated. Associated with each stream are its steady-state input temperature T_i, output temperature T_i^* and heat capacity flow rate W_i (average heat capacity multiplied by mass flow rate). There are also available N_{hu} heating utility streams and N_{cu} cooling utility streams. The synthesis problem is to create several optimum and suboptimum networks of units (exchangers, heaters and/or coolers) so that the specified stream outlet temperatures are reached.

The optimum and suboptimum networks should achieve or nearly achieve at least the following multiple objective criteria: (i) approaching a practical minimum loss in thermodynamic available energy during heat exchange among hot and cold process streams; (ii) minimizing the number of units; (iii) minimizing the investment cost of units; and (iv) minimizing the operating cost of utilities. Note that these criteria have been anticipated or utilized in part in reference nos. 2 to 6.

An important feature of this multiobjective synthesis problem is that some of the criteria may conflict with others. For example, minimizing the loss of available energy during the

0097–6156/83/0235–0161$06.00/0

heat exchange process requires maximizing the heat transfer area, which tends to maximize the network investment (6). Also, minimizing the number of units does not necessarily lead to the minimization of the investment cost of units. This follows because the investment cost of units depends not only on the number of units, but also on how this total area is distributed among the different units (2,4). Thus, in order to synthesize several optimum and suboptimum networks, a multistep evolutionary strategy is recommended in this work, with each step emphasizing one of the criteria.

In general, the investment costs for the ith exchanger, heater and cooler, denoted by C_{Ei}, C_{Hi} and C_{Ci}, respectively, can be correlated by the empirical expressions $C_{Ei} = aA_{Ei}^b$, $C_{Hi} = aA_{Hi}^b$ and $C_{Ci} = aA_{Ci}^b$, where A_{Ei}, A_{Hi} and A_{Ci} are, respectively, the heat transfer areas of the ith exchanger, heater and cooler. The total network investment and utility operating cost (J) to be minimized can be expressed as:

$$J = \delta \left(\sum_i aA_{Ei}^b + \sum_i aA_{Hi}^b + \sum_i aA_{Ci}^b \right) + \sum_k \sum_l u_k S_{ukl} \qquad (1)$$

where S_{ukl} represents the amount of heating or cooling utility stream S_{ukl} such as stream or water spent at the ith auxiliary heater or cooler per year, u_k denotes the annual operating cost of the utility S_{uk}, and δ is the annual rate of return on investment. For convenience, the following simplifying assumptions have been included in the systematic synthesis of heat exchanger networks: (i) the use of single-pass countercurrent shell-and-tube exchangers; (ii) no phase changes of process streams; (iii) equal values of effective heat transfer coefficients for exchanges between two process streams and between two process and utility streams; and (iv) temperature-independent heat capacity flow rates.

The Thermoeconomic Approach

The thermoeconomic approach of Pehler and Liu (1) is based on both thermodynamic and economic considerations of the network synthesis problem. It consists of four steps. The detailed descriptions of the first two steps can be found from the references cited below. In this paper, some emphasis is placed on the thermodynamic availability analysis of the third and fourth steps which include practical heuristic and evolutionary rules for the systematic synthesis of energy-optimum and minimum-cost networks.

The first step is to represent the network synthesis problem by a heat content diagram (4) or by the method of temperature interval and problem table (3, pp. 32-38; 5). The second step is to determine the minimum heating and cooling utility requirements by the method of temperature interval and problem or by the minimum area algorithm (4).

Step 3: Synthesis of an Energy-Optimum and Nearly Minimum-Cost Network by the Thermodynamic (Minimum-Availability-Loss) Matching Rule. The thermodynamic matching rule states that: "The hot process and utility streams, and cold process and utility streams are to be matched consecutively in a decreasing order of their stream temperatures." This rule is based on the results of a thermodynamic availability analysis of the network synthesis problem.

An important feature of this analysis can be illustrated by considering the heat exchange between a hot gaseous stream and another cold gaseous stream occurring within a differential element (length) of an adiabatic, single-pass shell-and-tube exchanger. Suppose that the temperature and pressure of the hot stream change from T_h and P_h to $T_h +dT_h$ and $P_h + dP_h$, respectively; and the temperature and pressure of the cold stream also change from T_c and P_c to $T_c + dT_c$ and $P_c + dP_c$, respectively. Within the differential element of the exchanger, the rate of heat exchange between hot and cold streams, denoted by dQ, is equal to the rate of change of the enthalpy of the hot stream, $-dH_h$, and to that of the cold stream, dH_c. It can then be shown that the rate of loss of available energy during heat exchange between hot and cold streams is (7):

$$\begin{aligned}
\text{Rate of loss} \\
\text{of available} \\
\text{energy}
\end{aligned} = T_o \left\{ \frac{|d\dot{Q}|}{T_c} - \frac{|d\dot{Q}|}{T_h} - \frac{\dot{V}_h dP_h}{T_h} + \frac{\dot{V}_c dP_c}{T_c} \right\}$$

$$= T_o \left\{ |d\dot{Q}| \ (\frac{T_h - T_c}{T_h T_c}) - [\frac{\dot{V}_h dP_h}{T_h} + \frac{\dot{V}_c dP_c}{T_c}] \right\} \quad (2)$$

In Eq. (2), T_o represents the temperature of the surroundings; and V_h and V_c denote the rates of change of the volume of hot and cold streams, respectively. Since dP_A and dP_B are always negative in magnitude, the term within the bracket of Eq. (2) is generally negative. Eq. (2) then suggests that the rate of loss of available energy during heat exchange is minimized when the first term within the equation can be made to approach zero. The latter implies that T_h should be made to approach T_c. In the limit, when T_h is maintained only infinitesimally greater than T_c, the loss of available energy during the heat exchange process is at a minimum value.

The thermodynamic matching rule can be implemented on a heat content diagram or temperature interval diagram as follows. Starting with the highest-temperature heat source (hot process and utility streams), each differential element of heat is transferred to the highest-temperature heat sink (cold process and utility streams). This process continues with the heat of the intermediate-temperature heat source being transferred to the intermediate-temperature heat sink, and ends when the heat of the lowest-temperature heat source is given up to the lowest-temperature heat sink.

The preceding analysis provides the thermodynamic basis of a similar stream matching rule described in Corollary 3 of reference no. 4. In the thermoeconomic approach, the thermodynamic matching rule is not only applied in the initial generation of an energy-optimum and nearly minimum-cost network, but also in the evolutionary synthesis of an energy-optimum and minimum-cost network.

Step 4: Evolutionary Synthesis of an Energy-Optimum and Minimum-Cost Network by Minimizing the Number of Units and Approaching a Practical Minimum Loss in Available Energy during Heat Exchange Through Systematic Merging, Shifting and/or Unsplitting of Units (1)

(a)   Determination of the most probable minimum (quasi-minimum) number of units (exchangers, heaters and coolers), N_{min}, according to (3):

$$N_{min} = N_h + N_c + N_{hu} + N_{cu} - 1 \qquad (3)$$

where N_h and N_{hu} are the numbers of hot process and utility streams, respectively; and N_c and N_{cu} are the numbers of cold process and utility streams, respectively.

(b)   Minimization of the number of units to reduce the network investment through systematic merging of units according to the basic idea described in reference no. 4. In particular, without increasing the total heat transfer area of units, the network investment can be reduced if: (i) several units can be combined into a single one, or (ii) a smaller number of units are to be used.

(c)   Application of the thermodynamic matching rule to the modified network with a fewer number of units. This entails systematic shifting and/or unsplitting the remaining units in order to reach a practical minimum loss in available energy during heat exchange.

For step 4 of the thermoeconomic approach, the following evolutionary rules have been presented in reference no. 1. These rules are to be applied sequentially in their numerical order; that is, Rule 1 should be applied before Rules 2 and 3, etc.

Rule 1. Delete, shift or merge units so as to reduce the number of units in a selected local subnetwork (i.e., a selected number of units in an initial network) and to minimize the required changes to adjacent subnetworks due to changes in heating or cooling load. When selecting candidates for unit modifications, choose initially a redundant heater or cooler exceeding the minimum heating or cooling requirement found in Step 2 and then a redundant exchanger with a small heat load exceeding the quasi-minimum number of units determined in Step 4a. Avoid modifying any single-exchanger match between two hot and cold streams. If a selected unit modification in a given subnetwork results in extensive structural modifications in adjacent or other subnetworks, then another candidate for unit modification should be evaluated.

Rule 2. Shift heaters or coolers to approach a practical minimum loss in available energy during heat exchange:
(a) When shifting a heater (cooler) between matches on a given cold (hot) process stream, always shift the heater (cooler) from the low-temperature (high-temperature) portion to the high-temperature (low-temperature) portion of the cold-stream (hot-stream) match.
(b) When merging two heaters (coolers) matched with two different cold (hot) process streams, always shift the heater (cooler) from one stream to the other so that the resulting merged heater (cooler) will have a higher (lower) arithmetic average of its input and output temperatures.

Rule 3. Reduce the number of units by deleting repeated matches between two hot and cold process streams in a given network. In particular, if a given network contains a local subnetwork in which a hot (cold) stream matches the same cold (hot) stream which it has matched before, delete one of these repeated matches.

Rule 4. Unsplit a given splitting network to minimize the number of units and reduce the loss in available energy during heat exchange. When unsplitting a given splitting network, always match the hot and cold process streams in the resulting network in a decreasing order of their arithmetic averages of input and output temperatures.

Thermoeconomic Aspects of Evolutionary Rules of Pehler and Liu

Thermodynamic and economic aspects of the preceding evolutionary rules can be illustrated by translating them into a set of basic on-diagram modifications (BODM). These modifications can be applied directly on a heat content diagram or temperature interval diagram for the systematic evolutionary synthesis of energy-optimum and minimum-cost networks. For convenience, the discussion below will be based on the network representation by a heat content diagram only.

BODM No. 1 (Figure 1). This modification illustrates the use of Rule 2a to guide the shift of a heater on a given cold process stream. The evolutionary changes shown in Figure 1 are as follows: (a) On S_{c1}, shift H_1 to the high-temperature portion, and (b) shift E_1 downward on S_{c1} to a position adjacent to E_2. Such an upwardly-directed heater shift results in a decrease of the loss in available energy during the heating process and makes the use of the heating utility more efficient thermodynamically. This heater shift is, of course, subjected to the constraint of the minimum approach temperature for the heater.

BODM No. 2 (Figure 2). This modification illustrates an application of Rule 2a to guide the shift of a cooler on a given hot process stream. As shown in Figure 2, the evolutionary changes involve: (a) shifting C_1 downward to the low-temperature portion of S_{h1}, and (b) shifting E_2 upward on S_{h1} to a position adjacent to E_1. Subjected to the constraint of the minimum approach temperature for the cooler, this cooler shift leads to a thermodynamically more efficient use of the cooling utility.

BODM No. 3 (Figure 3). This modification illustrates the use of Rule 2b to choose a heater to be shifted from one cold stream to the other in a given subnetwork. The evolutionary changes shown in Figure 3 are as follows: (a) shift H_1 from S_{c1} to S_{c2} and merge it with H_2 to form a composite exchanger $H_1 + H_2$; (b) on S_{c2}, decrease the heat load of E_2; and (c) on S_{h1}, enlarge E_1 by changing the stream splitting ratio of S_{h1}. As Rule 2b suggests, the choice of a heater to be shifted is determined by the composite heater formed which has the highest arithmetic average of its input and output temperatures. This implies that a heater is generally shifted from a cold stream with a smaller heat capacity flow rate to another cold stream with a larger heat capacity flow rate so as to result in a thermodynamically more efficient use of the heating utility. Such a heater shift requires load changes in adjacent exchangers in the selected subnetwork. In particular, E_1 must have its heat load increased and E_2 must have its heat load decreased. Fortunately, an attractive feature of having a splitting network like that shown in Figure 3 is that such load changes can be easily accomplished by changing the stream splitting ratio of S_{h1}. As a result, no structural changes outside of the selected subnetwork are required.

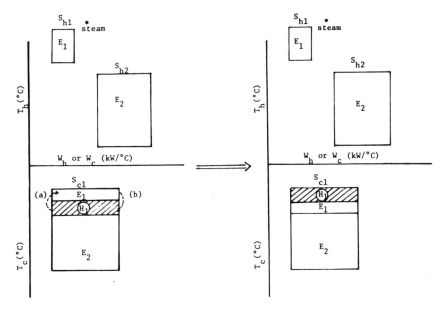

Figure 1.   BODM no. 1 corresponding to rule 2a.

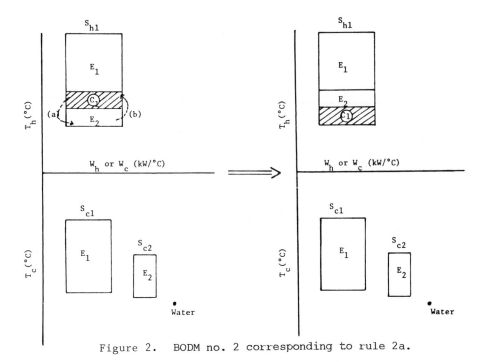

Figure 2.   BODM no. 2 corresponding to rule 2a.

BODM No. 4 (Figure 4). The modification illustrates an application of Rule 2b to choose a cooler to be shifted from one hot stream to the other in a given subnetwork. As suggested by Rule 2b, the choice of a cooler to be shifted is determined by the composite cooler formed which has the lowest arithmetic average of its input and output temperatures. This leads to a thermodynamically more efficient use of the cooling utility. The specific evolutionary changes shown in Figure 4 are as follows: (a) shift C_1 from S_{h1} to S_{h2} and merge it with C_2 to form a composite cooler $C_1 + C_2$; (b) on S_{h1}, enlarge E_1 to compensate for the load reduction on E_2 due to the shifted C_1 on S_{h2}; (c) on S_{c1}, enlarge E_1 and reduce E_3 which is matched with another hot stream in an adjacent subnetwork not shown in the figure; and (d) on S_{c2}, reduce E_2 by matching the low-temperature portion of S_{c2} with another hot stream in an adjacent subnetwork also not shown in the figure. Note that although BODM no. 4 is illustrated by a nonsplitting subnetwork in Figure 4, this BODM is also applicable to a splitting subnetwork.

BODM No. 5 (Figure 5). This modification illustrates the use of Rule 3 to reduce the number of units by deleting repeated matches between two hot and cold process streams in a given subnetwork. As shown in Figure 5, S_{h1} and S_{c1} are matched twice through E_1 and E_3. Thus, the evolutionary changes are: (a) on both S_{h1} and S_{c1}, E_3 is shifted upward through E_4 and merged with E_1; (b) on S_{h1}, E_4 is shifted downard to the low-temperature portion; and (c) on S_{c1}, E_2 is also shifted downward to the low-temperature portion.

The Evolutionary Development Method

The evolutionary development method of Linnhoff and Flower (2) utilizes the temperature interval diagram to represent an initially created network and also the concept of the freedom (F) of an exchanger. The latter has the physical dimension as a heat load (kW) and is related to the larger heat capacity flow rate of the two streams matched in an exchanger (CPL), the smallest actual temperature difference within the exchanger (ΔT_s) and the minimum approach temperature of the exchanger (ΔT_{min}) according to the expression

$$F = CPL (\Delta T_s - \Delta T_{min})$$ (4)

Linnhoff and Flower have claimed that the use of the exchanger freedom parameter allows one to evaluate the effects of shifting, merging or deleting heaters and coolers in a given network on the feasibility of the resulting modified network. In particular, these authors have proposed ten feasibility rules to guide the shifting, merging or deleting of heaters and

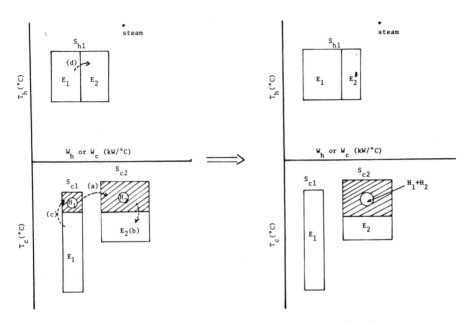

Figure 3.   BODM no. 3 corresponding to rule 2b.

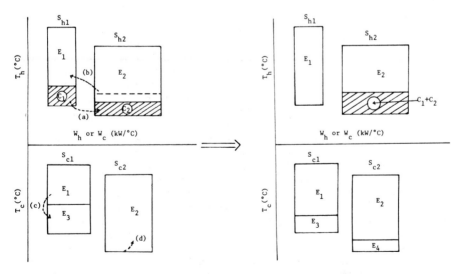

Figure 4.   BODM no. 4 corresponding to rule 2b.

coolers in a given network. However, no guidelines are given as
to which rule should be used in a given network. There are also
situations for which several of the feasibility rules can be
applied to a given network leading to the same changes in the
heat load and exchanger freedom parameter, but no advice is
given on the proper order of applications of these rules. Thus,
this lack of a definite strategy to apply the feasibility rules
represents a weak point of the ED method, particularly when
applying it to synthesis problems with six or more process
streams.

Thermoeconomic Analysis of Feasibility Rules of Linnhoff and
Flower

     As part of this work, the ten feasibility rules proposed by
Linnhoff and Flower have been systematically evaluated on the
heat content diagram with respect to their thermoeconomic
advantages and disadvantages for the evolutionary synthesis of
energy-optimum and minimum-cost networks. A specific goal was
to develop some definite recommendations for the applications of
these rules to properly guide the shifting, merging or deleting
of heaters or coolers in a given initial network.
     Figure 6 illustrates feasibility rule no. 6 of Linnhoff and
Flower, concerning the shifting of a cooler in a given
subnetwork. As shown in the figure, this rule begins with a
thermodynamically efficient cooler placement (i.e., on the
low-temperature portion of the hot stream) and then shifts the
cooler to a thermodynamically, less efficient position (i.e., on
the low-temperature portion of the matched cold stream). In the
resulting subnetwork, the shifted cooler actually serves as a
precooler for the cold stream before it enters the enlarged
exchanger with its heat load being increased from E to (E+A) kW.
Although there is an increase in the exchanger freedom parameter
of A kW through applying evolutionary rule no. 6, the
corresponding evolutionary change would not be generated by the
thermoeconomic approach. This follows because the latter would
not have allowed the cooler to be placed in its
thermodynamically inefficient position in the modified network.
     Figure 7 illustrates an inverse form of feasibility rule
no. 7 of Linnhoff and Flower, relating to the formation of an
exchanger by merging a heater and a cooler. This figure shows
that the cooler (heater) is initially placed on the
highest-temperature (lowest-temperature) portion of the hot
(cold) stream. After applying feasibility rule no. 7, a new
exchanger is formed which matches the highest-temperature
portion of the hot stream with the lowest-temperature portion of
the cold stream. The thermodynamic matching rule in Step 3 of
the thermoeconomic approach, suggests that both the placement of
the heater and cooler in the initial network and its subsequent
formation of a new exchanger according to the feasibility rule

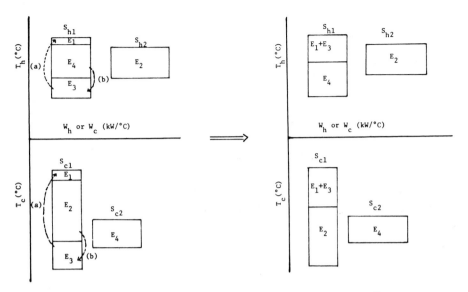

Figure 5.   BODM no. 5 corresponding to rule 3.

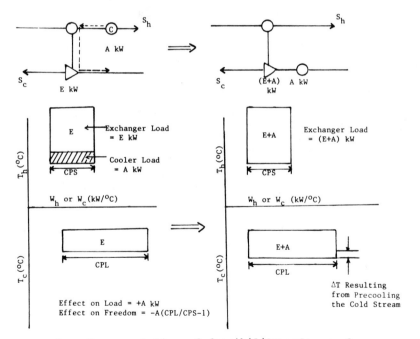

Figure 6.   Representation of feasibility rule no. 6
of Linnhoff and Flower ($\underline{2}$) by the heat content
diagram.

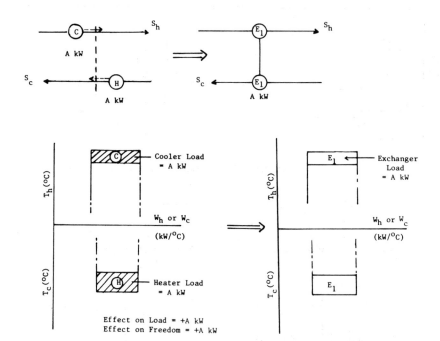

Figure 7. Representation of the inverse form of feasibility rule
no. 7 of Linnhoff and Flower (2) by the heat content diagram.

no. 7 represents a thermodynamically very inefficient exchange among hot and cold, process and utility streams. After evaluating all of the ten feasibility rules of Linnhoff and Flower in a similar fashion, a minimum set of three modified feasibility rules along with some definite recommendations for their applications have been identified and are discussed as follows.

Modified Feasibility Rule No. 1. Always shift a heater along the stream with the larger heat capacity flow rate (CPL), and a cooler along the stream with the smaller heat capacity flow rate (CPS) so that the shift leads to a thermodynamically more efficient use of the heating or cooling utility. Both shifts result in an increase of the exchanger freedom parameter, being equal to the load of the heater (A kW) or to the load of the cooler (A kW) multiplied by a factor, CPL/CPS. This modified feasibility rule includes the original feasibility rule nos. 1-2 of Linnhoff and Flower.

Modified Feasibility Rule No. 2. Consider the possibility of shifting a cooler (heater) from a hot (cold) stream to the lowest-temperature or highest-temperature portion of a cold (hot) stream so as to increase the approach temperature between the hot and cold streams due to the precooling or post-cooling of the cold stream (the preheating of post-heating of the hot stream). In particular, such a shift of the cooler (heater) should be considered if it leads to an increase in the approach temperature from an unacceptable low value to an acceptable high value so that the very large area of the exchanger resulting from the previously infeasible match becomes smaller. This modified feasibility rule includes the original feasibility rule nos. 3 to 6 of Linnhoff and Flower, and it is mainly related to the tradeoff between the network investment cost (determined by the approach temperature) and the utility operating cost. In other words, there may be an economic incentive to use an additional amount of heating or cooling utility to heat up or cool down a process stream. This situation may exist when the additional heating or cooling requirement increases the approach temperature of the exchanger in such a way that the reduction in the network investment cost effectively compensates for the increase in the utility operating cost. If such an incentive for precooling or post-cooling of the cold stream (preheating or post-heating of the hot stream) does not exist, the above-mentioned shift of the cooler (heater) should not be considered.

Modified Feasibility Rule No. 3. Consider the possibility of forming a new exchanger by merging a heater on the lowest-temperature portion of a hot stream with a cooler on the highest-temperature portion of a cold stream. Both heating and cooling utilities are originally placed in thermodynamically inefficient positions, which could, however, appear during the evolutionary changes of a given network. Forming a new exchanger from two thermodynamically inefficient heater and cooler results in an increase in the exchanger freedom parameter by A kW. This modified feasibility rule includes the inverse forms of the original feasibility rules no. 7-10 of Linnhoff and Flower. Note that in their forward forms, the last four feasibility rules of Linnhoff and Flower involve the thermodynamically irrational placement of a heater on a hot stream and a cooler on a cold stream in the initial network.

Based on the preceding analysis, it is evident that out of the ten feasibility rules of Linnhoff and Flower, only three of them (nos. 1 and 2, the inverse form of no. 8) may find some applications in the evolutionary improvement of a thermodynamically-based initial network such as that synthesized by steps 1 to 3 of the thermoeconomic approach. The remaining majority of the feasibility rules would rarely be applicable, as the placement of units in thermodynamically efficient positions can be assured in the generation of an initial network through the use of the thermodynamic matching rule. Consequently, only two of the feasibility rules (nos. 1 and 2) of Linnhoff and Flower have been adapted as Rule 2a in the thermoeconomic approach.

An Illustrative Example

The following example has been described in detail in Pehler and Liu ($\underline{1}$), in which it is shown that applications of the thermoeconomic approach have successfully generated optimum and suboptimum networks of five to ten process streams with much less time and effort compared to previous studies.

Figures 8a and 8b show, respectively, the heat content diagram and the grid (temperature interval) diagram for representing the final network (see Figure 8c) for a 7-stream problem (7SP1) reported by Masso and Rudd ($\underline{8}$) using a heuristic structuring method. The application of Step 2 of the thermoeconomic approach shows that the network represented by Figures 8a-8c uses more cooling utility for C_1 and C_2 (1281.2 kW) than the minimum amount required (1204.4 kW). The thermodynamic matching rule in Step 3 suggests that C_1 leads to a thermodynamically inefficient use of the cooling utility, as it is being placed on the intermediate-temperature portion of S_{h1}. Step 4a indicates that the network contains one more unit (H_1, 71.6 kW) than the quasi-minimum number of units (seven exchangers and coolers). Thus, the evolutionary synthesis of an improved network (Step 4b) can be done by deleting H_1 (Rule 1) and enlarging E_5 on S_{c4} to compensate for the elimination of the heating utility. E_4 on both S_{h1} and S_{c1} is reduced to accommodate for the increase in the heat

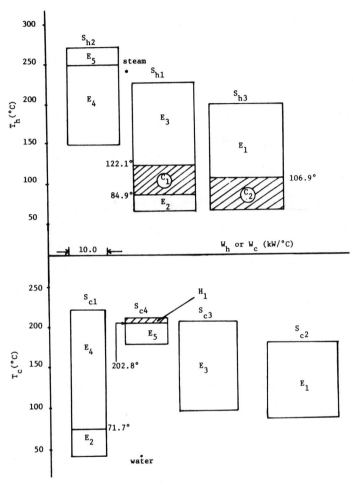

Figure 8a. Problem 7SP1: heat content diagram for the initial
network.

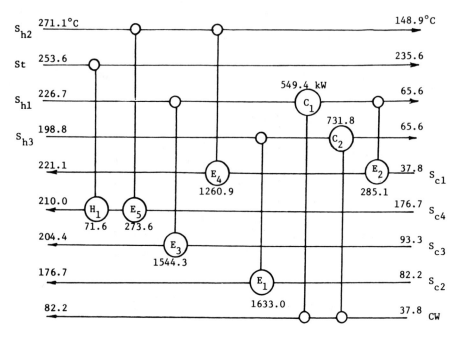

Figure 8b. Problem 7SP1: temperature interval diagram for the
initial network.

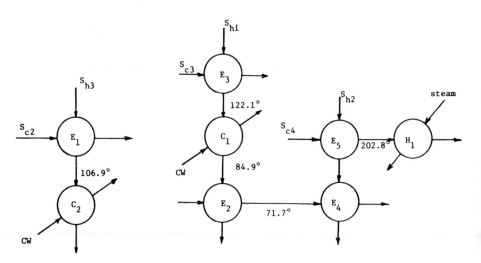

Figure 8c. Problem 7SP1: the initial network.

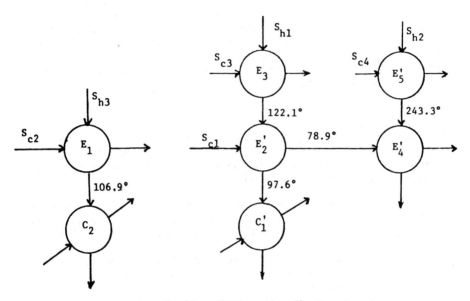

Figure 8d. Problem 7SP1: the final network.

load of E_5 on S_{h1}; and E_2 on both S_{h1} and S_{c1} is enlarged, which leads to a decrease in the heat load of C_1 on S_{h1}. Finally, the reduced C_1 on S_{h1} is shifted downward through the enlarged E_2 to the lowest-temperature portion of S_{h1} (Rule 2a), resulting in a thermodynamically efficient use of the cooling utility. Figure 8d shows the improved network, which is identical to the optimum network obtained by the evolutionary development (ED) method of Linnhoff and Flower (2).

Note that in applying the ED method to delete the unnecessary heater H_1 from the initial network shown in Figure 8a to 8c, Linnhoff and Flower had to consider whether this small heater could be shifted to a neighboring position of a cooler so that a fraction of the heat load could cancel that of the heater. The corresponding network calculations and manipulations were tedious.

Literature Cited

1. Pehler, F. A. and Liu, Y. A., "Studies in Chemical Process Design and Synthesis: VI. A Thermoeconomic Approach to the Evolutionary Synthesis of Heat Exchanger Networks," Chem. Eng. Commu., in press (1983).

2. Linnhoff, B. and Flower, J. R., "Synthesis of Heat Exchanger Networks: II. Evolutionary Generation of Networks with Various Criteria of Optimality," AIChE J., 24, 642 (1978).

3. Hohmann, E. C., "Optimal Networks for Heat Exchange," Ph.D. dissertation, University of Southern California, Los Angeles, California (1971).

4. Nishida, N., Liu, Y. A. and Lapidus, L., "Studies in Chemical Process Design and Synthesis: III. A Simple and Practical Approach to the Optimal Synthesis of Heat Exchanger Networks," AIChE J., 23, 77 (1977).

5. Linnhoff, B. and Flower, J. R., "Synthesis of Heat Exchanger Networks: I. Systematic Generation of Energy Optimal Newtorks," AIChE J., 24, 633 (1978).

6. Umeda, T., Itoh, J. and Shiroko, K., "Heat Exchange System Synthesis," Chem. Eng. Progr., 74, No. 7, 70 (1978).

7. Bett, K. E., Rowlinson, J. S. and Saville, G., Thermodynamics for Chemical Engineers, pp. 115-117, MIT Press, Cambridge, Massachusetts (1976).

8. Masso, A. H. and Rudd, D. F., "The Synthesis of System Designs: Heuristic Structuring," AIChE J., 15, 10 (1969).

RECEIVED August 29, 1983

Hierarchical Structure Analysis
Based on Energy and Exergy Transformation
of a Process System

MASARU ISHIDA

Research Laboratory of Resources Utilization, Tokyo Institute of Technology,
4259 Nagatsuta, Midori-ku, Yokohama, 227, Japan

Process systems are analyzed on the viewpoint of energy
and exergy transformation between processes. Since
processes including heat sources or sinks and work
sources or sinks are characterized thermodynamically by
vectors on a (ΔH, $T_0\Delta S$) plane called thermodynamic
compass, the first and second laws of thermodynamics
for a process system are associated with the direction
of the vector generated by summing up these vectors for
all processes. In order to disclose the hierarchical
structure of a process system, all processes in it are
distinguished into two groups, targets and coupled
processes, and six basic structures for process systems
--- singular system, binary system, multicoupler system,
multitarget system, looping multimediator system, and
accomodated system --- are discussed based on a new
diagram called Structured Process Energy-Exergy-flow
Diagram (SPEED). A few example process systems for
each basic structure are analyzed and the procedure
to disclose the hierarchical structure of complicated
process systems based on these basic structures is
discussed.

Exergy, or availability, is a quite powerful concept to be able to
deal in a unified manner with many kinds of energy, such as heat,
mechanical work, chemical energy, and so on ($\underline{1}$,$\underline{2}$,$\underline{3}$,$\underline{4}$).
Furthermore, the exergy analysis gives us the exergy destruction
in each portion in the process system. Since it is given as a
quantitative value denoting the deviation from the ideal process
system, we may judge based on it whether the modification of some
portion in the process system will lead significant energy saving.
    Up to this stage, the concept of exergy is applied as a tool
of the analysis of a process system. The next target then may be
to develop a methodology how we can use the concept of exergy to
synthesize a process system.

0097–6156/83/0235–0179$09.00/0

The main feature of a process system is to attain some target
process by utilizing other processes and transforming energy and
exergy among them. Accordingly the process system synthesis is
equivalent to constructing proper networks for such tranformation
of energy and exergy. Again the concept of exergy is quite
important to analyze various kinds of processes, such as thermal,
chemical, and mechanical ones, in a systematic manner.

The purpose of this paper is to offer a methodology to
analyze or synthesize the hierarchical structure of a process
system. In the first section, we outline the thermodynamics for a
process and a process system based on not only the first law of
thermodynamics but also the second law. There we will introduce
the concept of thermodynamic compass. In the second section, the
transformation of energy and exergy among processes is discussed.
It is shown that by applying thermodynamic compass, both physical
and chemical processes can be treated in a unified manner. In the
third section, six basic structures of a process system are
discussed. The network of the transformation of energy and exergy
among processes is represented by the SPEED (Structured Process
Energy-Exergy-flow Diagram). In the fourth section, we summarize
the methodology for analysis and synthesis of process systems.

Thermodynamics of a Process and a Process System

Figure 1 (a) shows the scheme of a process. The circle denotes
the control volume of the process and the arrow denotes the flow
of materials. For physical processes such as heating, cooling,
mixing, and separation, the material itself remains the same and
only its state is changed in the control volume. At the steady
state, the mole number for each component which enters the process
should be equal to that which leaves the process. For chemical
processes such as reactions, on the other hand, the material
itself is changed and the mole (or atomic) number for each element
should be kept constant.

Since the total enthalpy and entropy of the input and output
materials can be calculated, the enthalpy increase ΔH and the
entropy increase ΔS caused by a process can be obtained as
follows.

$$\Delta H = H_{out} - H_{in} \qquad\qquad (1)$$
$$\Delta S = S_{out} - S_{in} \qquad\qquad (2)$$

Although ΔH and ΔS for a process are defined based on the
input and output of the materials, the process may receive heat
Q_{in} or work W_{in} or may release heat Q_{out} or work W_{out}, as
schematically shown in Fig. 2.

Since ΔH and ΔS for each process can be specified, the
thermodynamic characteristics of the process may be represented by
a vector on the $(\Delta H, T_o\Delta S)$ plane shown in Fig. 1 (b). As we
discuss in later sections, the direction of this vector suggests

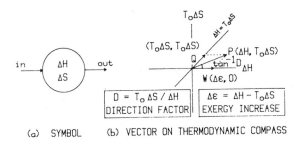

(a) SYMBOL   (b) VECTOR ON THERMODYNAMIC COMPASS

Figure 1. Representation of a process.

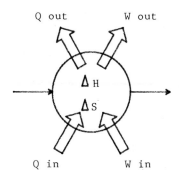

Figure 2. Flow of heat Q and work W.

the procedure to fulfill that process. Accordingly this diagram
is called the thermodynamic compass. The compass is an important
tool to find out proper direction in orienteering or
mountaineering. Similarly the thermodynamic compass will give us
a hint especially for the process system synthesis.

Since the direction of the vector is important, the tangent
of the vector is called the direction factor and denoted by D (5):

$$D = \frac{T_o \Delta S}{\Delta H} \tag{3}$$

Let us examine several typical processes and draw their
vectors on the thermodynamic compass.

Example Processes

Thermal processes. When n moles of a certain substance is heated
from T_1 to T_2, we have

$$\Delta H = \int_{T_1}^{T_2} n c_p dT \quad \text{and} \quad \Delta S = \int_{T_1}^{T_2} n(c_p/T)dT$$

When the heat capacity c_p is assumed to be constant between these
temperatures, the direction factor D is given as

$$D = \frac{T_o \Delta S}{\Delta H} = \frac{T_o(\ln T_2 - \ln T_1)}{T_2 - T_1} = \frac{T_o}{T_{1n}} \tag{4}$$

where T_{1n}, defined by $(T_2-T_1)/\ln(T_2/T_1)$, is the logarithmic mean
of T_1 and T_2.

Then the vectors of the heating and cooling processes are
obtained as shown in Fig. 3. It is seen that the vector of the
heating process always appears on the first quadrant of the
thermodynamic compass, whereas the cooling process on the third
quadrant, giving rise to the following four cases.

$$
\begin{aligned}
&T_2 > T_1 \text{ and } T_{1n} > T_o \quad \text{for heating} \\
&T_2 > T_1 \text{ and } T_{1n} < T_o \quad \text{for refrigerant} \\
&T_2 < T_1 \text{ and } T_{1n} > T_o \quad \text{for cooling (i.e., heat source)} \\
&T_2 < T_1 \text{ and } T_{1n} < T_o \quad \text{for refrigeration}
\end{aligned}
$$

It is found from Fig. 3 that the direction factor D for these
processes depends significantly on the logarithmic mean
temperature T_{1n} and that their vectors never appears on the second
or fourth quadrant.

Separation and mixing. For a separation process, the changes in
enthalpy and entropy are generally positive and negative,
respectively. So their direction factor becomes negative and its
absolute value is rather large, because the enthalpy change is
relatively small. For example, when 250 moles of aqueous solution
of 40 mol% methanol is distilled and separated into aqueous

solutions of 95.8 mol% methanol and 5.8 mol% methanol, we have

$$\{ 100 \text{ CH3OH} + 150 \text{ H2O} \}$$
$$= \{ 91 \text{ CH3OH} + 4 \text{ H2O} \} + \{ 9 \text{ CH3OH} + 146 \text{ H2O} \}$$
$$\Delta H = 0.18 \text{ kJ}, \quad T_0 \Delta S = -55.1 \text{ kJ}, \quad \Delta \varepsilon = 55.3 \text{ kJ}$$

and
$$D = -306$$

where the braces in the above formula mean that the stream within them consists in the same phase and composes a mixture.

Then the vector for the sepation process appears on the fourth quadrant as shown in Fig. 4.

Since the mixing is the reverse process of separation, the changes in enthalpy and entropy are usually negative and positive, respectively. Therefore the vector appears in the second quadrant on the thermodynamic compass.

Special processes without flow of materials

The processes discussed so far had inputs and outputs of materials. However, the following four processes which have neither input nor output are introduced to make the thermodynamics of a process system quite simple.

Heat source. When a heat reservoir which is able to release heat of the quantity Q at the temperature T shown in Fig. 5 (a) is regarded as a process, the following equation similar to Eq. (4) holds.

$$\Delta H = -Q, \quad \Delta S = -Q/T, \quad \text{and} \quad D = T_0/T \qquad (5-a)$$

Namely, the direction factor for a heat sink is inversely proportional to its temperature.

Heat sink. Similarly for a heat reservoir to absorb heat of the quantity Q at the temperature T shown in Fig. 5 (b), we obtain

$$\Delta H = Q, \quad \Delta S = Q/T, \quad \text{and} \quad D = T_0/T \qquad (5-b)$$

Accordingly the vectors for the heat source or sink are equal to those of thermal processes shown in Fig. 3, when the logarithmic mean temperature T_{ln} is replaced by the temperature T of the heat source or sink.

Work source. When a work reservoir which is able to release work of the quantity W shown in Fig. 6 is regarded as a process, the following equations are obtained.

$$\Delta H = -W, \quad \Delta S = 0, \quad \text{and} \quad D = 0 \qquad (6-a)$$

Work sink. Similarly for a work reservoir to absorb work of the quantity W shown in Fig. 6, we obtain

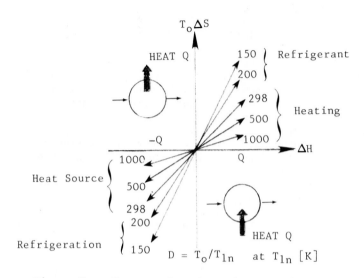

Figure 3.   Vectors for thermal processes.

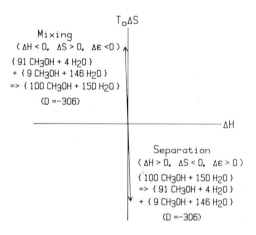

Figure 4.   Vectors for separation and mixing processes.

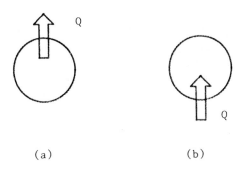

Figure 5. Heat source (a) and heat sink (b).

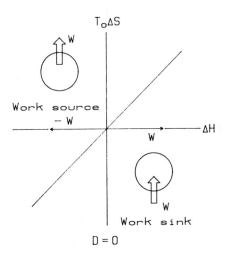

Figure 6. Vectors for work source and sink.

$$\Delta H = W, \quad \Delta S = 0, \quad \text{and} \quad D = 0 \tag{6-b}$$

The sources or sinks of electricity or light may also be regarded as special processes, but their vectors are equivalent to those of work sources or sinks.

Classification of processes on thermodynamic compass

We have examined the characteristics of the enthalpy increase ΔH and entropy increase ΔS for a process. Then let us consider the exergy increase $\Delta \varepsilon$ for the process.

When $\Delta \varepsilon$ is defined as the difference between the exergy of the outlet materials ε_{out} and that of the input ones ε_{in}, we have

$$\begin{aligned}
\Delta \varepsilon &= \varepsilon_{out} - \varepsilon_{in} \\
&= [(H-H_0) - T_0(S-S_0)]_{out} - [(H-H_0) - T_0(S-S_0)]_{in} \\
&= \Delta H - T_0 \Delta S \\
&= \Delta H \,(1 - D)
\end{aligned} \tag{7}$$

In deriving the above equation, the following relation at the reference state is applied.

$$[H_0]_{in} = [H_0]_{out} \quad \text{and} \quad [S_0]_{in} = [S_0]_{out} \tag{8}$$

When the concept of the exergy increase $\Delta \varepsilon$ is applied, the process vector on the thermodynamic compass shown in Fig. 1 (b) may be decomposed into two vectors Q and W. The vector Q on the diagonal line is equivalent to the heat sink at the reference temperature T_0, of which direction factor is unity. On the other hand, the vector W on the abscissa is equivalent to the work sink and has the magnitude of the exergy increase of the process, $\Delta \varepsilon$. In this way, the exergy increase $\Delta \varepsilon$ may easily be obtained graphically from the process vector on the thermodynamic compass. Therefore the plane of the thermodynamic compass may be divided into six regions by these three lines, $\Delta H = 0$ (ordinate), $\Delta S = 0$ (abscissa), and $\Delta \varepsilon = 0$ (diagonal line), as shown in Fig. 7. According to this division, the process may be classified into six types: heating, separation, refrigeration, heat source, mixing, and refrigerant. With respect to ΔH, the process of the refrigerant, heating, and separation types are energy-accepting (i.e., $\Delta H > 0$) and those of the refrigeration, heat source, and mixing types are energy-donating (i.e., $\Delta H < 0$). With respect to $\Delta \varepsilon$, on the other hand, the processes of the heating, separation, and refrigeration types are exergy-accepting (i.e., $\Delta \varepsilon > 0$) and those of the heat source, mixing, and refrigerant types are exergy-donating (i.e., $\Delta \varepsilon < 0$). Also the former is called endergonic and the latter exergonic.

Other typical processes

Before discussing about the characteristics of a process system consisting of plural processes, some other typical processes such as reactions and polytropic processes will be considered.

<u>Chemical reaction</u>. To perform detailed exergy analyses of chemical reactions, information about composition of both reactants and products is required. Then the changes in enthalpy, entropy, exergy, and the direction factor for the reaction process can be calculated. For primary dicussions for the reaction system synthesis, however, the standard direction factor $D°$ defined by the following equation may be utilized.

$$D° = T_0 \Delta S° / \Delta H \tag{9}$$

where $S°$ is the entropy of each component under unit pressure.

Figure 8 shows the vectors on the thermodynamic compass for typical endergonic reactions. It is found that for chemical reactions all three types can be observed. However, the most common one may be the heating type of which examples are: the reduction of metallic oxides and the decomposition of water to hydrogen and oxygen.

On the other hand, Fig. 9 shows the vectors for exergonic reactions. They often plays the role of donating exergy to other processes and their reactants especially with negative or small positive values for the direction factor such as ATP shown in Fig. 9 are called high-exergy compounds.

Since ΔH and ΔS are scarcely dependent on temperature T, the direction factor D for chemical reacions is almost constant over a wide range of temperature, as shown in Fig. 10. When it is assumed to be independent of temperature, its value is equal to the reciprocal of the dimensionless temperature at which the change in Gibbs free energy ΔG becomes zero as follows.

$$D = T_0 \Delta S / \Delta H = T_0 / T_{eq} \tag{10}$$

where

$$\Delta G = \Delta H - T \Delta S = 0 \quad \text{at } T = T_{eq}$$

By similar reason, the vectors for reactions in Figs. 8 and 9 are scarcely affected by the temperature.

On the other hand, as shown in Fig. 11, the direction factor D is affected by the pressure according to the following simple relation (<u>6</u>).

$$D \text{ at } P_2 - D \text{ at } P_1 = -\frac{\mathcal{R} T \Delta n}{\Delta H} \ln \frac{P_2}{P_1} \tag{11}$$

where Δn denotes the change in mole number by the reaction.

The advantage of the direction factor D for chemical reactions will be discussed later.

<u>Polytropic process</u>. Compression and expansion are important processes in chemical industry. The changes in enthalpy and

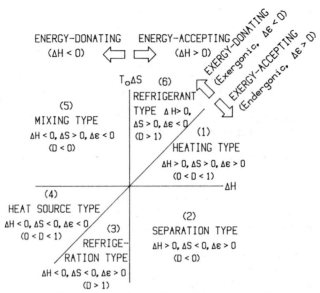

Figure 7.   Classification of processes on thermodynamic
compass.

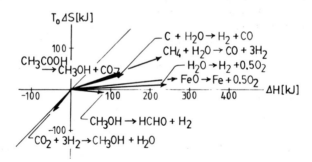

Figure 8.   Examples of endergonic reactions.

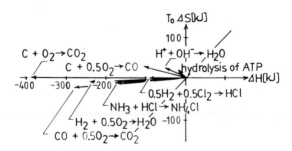

Figure 9.   Examples of exergonic reactions.

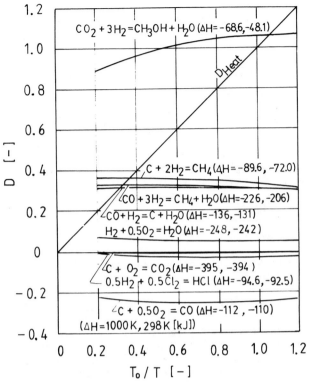

Figure 10.   Effect of temperature T on D.

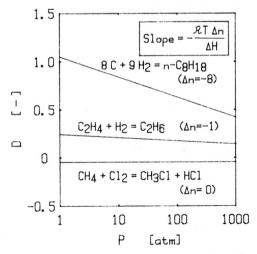

Figure 11.   Effect of pressure on D of reaction processes.

entropy of n moles of an ideal gas from the state at T_1 and P_1 to T_2 and P_2 can be obtained as follows.

$$\Delta H = n \int_{T_1}^{T_2} c_p dT = n c_p (T_2 - T_1)$$

$$\Delta S = n \left[\int_{T_1}^{T_2} \frac{c_p}{T} dT - \mathcal{R} \ln \frac{P_2}{P_1} \right] = n \ln(\frac{T_2}{T_1})^{c_p} (\frac{P_1}{P_2})^{\mathcal{R}}$$

When the polytropic relation, PV^m = constant, is assumed, the direction factor is given as follows (5):

$$D = \frac{T_0}{(c_p/c_m) T_{1n}} \qquad (12)$$

where m is the polytropic exponent and c_m is the polytropic heat capacity defined by

$$c_m = c_v (m - \gamma)/(m - 1)$$

Figure 12 shows the two process vectors. One is the case when a mole of nitrogen at 298.2 K and 1 atm is compressed up to 4 atm and the other is for the reverse case when the compressed nitrogen is expanded again to 1 atm. Hence the direction factor becomes positive for compression (i.e., $m > \gamma$) and negative for expansion (i.e., $m < \gamma$). It is zero for adiabatic compression or expansion (i.e., $m = \gamma$).

Thermodynamics of a process system

Let us consider a process system which consists of several processes such as heat exchange, separation, mixing, chemical reaction, and so on.

Usually for the process with input and/or output of heat and work shown in Fig. 2, the first law of thermodynamics is represented by the following equation.

$$\Delta H = Q_{in} + W_{in} - Q_{out} - W_{out} \qquad (13)$$

In the above equation, ΔH is given as a derivative value, while Q and W are given as absolute values. As far as simple process systems are concerned, such description works well. As the process system becomes complex, however, it sometimes leads to confusion. Then we may take two other approaches.

Approach based on absolute values. By introducing the enthalpy at the reference state, H_0, and by replacing $(H - H_0)$ by Θ, we have

$$[\Theta + Q + W]\text{input} = [\Theta + Q + W]\text{output} \qquad (14)$$

Θ is the enthalpy calculated by setting zero at the reference state similar to the concept of exergy ε.

On the other hand, the second law gives rise to the following equation concerning with the exergy.

$$[\Theta + Q (1- \frac{T_0}{T}) + W]\text{in}$$

$$= [\Theta + Q (1- \frac{T_0}{T}) + W]\text{out} + T_0 \, S_{irr} \qquad (15)$$

where $T_0 \, S_{irr}$ represents the exergy destruction caused by irreversibility and takes always a positive value.

When this approach is adopted, we must define the reference state definitely.

Approach based on derivative values. As stated previously, the terms of heat and work may be treated as derivative values by introduncing special processes without the flow of materials.

As a general case, let us consider a set of n processes enclosed in an adiabatic enclosure shown in Fig. 13 (a). Such a system is called a process system. In this paper, the boundary of process systems is represented by broken lines, as shown in Fig. 13.

The first and second laws of thermodynamics for such a process system is represented by

$$\Sigma \Delta H_i = 0 \qquad \text{(First law of thermodynamics)} \qquad (16)$$

$$\Sigma \Delta S_i \geq 0 \qquad \text{(Second law of thermodynamics)} \qquad (17)$$

The increase in entropy is caused by irreversibility in the system.

From these two equations, the exergy destruction in the system is obtained as

$$- \Sigma \Delta \varepsilon_i = - \Sigma (\Delta H_i - T_0 \Delta S_i) = T_0 \Sigma \Delta S_i \geq 0 \qquad (18)$$

From these equations it is concluded that the summation of all process vectors gives rise to a vertical vector of which the magnitude corresponds to the exergy destruction, as shown in Fig. 13 (b).

To compare this approach with the previous one based on absolute values, consider a process system in which a process with ΔH and ΔS is coupled with a heat source with Q_{in}, a heat sink with Q_{out}, a work source with W_{in}, and a work sink with W_{out}. Then Eq. (14) is obtained by substituting Eqs. (5) and (6) into Eq. (16). Similarly from Eq. (18) we have

$$[\Theta + Q \ (1- \frac{T_0}{T}) + W]_{in}$$

$$= [\Theta + Q \ (1-\frac{T_0}{T}) + W]_{out} + T_0 \Sigma \Delta S_i \qquad (19)$$

Hence, by introducing the special four kinds of processes, the term $T_0 \ S_{irr}$ in Eq. (15) is found to be equal to $T_0 \Sigma \Delta S_i$.

Energy and Exergy Transformation in a Process System

By adopting the derivative approach and applying the thermodynamic compass and the concept of a process system, we may demonstrate the characteristics of exergonic (i.e., exergy-donating) processes clearly. For example, a process of the heat source or mixing type may compose a process system when it is coupled with a heat sink at the reference temperature T_0, giving rise to positive exergy destruction, as shown in Fig. 14 (a). Similarly, a process of the refrigerant type makes a process system by being coupled with a heat source at T_0, as shown in Fig. 14 (b). Hence the exergonic process is defined as a process which is able to compose a process system when it is coupled with a heat sink or source at the reference temperature T_0. On the other hand, the endergonic process needs some exergy source besides the heat source or sink at T_0 to compose a process system, as will be discussed later.

By taking notice of the transformation of exergy among processes, we may deal with physical processes and chemical ones in unified manner. For example, Fig. 15 (a) compares a heat exchanger and a chemical reactor. For the former case, the fluid passing through the inner tube will gain heat, resulting in the temperature rise by α , whereas the fluid in the annulus gives exergy to the inner fluid. For the latter case, on the other hand, an endergonic reaction, $FeO + H_2$ => $Fe + H_2O$ takes place. These two systems seem to have nothing to do with each other. However, when the scheme of exergy transformation is examined as shown on the right side of Fig. 15, we may find that they have quite similar structure. Actually this endergonic reaction is of the heating type and hence its vector appears on the first quadrant as that of the heating process.

Nevertheless the criterion of the heat exchanger has been described by the term of temperature, whereas that of the chemical reaction system by the Gibbs free energy. The above fact suggests that the same criterion may be applied to both cases when systematic analysis of a process system is performed based on the concept of exergy transformation. This point will be discussed more quantitatively in the later section.

We have observed that the processes of the same type may be perfomed by the same coupler and hence have similar scheme for the exergy transformation. Then suppose three endergonic processes of

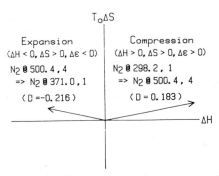

Figure 12. Vectors for polytropic compression and expansion.

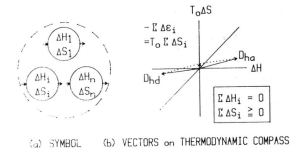

(a) SYMBOL   (b) VECTORS on THERMODYNAMIC COMPASS

Figure 13. Concept of a process system.

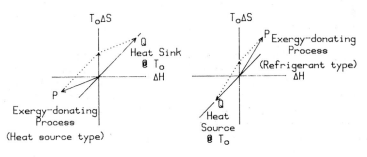

(a) Heat source type   (b) Refrigerant type

Figure 14. Coupling of an endergonic process with a heat source or sink at T_O.

different types shown in Fig. 16. Although the exergy increase of
all these processes is about 60 kJ, the procedures to realize them
are quite different from each other. Therefore not only the
exergy increase $\Delta \varepsilon$ but also the direction of the vector on the
thermodynamic compass D is found to be a key parameter. In the
next section, the relationship between the direction of the
process vector and various kinds of scheme of exergy
transformation in a process system will be discussed.

Basic structures of a process system

By representing each process by a vector on the thermodynamic
compass, any process system may be represented as a set of
vectors. When there are many processes in the system, however,
the interaction among the processes becomes quite complicated. To
clarify such complicated interactions, the concept of the
hierarchical structure of a process system will be introduced.
        It is shown in this section that there are six basic
structures and that even a complicated process system is composed
of the combination of these basic structures.

Singular system. It is found from Eqs. (16) and (17) that only a
process with $\Delta H = 0$ and $\Delta S \geq 0$ is able to constitute a process
system by itself. Then it can proceed spontaneously without
tranformation of exergy with other processes, as schematically
shown in Fig. 17 (b). Of couse its vector appears vertically on
the thermodynamic compass, as shown in Fig. 17 (a) and its length
corresponds to the magnitude of the exergy destruction in the
system.
        Figure 18 shows an example of a singular system when the
water-gas-shift reaction takes place adiabatically. The braces
mean the mixture and the quotation marks over the reaction formula
shows that this reaction is the target of this process system.
The temperature [K] and pressure [atm] of the input and output
streams are specified following the mark @. The adiabatic nature
of this process is disclosed by the value of ΔH. The exergy
destruction $T_0 \Sigma \Delta S_i$ is, therefore, given by changing sign of the
exergy increase of this adiabatic process. Adiabatic compression
and expansion are other examples of the singular system.

Binary system. The binary system composed of two processes is the
most basic system and it is quite frequently applied. Generally
one of the binary processes is the target and the other donates
exergy to it or accepts exergy from it. Hence the latter process
may be called a coupled process or simply coupler.
        When the processes 1 and 2 constitute a process system, the
following equation is derived from Eqs. (16) and (17).

$$\Delta H_1(D_1 - D_2) \geq 0 \qquad \text{(for a binary system)} \qquad (20)$$

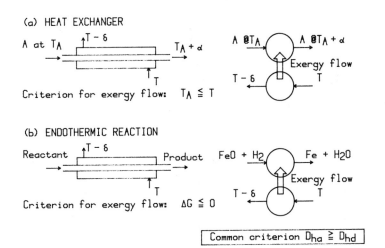

Figure 15. Analogy between physical processes and chemical processes.

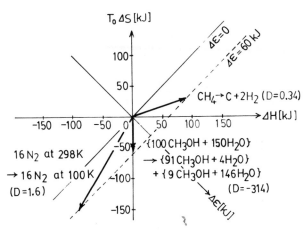

Figure 16. Three kinds of processes with almost equivalent amount of $\Delta \varepsilon$.

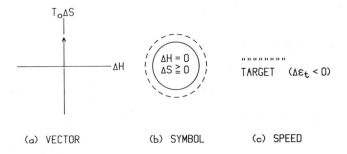

(a) VECTOR   (b) SYMBOL   (c) SPEED

Figure 17. Representation of a singular system.

$$\{ 5\, CO + 3\, H_2 + CO_2 + 4\, H_2O \} \, @\, 400,1$$
$$=> \{ 3\, CO + 5\, H_2 + 3\, CO_2 + 2\, H_2O \} \, @\, 585,1$$
$$\Delta H = 0, \quad \Delta \varepsilon = -27.4\, kJ, \quad D = Infinite$$
$$T_o \Sigma \Delta S_i = 27.4\, kJ$$

Figure 18. SPEED for an adiabatic reaction system.

When one of the binary processes is energy-accepting (i.e., $\Delta H > 0$), the other is always energy-donating (i.e., $\Delta H < 0$). By denoting the direction factors of both processes by D_{ha} and D_{hd}, respectively, the above equation reduces to

$$D_{ha} \geq D_{hd} \qquad \text{(for a binary system)} \qquad (21)$$

because the enthalpy increase for the energy-accepting process, ΔH_{ha}, is always positive. Eq. (20) or (21) is the necessary condition for the binary system to hold.

Let us apply the above criterion to two example cases shown in Fig. 15.

For a heat exchanger, Eq. (5) is substituted into Eq. (21), giving rise to the following criterion based on the mean temperature.

$$[T_{ln}]_{hd} \geq [T_{ln}]_{ha}$$

This is the criterion of the heat exchanger derived from Eq. (21).

With respect to chemical reactions, the most popular combination is the case when an endothermic reaction is coupled with a heat source as shown in Fig. 15 or an exothermic reaction is coupled with a heat sink. Since the direction factor for heat sources or sinks at the temperature T is given as T_0/T, the criterion, Eq. (20), is reduced to

$$\Delta H[D - (T_0/T)] \geq 0 \quad \text{(with heat source or sink at T)} \qquad (22)$$

or we may derive

$$D > T_0/T \quad \text{for an endothermic reaction}$$
and
$$D < T_0/T \quad \text{for an exothermic reaction}$$

where D is the direction factor for the reaction process.

Therefore, only from the value of D , we may estimate the temperature range in which the reaction may proceed. As a first-step approximation, we may use D at the reference temperature T_0, since D is almost independent of temperature, as shown in Fig. 10. This point may be an important advantage of the criterion based on D especially for reaction system synthesis.

When $-(T/T_0)$ is multiplied to both sides of Eq. (22), the left-hand side becomes equal to the increase in Gibbs free energy ΔG and we obtain the following relation.

$$\Delta G = \Delta H - T\Delta S = - T\Sigma \Delta S_i \leq 0$$
$$\text{(with heat sink or source at T)}$$

This criterion is very popular in chemistry to examine the equilibrium condition for reactions and it can easily be derived

from more general criterion, Eq. (21). For example, the criterion
based on Gibbs free energy is valid only for isothermal systems
but Eq. (21) may be applied also to nonisothermal systems. This
advantage is efficacious in the exergy analysis of a process
system, because almost all processes in process systems proceed
nonisothermally.

Figure 19 (a) shows that there are four combinations between
an endergonic process (denoted as 1) and an exergonic one (denoted
as 2). Namely, for a process of the heating type, we may combine
a process of the heat source or mixing type, as shown in Figs. 19
(a-i) and (a-ii). A process of the separation type needs a
process of the mixing type, as shown in Fig. 19 (a-iii). Hence,
as far as the binary system is concerned, a process of the other
types cannot be coupled with a process of the separation type.
Similarly, a process of the refrigeration type needs a process of
the refrigerant type, as shown in Fig. 19 (a-iv). In any of the
above processes the exergy transforms from the exergonic process
(Process 2) to the endergonic one (Process 1), as shown in Fig. 19
(b). When the target process is endergonic, an exergy donor may
become the coupler. On the other hand, when the target is
exergonic, an exergy acceptor will be coupled, as shown in Fig. 19
(c).

In order to represent the transformation of exergy among
processes and the hierarchical structure of a process system, we
will introduce the Structured Process Energy-Exergy-flow Diagram
which will be abbreviated as SPEED ($\underline{7},\underline{8}$). The SPEED is a
computer-oriented diagram which has functions like a process flow
diagram but can easily be read by a computer. It is proposed to
describe the scheme of transformation of exergy among processes
and to disclose the hierarchical structure of a process system.
Namely, the target process is declared definitely at the top. To
identify it as the target, the quotation mark is typed over its
formula, whereas the exergy donor and the exergy acceptor are
represented by setting a straight line and a ripple line,
respectively. In the SPEED, couplers such as donors or acceptors
are below the target in rank and the indent is set before
couplers, as shown in Fig. 19 (c). By introducing these rules,
all thermodynamic calculations can be performed by a computer.

On the other hand, when the binary processes are both
exergonic, there are only two combinations, as shown in Fig. 20
(a): One is the combination of the mixing and refrigerant types
and the other of the combination of the heat source and
refrigerant types. A process system to cool a certain substance
at 400 K to 300 K by a refrigerant at the temperature less than
the reference temperature T_0 is an example of the latter
combination. In such a process system, the summation of the
exergy decrease $-\sum\Delta\varepsilon$ for both processes yield the exergy
destruction in the process system. Since the coupler in such a
system usually plays the role of accelerating the rate of the
target process, it is called the accelerator and denoted by a
a double line in SPEED, as shown in Fig. 20 (c).

<u>Multicoupler system.</u> The target may be linked with plural
couplers such as exergy donors, acceptors, and/or accelerators, as
shown in Fig. 21 (a). Such a set of processes is called a
multicoupler system. An example is seen in the electrolysis of
water. From the value of the direction factor of the target
process, the temperature as high as 6000 K ($T_0/T = 0.05$) is
required when only a heat source is selected as exergy donor. On
the other hand, when an electricity source and a heat source at
the ambient temperature T_0 are selected as couplers, an ideal
process system with no exergy destruction may theoretically be
constructed as shown in Fig. 22. As seen in this figure, the
required quantity of electricity corresponds to the change in
Gibbs free energy Δ G of the target, and it is to be noted that
the heat source at T_0 supplies the rest of energy. In the SPEED,
the couplers which are linked to the same target are typed in the
same rank.

Another example of a multicoupler system is the separation
system. it was mentioned in Fig. 19 (a-iii) that the separation
type has only one combination, namely with the mixing type. it is
known that active transport of various kinds of ions may be
achieved by the help of the hydrolysis of ATP of the mixing type.
As shown in Fig. 23, we may synthesize a process of mixing type P
by combining two vectors of different types. For distillation,
for example, the process in the reboiler is of the heat source
type (vector i-a in Fig. 23), and that in the condenser is of the
heating type (i-b). Addition of these two vectors results in a
vector of of the mixing type, P. Other examples are addition of
the vector of the refrigerant type (ii-a) and that of the
refrigeration type (ii-b) or of the vector on the abscissa (iii-a,
an extreme case of the heat source type) and that of the heating
type (iii-b). The former may be observed for the case of
low-temperature separation and the latter for electrodialysis or
reverse osmosis.

When two processes a and b are combined, the resultant
process has the following properties.

$$\Delta H_{a+b} = \Delta H_a + \Delta H_b \tag{23}$$

$$D_{a+b} = \frac{\Delta H_a D_a + \Delta H_b D_b}{\Delta H_a + \Delta H_b} \tag{24}$$

The process system which contains several couplers for a
single target may be converted to a binary process system by
assembling all coupled processes into one coupler as schematically
shown in Fig. 21.

<u>Multitarget system.</u> Since a process is transition of the state of

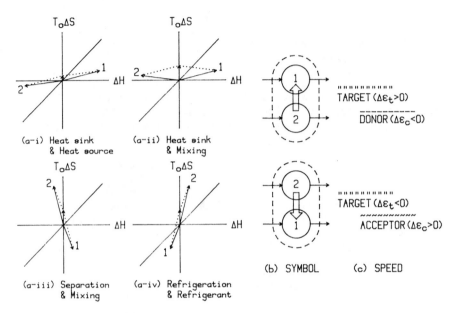

Figure 19. Combination of enderdonic process 1 and exergonic
process 2.

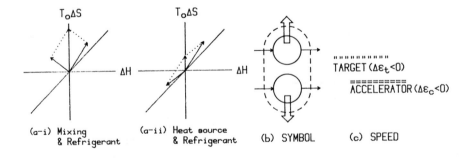

Figure 20. Combination of two exergonic processes.

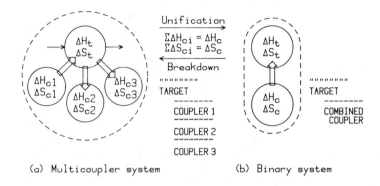

(a) Multicoupler system   (b) Binary system

Figure 21.   The scheme of a multi-coupler system and its relation to a binary system.

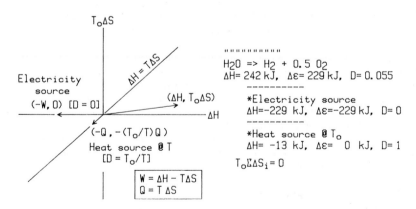

Figure 22.   Thermodynamic compass and SPEED for electrolysis of water.

materials, the target may be decomposed into series of subtargets, in which the output of a subtarget may become the input of the next subtarget, as shown in Fig. 24. In order to convert A to B, A is first reacted with M, giving rise to C. Then C is converted to D. And finally D gives B and M. Hence, the summation of changes in enthalpy and entropy of the subtargets becomes equilvalent to those of the original target.

$$\text{Target process} = \Sigma \text{ Subtarget process (i)} \qquad (25)$$

For example, for the target of decomposition of water into hydrogen and oxygen, we have subtargets such as

$$H_2O + Zn \quad => \quad ZnO + H_2$$
$$ZnO \quad => \quad Zn + 0.5\ O_2$$

On the thermodynamic compass, addition of the two subtarget vectors gives rise to the target vector as shown in Fig. 25.

When a target is decomposed into several subtargets, each subtarget and its coupled processes compose a process system of smaller scale, i.e., a subsystem, as shown by the broken lines in Fig. 24. For each subsystem, Eqs. (16) and (17) hold and we can obtain the exergy destruction in it, as shown in Fig. 26. Since the subtarget has the same characteristics as the target, the quotation mark is used also for subtargets in the SPEED.

By intoducing proper subtarget processes, each subsystem may be performed under milder conditions. In this example, the direction factor for the decomposition of zinc oxide is 0.16 and hence the temperature should be as high as 2000 K. But this temperature is much less than that of the target process, 6000 K.

Looping multimediator system. Even when there is only one target process, the process system may be decomposed to several subsystems by utilizing mediators. These mediators function as carriers of energy (enthalpy), entropy, and/or exergy. Figure 27 shows the scheme of the mediator loop. In this example, the loop consists of three mediators, $M_1 => M_2$, $M_2 => M_3$, and $M_3 => M_1$. The first is linked to the target process, and the others to the couplers. Mediators are denoted by a dotted line in the SPEED.

For a looping multimediator system, we have

$$\Sigma \text{ Mediator processes (i)} = 0 \qquad (26)$$

Hence, when all mediators are eliminated, it may be reduced to a multicoupler system.

When there are mediators in the system, we may decompose the system into subsystems which are also enclosed by broken lines in Fig. 27.

Also there are many hierarchical structures which have multitargets and looping multimediators simultaneously. An

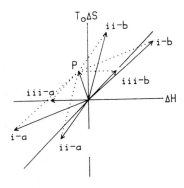

Figure 23. Synthesis of a mixing-type process P from two
processes a and b.

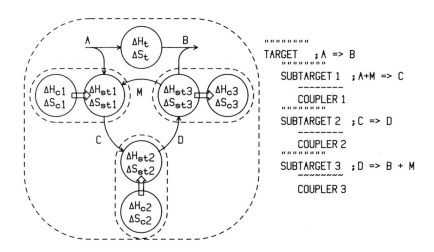

Figure 24. The scheme and SPEED for a multitarget system.

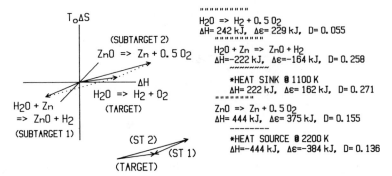

Figure 25. A multi-reaction system for decomposition of water.

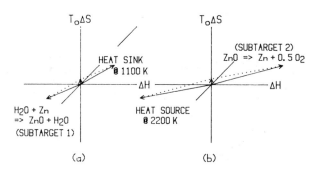

Figure 26. Thermodynamic compasses of two reaction subsystems for decomposition of water.

example is Carnot cycle of which process scheme is shown in Fig.
28, where four processes, isothermal expansion, adiabatic
expansion, isothermal compression, and adiabatic compression of
ideal gas form a mediator loop and transform thermal energy into
work. The process vectors for each subsystem in the Carnot cycle
are shown in Figs. 29 (a) through (d). In the isothermal
expansion, (a) in Fig. 29, the heat source gives entire energy to
the work sink and the entropy is stored in the expanded gas. In
the adiabatic expansion, (b) in Fig. 29, the mediator plays the
role of a donor to produce work. It accompanies no increase in
entropy but the decrease in pressure and temperature of the gas.
The next two stages are introduced to satisfy Eq. (26) by
consuming a part of work produced in the previous two stages.
Namely in the isothermal compression, (c) in Fig. 29, some work is
applied and the entropy which has been stored in the gas is
completely transferred to the heat sink at a lower temperature.
In the adiabatic compression, (d) in Fig. 29, work is applied
again to raise the pressure and temperature of the gas till the
initial state.

   As a whole, the Carnot cycle is found to consist of a
mediator loop shown in Fig. 30 (a) and a multicoupler system
composed of a net work sink as an overall target and the two
couplers: a heat source at high temperature and a heat sink at low
temperature, as shown in Fig. 30 (b). Hence the mediators act as
carrier of energy and entropy.

Accomodated process system. It is found in the previous sections
that the hierarchical network of the SPEED is constructed based on
subtargets and looping mediators. In actual process systems,
however, there may be subsystems to supply donor reactants or
exergy-donating utilities such as electricity and steam. Such
subsystems cannnot be classified into the previous groups because
the product of the target in the subsystem appears in the couplers
of the main process system. Therefore they are called the
accomodated process system and treated separately. Since each
accomodated system has a target, we may construct a SPEED for each
accomodated subsystem.

Summary of SPEED analysis

The exergy analysis of an existing process system by the SPEED and
the thermodynamic compass may be performed by the following steps.
1) Specify the target process.
2) List the coupled processes.
3) Disclose the system structure.
4) Output the exergy destruction.
   The first step is generally easy. Note however that the
target is given in the form of a process in the SPEED. In the
second step, main exergy donors, acceptors, and accelerators are
listed. In the third step we may find subtargets and/or looping

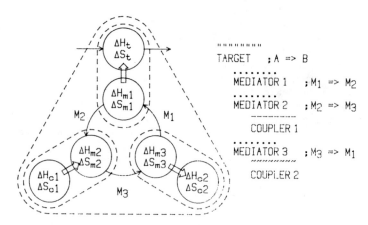

Figure 27. The scheme and SPEED for a looping multimediator system.

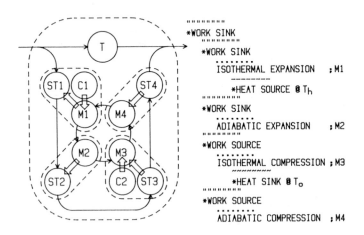

Figure 28. The scheme and SPEED for Carnot cycle system.

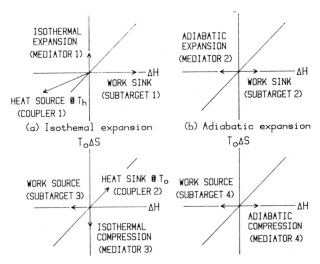

Figure 29. Thermodynamic compasses of four subsystems in Carnot cycle.

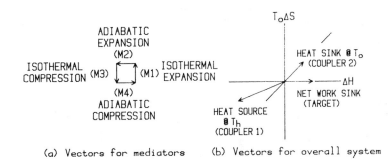

Figure 30. Mediator loop and thermodynamic compass for the whole Carnot cycle system.

mediators. The second and third steps are quite important,
especially when the process system becomes complex. During these
steps, the hierarchical scheme of energy and exergy transformation
among processes may be clarified. When the SPEED is completed,
the computer will read it and the exergy destruction in each
subsystem as well as the whole system may automatically be
outputted. Namely, in the SPEED, the thermodynamic calculations
are performed by the computer based on the compiled data base, and
the values of ΔH, $\Delta \varepsilon$, and D are outputted for each process and the
exergy destruction $T_0 \Sigma S_i$ for a process system (8).

On the other hand, synthesis is a problem to construct a
SPEED for a given target, i.e., to find appropriate subtargets,
mediators, exergy donors, exergy acceptors, and accelerators.
Since the SPEED is constructed based on top-down description, it
may be performed as follows.
1) Specify the target process. Jump to step 3).
2) Introduce appropriate subtargets or looping mediators to
   compose subprocesses.
3) Judge whether the target may be realized by appropriate
   couplers. If impossible, jump to step 2).
4) Take mixtures into consideration, if necessary, and solve the
   separation and heat exchange tasks.
5) Output the exergy destruction
6) Modify the inlet or outlet conditions of the process and/or the
   system structure if necessary.

Conclusion

1) On the thermodynamic compass, processes are classified into six
   types --- heating, separation, refrigeration, heat source,
   mixing, and refrigerant --- from the viewpoint of thermodynamics. This classification can be applied to both physical
   and chemical processes.
2) The direction facctor D (=$T_0\Delta S/\Delta H$) is given as T_0/T_{1n} for
   thermal processes, T_0/T_{eq} for chemical reactions, and
   $T_0[(c_p/c_m)T_{1n}]$ for polytropic processes. Based on it,
   a new criterion for processes to constitute a process system is
   derived.
3) Six basic system structures --- singular system, binary system,
   multicoupler system, multitarget system, looping multimediator
   system, and accomodated system --- are introduced from the
   viewpoint of exergy transformation and a top-down hierarchical
   description of the whole process system based on the SPEED is
   demonstrated.

Appendix 1   Efficiency of exergy transformation

We have discussed two kinds of approaches based on absolute values and on derivative ones. Both approaches may also be applied to defining the efficiency of the exergy transformation in the process system.

Based on absolute values. When the input and output of materials, heat, and work for the whole process is considered, the efficiency of the exergy transformation from all inputs to the objective outputs is obtained as follows.

$$\eta_\varepsilon = \frac{\text{Exergy of objective outputs}}{\text{Exergy of all inputs}}$$

$$= \frac{\text{Exergy of objective outputs}}{\text{Exergy of objective outputs} + \text{Exergy of wastes} + \text{Exergy destruction}} \tag{27}$$

When we adopt this definition, the objective outputs must be declared explicitely.

Since all terms in Eq. (27) are positive, the value of η_ε ranges from zero to unity. For ideal (i.e. reversible) process systems, η_ε is given as unity.

Based on derivative values. When the exergy transformation among the target process and the coupled processes is taken into consideration, the efficiency of the exergy transformation is given as

$$\eta_{\Delta\varepsilon} = \frac{\text{Exergy increase in the target process}}{\text{Exergy decrease in the coupled processes}}$$

$$= \frac{\text{Exergy increase in the target process}}{\text{Exergy increase in the target process} + \text{Exergy destruction}} \tag{28}$$

In this case, the target process should be specified clearly. For ideal process systems, $\eta_{\Delta\varepsilon}$ is unity. When the numerater is negative, however, it becomes greater than unity. This is because $\eta_{\Delta\varepsilon}$ has the physical meaning only when the target is endergonic.

In any way, since there are some arbitrariness in choosing either objective outputs or target processes, the value of the efficiency must be interpreted carefully.

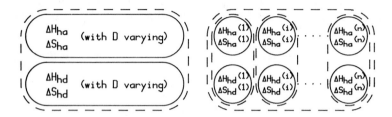

(a) System of distributed-   (b) System of multistaged lumped-
    parameter processes   parameter processes

Figure A1. Decomposition of distributed-parameter process
into multistaged lumped-parameter process.

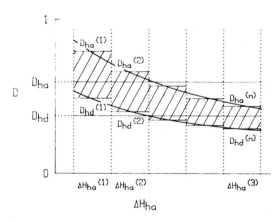

Figure A2. Energy - Direction factor diagram.

Appendix 2   Distributed parameter analysis

In this paper, the direction factor D is calculated based on the changes in enthalpy and entropy between inputs and outputs. Namely we have regarded D as a lumped parameter.

We may take another viewpoint by taking into consideration the variation of the direction factor in a process. A simple method to analyze the effect of the distribution of D to decompose the process into multistaged lumped-parameter processes in such a manner that each subsystem may satisfy Eqs. (16) and (17), as shown in Fig. A1. Therefore we have the following equation for the i-th subsystem.

$$D_{ha}^{(i)} \geqq D_{hd}^{(i)} \tag{29}$$

And the exergy destruction in it is given by

$$T_0 \Sigma \Delta S_i^{(i)} = \Delta H_{ha}^{(i)} [D_{ha}^{(i)} - D_{hd}^{(i)}] \tag{30}$$

Then the exergy destruction for any process system may be obtained as the shaded area on the energy – direction factor diagram shown in Fig. A2. When the number of subprocesses is increased, the width of each $\Delta H_{ha}^{(i)}$ is decreased, resulting in continuous change in $D_{ha}^{(i)}$ and $D_{hd}^{(i)}$. Of course the exergy destruction obtained in this method is the same as that in the text (9), but Eq. (29) becomes now the sufficient condition for a process system to hold.

Literature Cited

1) Denbigh, K.G. Chem. Eng. Sci., 1956, 6, 1
2) Gaggioli, R.A. Chem. Eng. Sci., 1961, 16, 87; 1962, 17, 523
3) Riekert, L. Chem. Eng. Sci., 1973, 29, 1613
4) Gaggioli, R.A. and Petit P.J. Chem. Tech. 1977, 7, 8
5) Oaki, H. and Ishida M. J. Chem. Eng. Japan 1982, 15, 51
6) Ishida, M., Oaki, H. and Suzuki, T. Kagaku-Kogaku 1982, 46, 175
7) Oaki, H., Ishida, M., and Ikawa T. J. Japan Petrol. Inst. 1981, 24, 36
8) Ishida, M. and Oaki H. AIChE 91st National Meeting, Detroit 1981
9) Ishida, M. and Kawamura, K. Ind. Eng. Chem. Process Des. Dev. 1982, 21, 690-695

RECEIVED July 7, 1983

THERMOECONOMIC ANALYSES

Strategic Use of Thermoeconomics for System Improvement

Y. EL-SAYED and M. TRIBUS

Center for Advanced Engineering Study, Massachusetts Institute of Technology, Cambridge, MA 02139

Second Law analysis combined with a cost consideration helps to understand the economics of lost work in an energy intensive system. This understanding may guide to innovative conceptual system designs. Familiar examples in the steady state are used for illustrative purposes.

A process engineer creating a conceptual design for an energy intensive system is expected to consider the extremization of two leading criteria: an energy efficiency criterion and a cost criterion which includes competing material demands. Both are usually subject to constraints imposed for technical, environmental, economic or legal reasons.

The options open to meet these criteria include:
- Choose another operating point (different pressure, flow rate, ...)
- Redesign the flow chart (introduce new device, add or remove items of equipment, provide by-pass flows, ...)
- Redefine the purpose (consider a dual purpose, find a use for a wasted stream, ...)
- Modify environment (relax or control a boundary constraint, negotiate a legal constraint, ...)
- Identify areas for R&D implementations

With the help of computer programs and suitable strategy for their use, it is possible to examine a great number of alternative solutions energy-wise and cost-wise and to enhance the evolution to competitive alternatives.

A good strategy takes into account the evolutionary nature of design concepts, the appropriate level of detail of describing the system and the degree of accuracy of available and assumed data for properties, performance and cost.

The purpose of this study is to develop such a strategy

0097–6156/83/0235–0215$06.75/0
© 1983 American Chemical Society

from a solid theoretical base and to formalize what many good
engineers appear to do informally.

The basic idea is to treat a system or a plant as if it is
embedded in two environments:

1) A physical environment described by a reference pressure
p_o, reference temperature T_0, and a set of reference
chemical potentials $[\mu_{c,o}]$ (or alternatively reference
compositions $[x_{c,o}]$)

2) An economic environment described by a set of reference
prices $[c_i]$ including that of capital

With the system embedded in the physical environment all
materials and energies of interest are evaluated according to
their work potentials (exergies). The criterion for the system
is a work measure. A typical function rated per unit time is

$$\phi_{o,e} = \Sigma EI - \Sigma EP \tag{1}$$

where the first term is the sum of all input exergies, the
second term is the sum of all useful product exergies, and the
function is to be minimized. The criterion may also take an
efficiency form to be maximized.

With the system embedded in the economic environment, all
the energies and materials of interest are evaluated according
to their economic potentials (costs). The criterion function
is a cost measure. A typical function rated per unit time is

$$\phi_{o,c} = \Sigma c_f F + \Sigma z - \Sigma c_p P \tag{2}$$

where the first term is the cost of major feed stocks, the
second is the cost of processing devices of competing material
content and the third is the value of useful products. The
function is to be minimized.

Thermoeconomics makes the connection between these two
evaluations visible throughout the system.

The idea of coupling physical and cost streams is not new.
The earliest example of which we are aware is due to Manson
Benedict in an unpublished set of notes in 1949 reported by
Gyftopoulis ($\underline{1}$). Other examples have been reported ($\underline{10}$, $\underline{11}$, $\underline{12}$).

Potential Work and Lost Work Functions

The irreversibility accompanying a steady flow process may be
computed for any zone by noting the fluxes of entropy into and
out of the zone. Leaving aside the problems posed by semi-
permeable membranes, which introduce ambiguities into the meanings
of heat and work, (2), equation (3) provides such a balance:

$$S^{cr} = \sum_e m_e s_e - \sum_i m_i s_i - \sum_b Q_b / T_b \tag{3}$$

As de-Nevers has pointed out (3), equation (3) suffices to
compute the "lost work", $T_0 S^{cr}$, of the process in the zone,

requiring only knowledge of T_0, the reference temperature and of absolute entropies, which are usually computed anyway as part of cycle calculations.

It is easy to apply equation (3) to all the system or to a subsystem. However, it is also desirable to evaluate the lost work which accompanies the discharge of a stream into a reference environment or alternatively the maximum work which can be obtained if the stream is brought reversibly to the state of the reference environment. The function "Essergy" equation (4), derived and named by Evans (4) leads to such evaluations

$$E = U + p_0 V - T_0 S - \sum_c \mu_{c,o} N_c \qquad (4)$$

(see nomenclature). Equation (4) requires in addition to the knowledge of T_0, the knowledge of a reference pressure p_0 and reference compositions $[x_{c,o}]$. When equation (4) is applied to a flowing stream equation (5) for the "Flow Essergy" is obtained

$$E^f = H - T_0 S - \sum_c \mu_{c,o} N_c \qquad (5)$$

There seems to be an agreement to name equation (5) "Exergy". The lost work (called also dissipation), in a system or a subsystem may also be computed from an "exergy balance" according to equation (6).

$$T_0 S^{cr} = \sum E^f_{in} - \sum E^f_{out} + \sum E^Q_{in} - \sum E^Q_{out} + \sum E^W_{in} - \sum E^W_{out} - dE_{stored}/dt \qquad (6)$$

In using equation (6) work flux is taken as equal to exergic flux. Heat flux is multiplied by the Carnot ratio to obtain its exergic flux, i.e.

$$E^Q = Q(T-T_0)/T \qquad (6a)$$

The dissipation in a plant is equal to the sum of the dissipations in the separate zones of the plant. Therefore, if the total dissipation in the plant is, $T_0 S^{cr}$, it may be decomposed into:

$$T_0 S^{cr} = \sum_k \sigma_k \qquad (6b)$$

where, σ_k, represents the dissipation in zone k.

The Costing of Exergies and Dissipations

There are mainly two reasons for attaching a price tag to an exergy or to a dissipation somewhere in the system whether the criterion is an energy efficiency criterion such as equation (1) or a cost criterion such as equation (2). The reasons are:

- To obtain a distribution throughout the system of the items involved in the criterion function

- To change the distribution in a direction extremizing
  the criterion function

The first reason results in a set of direct prices $[\gamma]$. The
second reason results in a set of differential prices $[\delta]$. The
two kinds of prices permit two different modes of analysis:

- Thermoeconomic accounting using algebraically determined
  prices permitting comparisons of subsystems and their
  costing items as though they were relatively independent.
- Thermoeconomic optimization using differentially derived
  prices, permitting the analysis of the system's local
  and global responses to well specified small changes in
  the state of the system, and leading to sensitivity
  analysis and optimization techniques.

Two subsets of the differential prices $[\delta]$ are most convenient in directing to improved values of the criterion function:
Marginal prices $[\Theta]$ and shadow prices $[\lambda]$. A marginal price Θ
is the change of the criterion function per unit change of a
free variable. A shadow price λ is the change per unit change
of a dependent variable. Exergy and dissipation prices follow
by the chain rule.

For example considering the energy criterion, equation (1),
for a single exergy input EI and a single fixed produce EP, all
direct prices are unity, i.e.

$$[\gamma] = [1] \tag{7a}$$

and the differential price of a dissipation σ_j due to a change
of a free variable y_k is

$$\Theta_{j,k} = (\partial EI/\partial y_k)/(\partial \sigma_j/\partial y_k) \tag{7b}$$

Performance-Cost Modeling

The computation of exergies and dissipations and their costing
requires a mathematical description of the system's performance
and cost.

A suitable level of detail for the purpose of comparing
alternative design concepts is the description of the performance
and the cost of each subsystem by overall descriptors. A more
detailed description may be achieved by subdividing the subsystems
into smaller subsystems.

Essential relations describing each subsystem are overall
equations for mass balance, for energy balance, for performance,
and for costing in terms of performance. Presently available
cost trends (17) in terms of capacity parameters (e.g. area,
mass rate, power, ...) are suitable costing equations to start
with. They may be implemented to include the influence of
variables such as pressure, temperature or efficiency whenever
sufficient data are available.

Thermoeconomic Accounting

Thermoeconomic accounting usually involves a cost criterion. In
this mode of analysis the effect of the economic environment is
traced by following each input stream and assigning monetary
values in accordance with the exergic content. For example, the
pressure portion of the exergy may be followed from a pump to
the frictional dissipation in a heat exchanger. The cost to
produce the fluid mechanical exergy at the pump can be assigned
to the pressure drop process. This assignment makes it evident
at what cost per unit of pressure loss it pays to change the
heat exchanger. The exergy influx with stream "i" may be
apportioned to either an exiting useful stream ("j"), an exiting
wasted stream ("ℓ") or a dissipation process ("k"). In the
accounting mode the engineer simply studies the process diagram
and assigns values to the x's in equation (8):

$$EI_i = \Sigma_j x_{ij} EP_j + \Sigma_\ell x_{i\ell} ER_\ell + \Sigma_k x_{ik} \sigma_k \tag{8}$$

The prices of exergies [EP] and [ER] and of dissipations [σ] are
obtained by multiplying each input EI_i by its boundary price and
summing over all inputs. Consider for simplicity a plant with
only one input and one product stream, with the objective to
reduce the cost as much as possible. For such a system the
objective function is:

$$\Phi_o = \alpha EI + \Sigma_k z_k - \beta EP \tag{9}$$

α and β are the boundary prices for input and product exergies.
If the exergic output is fixed, EI, may be eliminated between (8)
and (9) to give:

$$\Phi_o = \Sigma_k (\alpha\sigma_k + z_k) + \Sigma_\ell \alpha ER_\ell + (\alpha-\beta) EP \tag{10}$$

On the other hand, if the input were fixed, as in rationing, the
value of EP would have been eliminated to give:

$$\Phi_o = \Sigma_k (\beta\alpha_k + z_k) + \Sigma_\ell \beta ER_\ell + (\alpha-\beta) EI \tag{11}$$

The first terms in equations (10) and (11) are the cost of
amortization and dissipation in each zone. In (10) the dissipation is "priced" at the cost of the extra fuel required to make
up for the losses. In (11) the price reflects lost revenue
through loss in production at fixed input.

When there are several inputs and outputs, the assignment
of the x's in equation (8) may become arbitrary. When the prices
per unit of exergy in the input streams are not too different,
it is useful to define an _average_ price α^* and use this average
in the calculations. This kind of accounting allows the
comparison of the cost of dissipators with the cost of their
dissipations.

Another accounting approach is proposed by Gaggioli (11) by which the cost of input energy and the cost of processing devices are allocated to a set of useful product exergies within the system. The price of a dissipation is the price of its immediate exergy input. This kind of accounting permits the comparison of alternative subsystems producing the same product exergy. The accounting approach is the simplest to use and often reveals opportunities for improvement which are not otherwise obvious. The accounting method, however, does not use knowledge of system behaviour. It does not reveal how a capital investment in one part of the plant may affect the entropy generation in another. In the accounting method, entropy generation is a number not a function. Thermoeconomic accounting is demonstrated in example 1.

Thermoeconomic Optimization

Direct prices do not take into account the effect a decision in one part of a plant may have on the irreversibilities in another. Marginal and shadow prices do this but are more complicated to compute. They depend upon the system of equations (and their first derivatives with respect to the variables of interest) rather than upon only the states of various zones. The mathematical description of a thermodynamic process requires the specification of a set of "equations of constraint", represented here by the set, $[\Phi_j=0]$. The thermodynamic performance and stream variables are divided into two sets, state and decision variables, represented by $[x_j]$ and $[y_k]$, and each of the defining functions, $[\Phi_j]$, is expressed in terms of these variables. If the objective function, Φ_o, (whether it is an energy objective or a cost objective) is similarly expressed, a Lagrangian may be defined according to:

$$L=\Phi_o+\sum_j \lambda_j \Phi_j \qquad (12)$$

The way Lagrange's method of undetermined multipliers is interpreted here is not conventional. The approach is described in Appendix A. To guarantee that L is independent of the set $[x_j]$, set:

$$\partial L/\partial x_j=\sum_i \lambda_i b_{ij}+b_{oj}=0 \qquad (13)$$

The solution of the set of linear equations represented by (13) determines the shadow prices. In terms of these prices the marginal prices are given by:

$$\theta_k=\partial L/\partial y_k=\sum_i \lambda_i a_{ik}+a_{ok} \qquad (14)$$

When the functions Φ_o and $[\Phi_j]$ are analytic, the optimum is defined by $\theta_k=0$.

As proposed by Fax and Mills $(\underline{6})$, it is easier to keep track of the various derivatives if they are put into a matrix as shown in Table (1). In the general case, the shadow and marginal prices may be found from:

$$[\lambda_i] = -[b_{oj}][b_{ij}]^{-1}$$

$$[\theta_k] = [a_{ok}] - [b_{oj}][b_{ij}]^{-1}[a_{ik}] \tag{15}$$

If the basis matrix can be diagonalized, the marginal and shadow prices may be found in a direct way.

The matrices indicated in equation (15) are known as Jacobians. The properties of Jacobians have been of interest to thermodynamicists for many years and may be used to find an interesting interpretation of the meanings of the shadow prices. Using the methods described in $(\underline{7},\underline{8})$, define:

$$J[A] = J\left[\frac{\Phi_1, \ldots \Phi_{i-1}, \Phi_o, \Phi_{i+1}, \ldots \Phi_n}{x_1, \ldots x_{i-1}, x_i, x_{i+1}, \ldots x_n}\right]$$

$$J[B] = J\left[\frac{\Phi_1, \ldots \quad \Phi_n}{x_1, \ldots \quad x_n}\right]$$

$$J[C] = J\left[\frac{\Phi_1, \ldots \quad \ldots \Phi_n, \Phi_o}{x_1, \ldots \quad \ldots x_n, y_k}\right] \tag{16}$$

With these definitions, the shadow and marginal prices are found to be:

$$\lambda_i = -J[A]/J[B] = -J\left[\frac{\Phi_1, \ldots \Phi_o, \ldots \Phi_n}{\Phi_1, \ldots \Phi_i, \ldots \Phi_n}\right] = -(\partial\Phi_o/\partial\Phi_i)_{[\Phi_{j \neq i}=0]} \tag{17}$$

$$\theta_k = J[C]/J[B] = J\left[\frac{\Phi_1, \ldots \Phi_n, \Phi_o}{\Phi_1, \ldots \Phi_n, y_k}\right] = (\partial\Phi_o/\partial y_k)_{[\Phi_j=o]} \tag{18}$$

When the state variables satisfy the equations of constraint, the Lagrangian is equal to the objective function. Since the Lagrangian is then independent of the state variables, there are only three ways in which the objective function can be changed:
1. Change a decision variable (i.e., change the operating point)
2. Alter a constraint
3. Change something in the reference environments (physical or economic)

Table 1.	System Matrix $(n+1)(n+f)$		
Constraints Φ_i	Derivatives		
$(\Phi_i=\Phi_i-x_i=0)$	$\partial\Phi_i/\partial x_j=b_{ij}$		$\partial\Phi_i/\partial y_k=a_{ik}$
	$x_1\cdots \ \cdots x_j\cdots \ \cdots x_n$		$y_1\cdots \ \cdots y_k\cdots \ \cdots y_f$
Φ_1	$b_{11}\cdots \ \cdots b_{1j}\cdots \ \cdots b_{1n}$		$a_{11}\cdots \ \cdots a_{1k}\cdots \ \cdots a_{1f}$
Φ_i	$b_{i1}\cdots \ \cdots b_{ij}\cdots \ \cdots b_{in}$		$a_{io}\cdots \ \cdots a_{ik}\cdots \ \cdots a_{if}$
Φ_n	$b_{n1}\cdots \ \cdots b_{nj}\cdots \ \cdots b_{nn}$		$a_{n1}\cdots \ \cdots a_{nk}\cdots \ \cdots a_{nf}$
Objective			
Φ_o	$b_{o1}\cdots \ \cdots b_{oj}\cdots \ \cdots b_{on}$		$a_{o1}\cdots \ \cdots a_{ok}\cdots \ \cdots a_{of}$

For fixed environments, we have:

$$\delta\Phi_o=\sum_k(\partial\Phi_o/\partial y_k)dy_k+\sum_j(\partial\Phi_o/\partial\Phi_j)\delta\Phi_j \tag{19}$$

But in view of the previous equations,

$$\delta\Phi_o=\sum_k\theta_k dy_k-\sum_j\lambda_j\delta\Phi_j \tag{20}$$

Noting that $\Phi_j=\Phi'_j-x_j$, we have:

$$\delta\Phi_o=\sum_k\theta_k dy_k+\sum_j\lambda_j\delta x_j \tag{21}$$

We have found that shadow prices are easier to interpret if they are expressed as multiples of exergy prices by using the equation:

$$\Lambda_j=\lambda_j\partial x_j/\partial E_j \tag{22}$$

Marginal prices are more informative if expressed in non-dimensional form, Θ_k, using:

$$\Theta_k=\theta_k y_k/\Phi_o \tag{23}$$

Once the sets of prices, $[\theta_k]$, $[\lambda_i]$, are known, the prices associated with a change in an exergy or a dissipation may be obtained from the corresponding θ_k or λ_i by the chain rule. If, for example, an exergy stream E_s, has y_k and x_i as variables, then the prices associated with the changes of E_s through the changes in y_k and x_i respectively are:

$$\theta_s = \theta_k \partial y_k / \partial E_s \qquad\qquad (24)$$

$$\lambda_s = \lambda_i \partial x_i / \partial E_s \qquad\qquad (25)$$

Both modes of thermoeconomic analysis, accounting and optimization, are illustrated in example 2, considering a cost criterion function. The case of an energy criterion function may be included as a special case.

The Derivation of costing equations

When the cost of a device is best evaluated in terms of its geometrical variables, costing equations can be derived in terms of essential performance variables determining the performance of the device in a system. Geometrical and performance variables are generally related by design relations. The variables and the design relations determine together the design degrees of freedom. Three cases arise
 1) Essential performance variables determining the performance of the device in a system are equal to the design degrees of freedom. In this case the problem is a problem of exchange of variables.
 2) Essential performance variables are less than the design degrees of freedom. In this case, the cost is minimized with respect to the excess degrees of freedom and the envelope of minimum costs is the costing equation.
 3) Essential performance variables are more than the design degrees of freedom. In this case additional constraints are imposed on the essential performance variables.
 Since the second case arises often in practice, it is treated in the third of the example problems.

A Thermoeconomic Strategy

In conceptual designs the concern is as much with creativity as it is with analysis. Creative approaches are apt to introduce significant changes in flow charts, thereby altering the analyses. As options are discovered or created, changes in plans occur. Therefore a strategy, rather than a plan is required. Figure 1 presents the flow chart for a strategy which conforms to the general outline of a design process as described by Rosenstein (9).
 To start the process, the engineer defines as accurately as possible the inputs, outputs and energy and cost objective functions. Then the reference environments, both physical and economical are defined. These are not necessarily straight-forward steps, as for example when dealing with high rates of

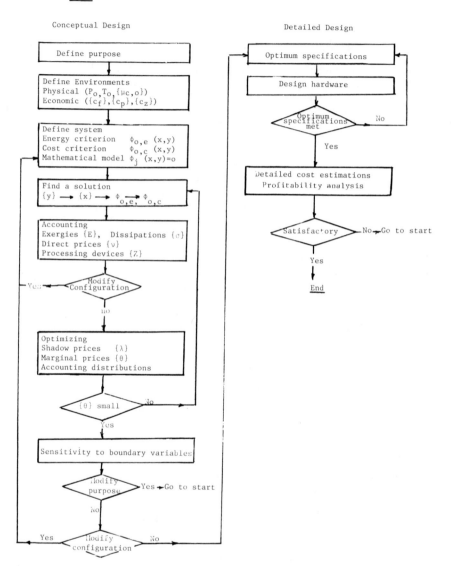

Figure 1.   System matrix (n+1) (n+f).

inflation or defining the chemical potential of a rare substance in the environment.

The next step is to develop a mathematical description of the system. From the perspective of mathematical analysis, the set of equations describing the constraints is the system. An algorithm which permits the computation of the performance of the system (i.e., solves the equations of constraint simultaneously) is a means to satisfactory solutions.

If the description of the system requires m variables, and there are n equations of constraint, there will be (m-n) degrees of freedom. The designer may choose these as "decision variables". The remaining n variables are called "state variables" and are found using the algorithm.

As indicated in the strategy, after the system has been defined, it is necessary to find at least one feasible solution to the set of equations, i.e., to demonstrate that the plant has at least one operating point. For that operating point the energy and cost objective functions may be computed. According to the strategy, accounting methods are first employed in tracing the exergic flows from the various inlets to the dissipative processes. This step leads to a comparison between the capital costs and exergy destruction costs in each zone. While such a picture does not reveal the system's behaviour, and therefore is not always accurate (as for example when a dissipative process merely gets displaced from one part of the system to another), nevertheless, because this is the least costly form of analysis and it often suggests fruitful avenues to pursue, from a strategic perspective, it is the most attractive first step. It may also be the only possible analysis in the absence of sufficient data. See the first example at the end of this paper for a case in which the analysis reveals a useful set of opportunities. Other examples are available.

Based upon this analysis the designer may decide either to change the configuration or to change the operating point.

As indicated in the strategy, if the flow chart is not to be changed, the next step is to compute the shadow, $[\lambda_i]$, and marginal, $[\theta_k]$, prices. The marginal prices form a vector which points to direction of an improvement in the existing design by changing the decision variables. (This step is essentially the same as conventional optimization.) Alternatively, the designer may use the shadow prices to investigate, approximately, the benefit of a sought configuration modification.

In each iteration with a new marginal price vector, the distributions by thermoeconomic accounting are always accessible if needed.

These elements of the strategy and the technique of developing the costing equations are illustrated in the following three examples.

EXAMPLE 1:   Manufacture of an Industrial Chemical
      At the time the analysis was undertaken, information on the
system performance was available for only one operating point and
cost information consisted of initial installation costs plus
estimates of life times. Equipment costs were not related to
operating conditions. Under these circumstances only the
accounting approach could be used. The available information
is summarized in Table II.   (For more details, see reference
10)
      The exergy flow diagram, Figure 2,   shows the results of
the exergy balances on the various zones of the system. The
reference state of zero exergy was taken as P_o=1 atm, T_o=528 R,
mass fraction water=0.3, mass fraction sodium palmitate=0.7.
The feed condition is a convenient reference state to take when
the analyst is not interested in potential uses of the product
in other environments. The evaluation of the x's in equation
(8) is relatively simple since all exergy inputs by heat may be
assumed to be converted to fluid thermal exergies or to be
dissipated in entropy generation. There is, however, some
ambiguity in the flash tank where the input mechanical and
thermal exergies are converted to chemical exergy, thermal
exergy and dissipation. Since the costs of the input exergies
are not too different from one another, it is convenient and
reasonable to assign an average price to the dissipation and
consider that the input exergies contribute in proportion to
their presence. The resulting direct prices, [γ], dissipation
values, [$\gamma\sigma$], and amortizations, [z], are given in Table III.
      The data suggests that improvements should be made as
follows:
      Substitute a boiling type heat exchanger (new zone 3) in
      place of the separate heater and flash tank (zones 3 and
      4).
      Use the latent heat of the exhaust steam in the preheater
      (zone 1).
      Substitute ordinary cooling water at 68 F for the brine
      in zone 7, using a larger heat exchanger with much
      greater agitation.
      Table III   presents the results of the analysis and a
comparison of three cases:
      The perfectly reversible process
      The original system
      The modified system

EXAMPLE 2: A Simple Open Cycle Gas Turbine
      This example treats a simple open-cycle gas turbine for
which the cost objective function, equations of constraint and
costing equations are all available in analytic form. Figure
3   shows these functions along with the fixed and variable
decision variables. Since the set of equations is diagonalized,

Table II.   Available data, example 1

Matter Streams	Thermodynamic Variables				Energy Streams	Thermo. Var.		Zones	Cost Variables		
	m lb/hr	P psig	T F	x_{H_2O} (mass)		T F	Rate kw		Capital k$	Maintenance k$/yr	Life yrs
External											
F_1	5000	35	68	0.3	W_1	–	4.5	Z_1	20	0.2	20
F_2	20	0	180	–	W_2	–	0.67	Z_2	15	3.5	14
P_1	4500	0	158	0.22	W_3	–	6.0	Z_3	46	0.2	25
R_1	520	0	217	1.0	W_4	–	9.0	Z_4	12	0.0	20
Internal											
S_1	5000	30	180	0.3	Q_1	365	151.0	Z_5	13	0.3	14
S_2	5000	570	182	0.3	Q_2	460	183.0	Z_6	45	7.5	14
S_3	5000	350	353	0.3	Q_3	14-18	-92.4	Z_7	18	0.5	14
S_4	4480	0	212	0.22	Reject J_1	180	2.25				
S_5	4480	90	212	0.22	J_2	212-217	4.27				
S_6	4500	50	212	0.22	J_3	212	0.34				
					J_4	212	5.9				

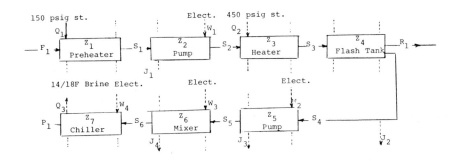

Input Energy Prices

150 psig steam	2.7 $/1000 lb
450 psig steam	3.05 $/1000 lb
Electrical energy	4.37 ¢/kwhr
14/18 F Brine	1.9 ¢/kwhr cooling load

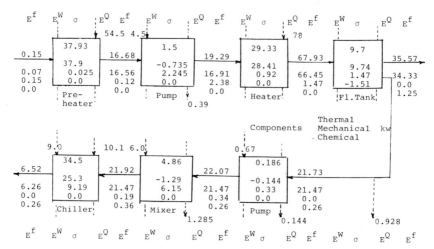

Figure 2.   Exergy flow diagram, example 1.

Table III.   Dissipation vs. amortization, example 1

Current System		Modified		Current System				Modified System			
{EI}@{α} Stream		{EI}	@ {α}	{σ, ER, EJ}@{γ}		{γσ,..}	vs{z}	{σ,...}	@{γ}	{γσ,..}	}vs{z}
	kw	¢/kwhr	kw	¢/kw hr	kw	¢/kw hr	¢/hr	¢/hr	kw	¢/kw hr	¢/hr	¢/hr

Stream	kw	¢/kwhr	kw	¢/kw hr		kw	¢/kw hr	¢/hr	¢/hr	kw	¢/kw hr	¢/hr	¢/hr
W_1	4.5	4.37	4.5	4.37	$σ_1$	37.93	3.00	114	20	19	3.1	59	54
W_2	.67	4.37	0.67	4.37	$σ_2$	1.5	4.37	6.56	21.42	1.5	4.37	6.56	21.42
W_3	6.0	4.37	6.0	4.37	$σ_3$	29.33	3.24	94.94	36.8	16.92	3.13	53	73.6
W_4	9.0	4.37	9.0	4.37	$σ_4$	9.7	3.39	32.88	12	-	-	-	-
Q_1	54.5	3.0	1.26	3.00	$σ_5$	0.186	4.37	0.81	18.58	1.86	4.37	0.81	18.58
Q_2	78	3.2	55.86	3.00	$σ_6$	0.486	4.37	21.24	64.28	4.86	4.37	2.24	64.28
Q_3	10.1	17.4	0	0	$σ_7$	34.5	7.62	262.8	25.72	23.5	3.535	83.0	25.72
F_1	∿ 0	-	-	-	ER_1	35.57	3.21	114.2	0	4.1	3.11	12.75	4.0
F_2	∿ 0	-	-	-	EJ_1	0.39	4.37	1.7	0	0.39	4.37	1.7	0
					EJ_2	0.928	3.2	2.97	0	0.927	3.0	2.78	0
					EJ_3	0.144	4.37	0.63	0	0.144	4.37	0.63	0
					EJ_4	1.29	4.37	5.62	0	1.29	4.37	5.62	0
					EJ_5	-	-	-	-	0.29	3.0	0.86	0

Case		ΣEI kw	α* ¢/kwhr	ΣαEI ¢/hr	Σz ¢/hr	ΣαEI+Σz ¢/hr
Least exergy requirement		1.5	4.37	6.6	∞	∞
	current	162.8	4.16	677.2	198.8	876
	modified	77.3	3.36	259.7	261.6	521.3

Net Values

the basis state variables, $[x_i]$, shadow prices, $[\lambda_j]$, and
marginal prices, $[\theta_k]$, may all be found explicitly, as indicated
in Figure 4.

The heat exchanger costing equation is the costing equation
derived in Example 3. The other costing equations were suggested
to us by experienced engineers and are illustrative only.

The first step in the strategy uses the accounting method,
with the results shown in Table IV.   In general, since the
capital costs are small compared to the irreversibility costs,
it appears to be advantageous to increase the capital investments
in each zone.

The marginal prices support this conclusion. The incremental
changes in decision variables, Δy_k, required to improve Φ_o
result in increased capital investment. Table V shows the
results of a stepwise change.

A configuration modification may be simulated by a modified
subset of the constraining equations. The shadow prices adjoint
to the modified constraints give the approximate redistribution
of the objective function resulting from the modification. The
approximation is due to the assumption that the shadow prices
remain constant. One change, suggested by the accounting method,
is to add a heat exchanger to recover the exergy wasted in the
exhaust stream.

For a given set of decisions, $[y_k]$,

$$d\Phi_o = -\Sigma \lambda_j d\Phi_j \tag{E2-1}$$

Adding a heat exchanger to the cycle will require a change in
constraints Φ_1, Φ_5, Φ_7 and Φ_8. For a given set of expressions,
$[\Phi'_j]$, these changes are represented, as shown in Table VI,
by:

$$\Delta\Phi_o = \lambda_1\Delta x_1 + \lambda_5\Delta x_5 + \lambda_7\Delta x_7 + \lambda_8\Delta x_8 \tag{E2-2}$$

or

$$\Delta\Phi_o = \lambda_1\Delta P_{cold} + \lambda_5\Delta P_{hot} + \lambda_7\Delta EI + \lambda_8 Z_x \tag{E2-3}$$

If there is to be a net improvement, this change should be negative, i.e., the cost of the exchanger, (Z_x), should not exceed
the value of the energy saving, $(\lambda_7\Delta EI)$, reduced by the cost of
the pressure losses on the cold and hot sides of the exchanger
$(\lambda_1\Delta P_{cold}, \lambda_5\Delta P_{hot})$.

For example, a heat exchanger with $\eta=0.6$, $\Delta P_{hot}=0.05$ atm,
$\Delta P_{cold}=0.05$ atm, should not cost more than 211 k\$/yr.

Examination of costing equation reveals that exchangers
in this range should cost about 3.6 k\$/yr., so that there is
obviously an opportunity to improve the system through addition
of a heat exchanger. The redistribution of cost given in Table VI
shows that the pressure loss penalty is high on the hot

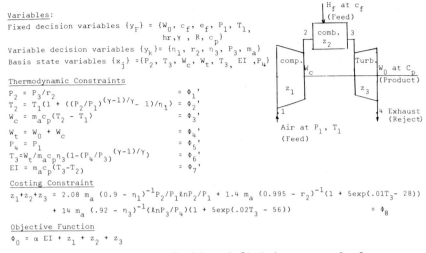

Variables:

Fixed decision variables $\{y_F\} = \{W_0, c_f, e_f, P_1, T_1, hr, \gamma, R, c_p\}$

Variable decision variables $\{y_k\} = \{n_1, r_2, n_3, P_3, m_a\}$

Basis state variables $\{x_j\} = \{P_2, T_3, W_c, W_t, T_3, EI, P_4\}$

Thermodynamic Constraints

$$P_2 = P_3/r_2 \qquad\qquad = \Phi_1{}'$$
$$T_2 = T_1(1 + ((P_2/P_1)^{(\gamma-1)/\gamma} - 1)/n_1) = \Phi_2{}'$$
$$W_c = m_a c_p (T_2 - T_1) \qquad = \Phi_3{}'$$
$$W_t = W_0 + W_c \qquad\qquad = \Phi_4{}'$$
$$P_4 = P_1 \qquad\qquad\qquad = \Phi_5{}'$$
$$T_3 = W_t/m_a c_p n_3 (1-(P_4/P_3)^{(\gamma-1)/\gamma}) = \Phi_6{}'$$
$$EI = m_a c_p (T_3 - T_2) \qquad = \Phi_7{}'$$

Costing Constraint

$$z_1 + z_2 + z_3 = 2.08 \, m_a (0.9 - n_1)^{-1} P_2/P_1 \ell n P_2/P_1 + 1.4 \, m_a (0.995 - r_2)^{-1}(1 + 5\exp(.01T_3 - 28))$$
$$+ 14 \, m_a (.92 - n_3)^{-1}(\ell n P_3/P_4)(1 + 5\exp(.02T_3 - 56)) \qquad = \Phi_8$$

Objective Function

$$\Phi_0 = \alpha \, EI + z_1 + z_2 + z_3$$

Figure 3.   Problem definition, example 2.

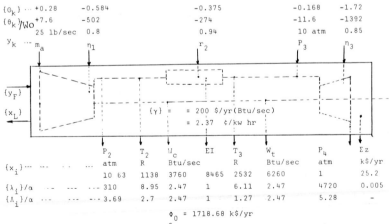

Direct Prices   Shadow Prices   Marginal Prices

$\{\gamma\} = \alpha$

$$\lambda_8 = 1$$
$$\lambda_7 = \alpha$$
$$\lambda_6 = \partial/\partial T_3 \; (z_2 + z_3 + \lambda_7 \Phi_7{}')$$
$$\lambda_5 = \partial/\partial P_4 \; (\lambda_6 \Phi_6{}' + z_3)$$
$$\lambda_4 = \partial/\partial W_t \; (\lambda_6 \Phi_6{}')$$
$$\lambda_3 = \partial/\partial W_c \; (\lambda_4 \Phi_4{}')$$
$$\lambda_2 = \partial/\partial T_2 \; (\lambda_3 \Phi_3{}' + \lambda_7 \Phi_7{}')$$
$$\lambda_1 = \partial/\partial P_2 \; (z_1 + \lambda_2 \Phi_2{}')$$

$$\theta_1 = \partial/\partial n_1 \; (z_1 + \lambda_2 \Phi_2{}')$$
$$\theta_2 = \partial/\partial r_2 \; (z_2 + \lambda_1 \Phi_1{}')$$
$$\theta_3 = \partial/\partial n_3 \; (z_3 + \lambda_6 \Phi_6{}')$$
$$\theta_4 = \partial/\partial P_3 \; (z_3 + \lambda_1 \Phi_1{}' + \lambda_6 \Phi_6{}')$$
$$\theta_5 = \partial/\partial m_a \; (z_1 + z_2 + z_3 + \lambda_3 \Phi_3{}' + \lambda_6 \Phi_6{}' + \lambda_7 \Phi_7{}')$$

$\{y_F\}$... $W_0 = 2500$ Btu/sec., $c_f = \$36/$Barrel, $e_f = 16000$ Btu/lb, $P_1 = 1$atm, $T_1 = 520$R, $hr = 8000$ hrs/yr, $c_p = .243$, $R = 0.07$ Btu/lbR, $\gamma = 1.4$

$\{\theta_k\}$	+0.28	-0.584			-0.375			-0.168	-1.72
$\{\theta_k\}/W_0$	+7.6	-502			-274			-11.6	-1392
	25 lb/sec	0.8			0.94			10 atm	0.85

y_k ... m_a n_1 r_2 P_3 n_3

$\{y_F\}$

$\{x_L\}$

$\{\gamma\} = = 200 \$/yr(Btu/sec) = 2.37 \, \cent/kw \, hr$

	P_2 atm	T_2 R	W_c Btu/sec	EI	T_3 R	W_t Btu/sec	P_4 atm	Σz k\$/yr
$\{x_i\}$	10 63	1138	3760	8465	2532	6260	1	25.2
$\{\lambda_i\}/\alpha$	310	8.95	2.47	1	6.11	2.47	4720	0.005
$\{\Lambda_i\}/\alpha$	3.69	2.7	2.47	1	1.27	2.47	5.28	-

$$\Phi_0 = 1718.68 \text{ k\$/yr}$$

$\{x_L\}$... $T_4 = 1501$ R, $n_0 = 0.295$, $\{E^f\} = \{0, 3300, 9200, 2600$ Btu/sec$\}$
$\{\sigma, ER\} = \{300, 2500, 400, 2600$ Btu/sec$\}$, $\{z\} = \{13, 0.8, 11.8$ k\$/yr$\}$

Figure 4.   Prices and numerical illustration, example 2.

Table IV.   Results of the Accounting Analysis, example 2

	Compressor	Combustor	Turbine	Exhaust
$\gamma\sigma$ k$/yr	60	500	80	560
z k$/yr	13.1	0.85	11.8	0

Table V.   Improving using marginal prices, example 2

Decision Variable	y_k	Θ_k	Δy_k	New y_k	New Θ_k
y_1: η_1 (compressor)	0.8	-0.584	+0.085	0.885	+3.14
y_2: r_2 (combustor)	0.94	-0.375	+0.04	0.98	-0.168
y_3: η_3 (turbine)	0.85	-1.72	+0.06	0.91	+3.77
y_4: P_3 atm	10	-0.168	+1.0	11	-0.088
y_5: m_a lb/sec	25	+0.28	-2.2	22.5	+0.254

Corresponding changes in ϕ_0 k$/yr

Variable	Value	Change	New Value
ϕ_0	1718.68	-232.78	1485.9
α EI	1693.0	-369.5	1323.5
z_1	13.1	+ 71.6	84.7
z_2	0.85	+ 1.31	2.16
z_3	11.8	+ 63.7	75.5

Table VI.   Evaluating regeneration using shadow prices, example 2

Affected ϕ_j	x_j	Δx_j (estimated individually	λ_j x 10^{-3}	$\lambda_j \Delta x_j$ k$/yr
ϕ_1	P_2	ΔP_{cold} = 0.05 atm	62	3.1
ϕ_4	P_4	ΔP_{hot} = 0.05 atm	944	47.2
ϕ_7	E_1	$-\eta_x m_a c_p (T_4 - T_2) = -1323$ Btu/sec	0.2	-264.6
		(η_x = 0.6)		
8	Σz	z_x = 3.64 k$/yr	1	3.6
				Σ = -210.7

$\Delta\phi_0$ k$/yr using Shadow Prices = $\Sigma \lambda_j \Delta x_j$ = -210.7

$\Delta\phi_0$ k$/yr without using Shadow Prices   = -236

side, indicating more saving may be achievable by reducing the
hot side pressure losses on the expense more costly exchanger.

It may be noticed that the analysis considering the energy
criterion $\Phi_{0\,e}$ is similar but with $\alpha=1$ and $z_1=z_2=z_3=0$. However,
the improved point for a given configuration is not an interior
point, as in the case of a cost criterion.

EXAMPLE 3

Translating a Cost Equation for a Heat Exchanger to an Optimized
Costing Equation.
Table VII describes the class of heat exchanger chosen and
summarizes the design equations and the capital cost equation .
The cost of the heat exchanger, expressed in terms of geometric
variables, (measured in feet) is given by:

$$Z=1150D^{1.05}L^{0.3}P^{0.75}(NN_i)^{0.975}N_e^{0.1}N_s N_p \quad \$$$

As indicated in Table VII, there are 32 variables constrained
by 23 equations, leaving 9 degrees of freedom. On the other
hand, the cycle analysis, requires a costing equation using four
variables, i.e.,

$$Z=Z(M, PL_i, PL_e, \eta)$$

Therefore the heat exchanger cost should be minimized with
respect to the remaining 5 degrees of freedom. Table VII shows
the set of equations as diagonalized, i.e., ready to be solved
without iterations. Since three of the four variables in the
costing equation are in the basis state variables (as the equations are displayed in Table VII), it is necessary to rearrange
the equations to make these variables part of the decision set.
When the equations are rearranged to exchange these basis
variables for decision variables, as shown in Table VIII, it
is no longer possible to solve for the basis set explicitly.
Instead iterative methods must be used. Because the variables
representing friction factors and unit thermal conductances
vary slowly, the system of equations converges rapidly if trial
values of these variables are inserted as part of the iterative
process. In this way the envelope of least cost designs is found
as a function of the decision set, [M, PL_i, PL_e, η]. In constructing the set, a systematic search was made in which 64
optimal designs were used to define the envelope of least cost
solutions, as given in Table IX. A final costing equation
(for $\eta \leq 0.6$ is:

$$Z=0.02m^{1.06}(\eta/1-\eta)^{0.5}\Delta P_i^{-0.38}\Delta P_e^{-0.11} \quad \$$$

Table VII.   Design and cost relations of a shell-and-tube heat exchanger, example 3

D = Outside tube diameter	L = Tube length	N = Tube/pass	N_i = Tube passes
N_s = Shells in a series	N_p = Shells in parallel	P = Pitch	N_e = Shell passes

Variable	x_i	Design Relation ϕ_i
1. H.T. Area/Shell	a	$= \pi\, D\, L\, N\, N_i$
2. Total H.T. area	A	$= a\, N_s\, N_p$
3. Internal Diam.	D_i	$= c_1\, D$
4. Tube side flow area	A_i	$= (\pi/4)D_i^2\, N\, N_p$
5. Tube side velocity	V_i	$= M/\rho A_i$
6. Shell diam.	D_s	$\simeq ((4/\pi)N\, N_i)^{0.5}P$
7. Shell side flow area	A_e	$= D_s L(1-D/P)N_p/N_e$
8. Shell side velocity	V_e	$= M/\rho A_e$
9. Tube side Reynolds	R_i	$= \rho V_i D_i/\mu$
10. Shell side Reynolds	R_e	$= \rho V_e D/\mu$
11. Tube direction reverse	F_d	$= (0.1 + 0.125(N_i - 1)/N_i)D_i/L$
12. Tube side friction	F_i	$= 0.046R_i^{-.2}+ F_d$ or $= 16R_i^{-1}$ $R_i < 3000$
13. Shell side friction	F_e	$= 0.66R^{-.15}$or $1.5Re^{-.6}$ $R_i < 2000$
14. Tube side H.T. coeft.	H_i	$= 0.023(k/D_i)R_i^{0.8}P_r^{0.33}$ or $= 1.86(k/D_i)R_i^{0.33}P_r^{0.33}$
		$R_i \quad 3000$
15. Shell side H.T. coeft.	H_e	$= 0.26(k/D)R_e^{0.6}P_r^{0.33}$ or $= 0.75(k/D)R_e^{0.4}P_r^{0.33}$ $R_e< 2000$
16. Tube side pressure loss	PL_i	$= 4F_i LN_i N_s \rho V_i^2/2gD_i$
17. Number of shell rows	N_r	$= D_s/P$
18. Shell side pressure loss	PL_e	$= 4F_e N_e r_s N_e\, V_e^2/2g$
19. Overall H.T. coeft.	U	$= (1/H_i + 1/H_e + 1/H_d)^{-1}$ $(H_d$ dirt coeft.)
20. Number of H.T. units	NTU	$= UA/M\, c_p$
21. Effectiveness (for 1-2 and cross flow types)	η'	$\simeq 0.6(1 - Exp(-1.5\ NTU))$
22. Counter flow factor	F_c	$\simeq 1.58(1 - Exp(-.5(1 + N_s + I(N_e - 1))))$ $I=0$ if $N_i \geq 2$
23. Effectiveness	η	$= \eta'F_c$

Assumptions: equal shell and tube side mass rates, constant properties and square in line tube arrangement

Capital Cost $		ϕ_0
01. Cost/shell	z'	$\simeq 527A^{0.6}D_s^{0.75}\, D^{0.45}L^{-0.3}N_e^{0.1}$
02. Cost $(D,L,P,\{N\})$	z	$\simeq 1150D^{1.05}L^{0.3}P^{0.75}(NN_i)^{0.975}N_e^{0.1}N_s N_p$
03. Applicability		. $3'<L<30'$, $3/8''<D<1.5''$, $1.25D<P<1.5D$, $250<a<5000$ Ft^2
		. Fixed tube sheet type
		. Material 316 SS
		. Operating pressures <300 psig

(References 13, 14, 15, 16)

Table VIII. Decision and state variables for minimizing cost, example 3

Column groups: $\{x_i\}$ = columns $x_1\,x_2\,x_3\,N\,L\,P\,R_i\,R_e\,F_i\,F_e\,H_i\,H_e\,U$; $\{y_k\}$ = columns $D\,N_p\,N_i\,N_s\,N_e$; $\{y_F\}$ = columns $PL_i\,PL_e\,\eta\,M$.

Design Relations $\{x_i\}=\{\Phi_i'\}$	x_1	x_2	x_3	N	L	P	R_i	R_e	F_i	F_e	H_i	H_e	U	D	N_p	N_i	N_s	N_e	PL_i	PL_e	η	M
$x=0.95(1-\mathrm{Exp}(-0.5(L+N_s+I(N_e-1))))$																						
$\quad I=0$ if $N_i\geq 2$																						
$x_1=LN=(Mc_p/1.5\pi DN_sN_pN_iU)\ln(x/x-\eta)$													•	•	•	•	•	•			•	•
$x_2=N^2/L=4^3F_i-N_iN_sM^2/2\pi^2\rho gN_p^2c_1^5D^5PL_i$				•					•					•	•	•	•		•			•
$x_3=N^{.5}(L(P-D))^2=\pi^{0.5}F_eN_sN_e^3M^2/\rho gN_i^2N_p^2PL_e$				•						•					•	•	•	•		•		•
$x_4=N=(x_1x_2)^{0.33}$	•	•		•																		
$x_5=L=x_1/N$	•			•	•																	
$x_6=P=(x_3^{0.5}/LN^{0.25})+D$			•	•	•	•								•								
$x_7=R_i=4M/\pi\mu NN_pD$				•			•							•	•							•
$x_8=R_e=\pi^{0.5}N_eDM/2\mu N_p(NN_i)^{0.5}L(P-D)$				•	•	•		•						•	•	•		•				•
$x_9=F_i=0.046R_i^{-0.2}+Rd(N_i,D,L)$ (or Laminar)					•		•		•					•		•						
$x_{10}=F_e=0.66R_e^{-0.15}$ (or Laminar eqn)								•		•												
$x_{11}=H_i=0.023(k/c_1D)R_i^{0.8}P_r^{0.33}$ (or Laminar)							•				•			•								
$x_{12}=H_e=0.26(k/D)R_e^{0.6}P_r^{0.33}$ (or Laminar eqn)								•				•		•								
$x_{13}=U=(1/H_i+1/H_e+1/H_d)^{-1}$											•	•	•									

Table IX. The costing equation, example 3

$$Z=km^{n1}(\eta/1-\eta)^{n2}(PL_i)^{n3}(PL_e)^{n4}\quad \$$$

m lb/hr, PL_i, PL_e in atm.

n,k	range	range of corresponding variable	Trend	Weighted average
n_1	1.038-1.085	0.1-1000 lb/sec	no trend	1.06
n_2 $\ \eta\leq0.6$	0.495-0.51	0.1-0.6	no trend	0.5
$\eta>0.6$	0.994-1.027	0.6-0.9	no trend	1.0
n_3	(-0.336)-(-0.4)	0.1-1000 psf	higher $|n|$ at lower PL_i	-0.38
n_4	(-0.082)-(-0.154)	0.1-1000 psf	higher $|n|$ at lower PL_e	-0.11
k	0.0145-.026		no trend	0.02

Conclusions

The examples given here represent a small sample of potential
applications. In test comparisons with some practicing engineers
it was found that the strategy was useful in directing their
attention to possible improvements that had not been obvious.
Of course, once they were pointed out, they became obvious, even
without analysis. We believe that the technique should be
extended to take into account time varying inputs and outputs
and systems in which various parameters are uncertain. The
methods should be incorporated with the techniques of decision
analysis, for example.

Nomenclature

$a_{ik}=\partial\Phi_i/\partial y_k$
$a_{ok}=\partial\Phi_o/\partial y_k$
$b_{ij}=\partial\Phi_i/\partial x_j$
$b_{oj}=\partial\Phi_o/\partial x_j$

c_f=market price, unit feed
c_p=market price, unit product
e_f= exergic content, unit feed
e_p= exergic content, unit product
$\alpha=c_f/e_f$= exergy based feed price
α^*=price averaged over all inlet streams
$\beta=c_p/e_p$= exergy based price per unit product
γ_k=unit price, exergy dissipation in zone z
δ=a differential price
λ_j=shadow price for Φ_j
Λ_j= exergy based shadow price for Φ_j
θ_k=marginal price (sensitivity) of y_k
Θ_k=dimensionless marginal price of y_k
Z=capital cost
z=annualized capital cost
E=essergy=$U+P_oV-T_oS-\Sigma\mu_{c,o}N_c$
E^f=flow essergy=exergy=$H-T_oS-\Sigma\mu_{c,o}N_c$, EI input, EP output, ER
    reject
E^Q=heat exergy=W
σ_k=dissipation (lost work) zone k
K=cost factor
m=mass flux, m_a=air mass flux
P=pressure
Q=heat flux
r=pressure ratio
S=Entropy flux
s=specific entropy
S^{cr}=rate of entropy creation
T=absolute temperature

t=time
V=variable
W=work flux
μ_c=chemical potential, species c
η=adiabatic efficiency
=thermal effectiveness
=exergy based efficiency
x_c=mass fraction, species c
x_{ij}=fraction of exergic content of stream i going to j
y_k=decision variable
x_i=basis state variable
Y_F=design parameter (fixed decision)
x_L=state variable, not in basis

Functions

Objective function: $\Phi_o=\Phi_o([x_i],[y_k])$
Constraint expression: $\Phi'_j([x_i],[y_k])$
Equation of constraint $\Phi_j=x_j-\Phi'_j=0$

Subscripts

o,e,i,b=reference state, exhaust, inlet, boundary
t,s,p=thermodynamic, stream, performance
d,g,r,=design, geometric, rate
c,m,a=cost, manufacture, amortization
i=1,2,...I feeds
j=1,2,...n constraints, basis variables
j=1,2,...J products
k=1,2,...K zones
k=1,2,...F degrees of freedom=number of decision variables

Acknowledgments
     The work described in this report was funded by a grant
from the Department of Energy, whose support is gratefully
acknowledged.

Literature Cited

1. Gyftopoulos, Elias, Lazaridis, Lozaros J. and Widner,
    Thomas, Potential Fuel Effectiveness in Industry, Ballinger
    Publishing Company, Cambridge, MA 1974
2. Tribus, M, Thermostatistics and Thermodynamics, D. van
    Nostrand Co., Inc. 1961. pp 605, 607
3. De-Nevers, Joel and Seader, J. D. "Lost Work: A Measure of
    Thermodynamic Efficiency", Energy, Vol. 5, No. 8-9, August-
    Sept. 1980, pp 7859-769

4. Evans, R. B., "A Proof that Essergy is the Only Consistent Measure of Potential Work for Chemical Systems", PhD thesis, Dartmouth College, NH, 1969

5. Tribus, M., El-Sayed, Y. "Thermoeconomic of Energy Systems", Final Report, DoE Agreement, DE-ACO-79ER10518.A000, Center for Advanced Engineering Study, MIT, Cambridge, MA 1982

6. Fax, D. H. and Mills, R. R., "Generalized Optimal Heat Exchanger Design", Trans. ASME, April 1957, p 653

7. reference 2, pp 456-460

8. Tribus, M. and El-Sayed, Y., Progress Report, DoE Grant, DoE Agreement, DE-ACO2-79ER10518.A000, June 1980, MIT, Cambridge, MA

9. Rosenstein, A. B., Rathbone, R. R., Schneerer, W. F., "Engineering Communications", Prentice Hall, 1964

10. Tribus, M. and El-Sayed, Y., "Thermoeconomic Analysis of an Industrial Process", report, Center for Advanced Engineering Study, MIT, Cambridge, MA

11. Gaggioli, R., "A Thermodynamic-Economic Analysis of the Synthane Process", Final Report for US DoE, Marquette University, Milwaukee, Wisconsin, Nov. 1978

12. Evans, R. B., Tribus, M., Thermoeconomics of Saline Water Conversion", I&EC Process Design and Development, Vol. 4, April, 1965, pp 195-206

13. Peters, M., Timmerhaous, K., "Plant Design and Economics for Chemical Engineers", McGraw Hill, 1968, Chapter 14

14. Welty, J. R., Engineering Heat Transfer, Wiley International, 1974

15. Popper, H., "Modern Cost-Engineering Techniques", McGraw Hill, 1970

16. Garrison, C. M. and Steinmetz, F. J., "Energy Optimization of Interchangers", Heat Transfer--Orlando, 1980, pp 301-309, AICHE Symposium Series, Vol. 76, No. 199

17. Humphreys, Kenneth and Katell, Sidney, "Basic Cost Engineering", Marcel Dekker, 1981

Appendix A

Lagrange's Method Revisited

The mathematical description of a system is determined by a set of equations of constraint, $[\Phi_j=0]$, $j=1,2,3,\ldots n$ The expressions, Φ_j, are functions of two sets of variables:

Decision variables, y_k, which, within limits, may be set by the designer, and

State variables, x_j, which are determined by the equations of constraint, once the decision variables have been chosen. If the total number of variables is m, the number of equations of constraint is n, the degrees of freedom will be $f=m-n$, equal to the number of decision variables.

If the performance of the system is defined by an "objective function", denoted by Φ_0, the task of the designer may be stated as follows:

Extremize $\Phi_0 = \Phi_0([x_i], [y_k])$ (A-1)

Subject to $\Phi_j = \Phi_j([x_i], [y_k]) = 0$ (A-2)

If it were possible to do so, the equations of constraint would be solved for the individual values of x_i, and these would be substituted into the objective function to produce a new function which did not depend upon the set $[x_i]$. Call this new function $L([y_k])$. The extremum could then be found by differentiating the function L with respect to each of the independent decision variables and setting each derivative equal to zero.

Unfortunately, in most instances it is not possible to solve the set of equations of constraint explicitly for the state variables. A method for constructing the function, L, called the Lagrangian, is due to Lagrange. According to this method, L is defined by:

$$L = \Phi_0 + \Sigma \lambda_j \Phi_j$$ (A-3)

In the above equation λ is called a Lagrangian multiplier and is considered to be a function of the sets $[y_k]$ and $[x_i]$.

$$\lambda_j = \lambda_j([x_i], [y_k])$$ (A-4)

To guarantee that L, as defined above, will not be a function of $[x_i]$, differentiate L with respect to each x_i, and set the derivative equal to zero.

$$\partial L / \partial x_i = (\partial \Phi_0 / \partial x_i) + \Sigma_j \lambda_j (\partial \Phi_j / \partial x_i) + \Sigma_j \Phi_j (\partial \lambda_j / \partial x_i) = 0$$ (A-5)

If the set $[\Phi_j = 0]$ is satisfied, the set of derivatives in the equation above should <u>not</u> be interpreted as determining an extremum, but rather as guaranteeing that L does not depend upon $[x_i]$.

With $[\Phi_j = 0]$ satisfied, Equation (A-5) becomes

$$\partial L / \partial x_i = (\partial \Phi_0 / \partial x_i) + \Sigma_j \lambda_j (\partial \Phi_j / \partial x_i) = 0$$ (A-6)

Equation (A-6) is linear in the Lagrangian multipliers and can be solved as indicated in equation 13.

The first derivative of L with respect to y_k, <u>provided</u> the set $[\Phi_j = 0]$ is satisfied, is the derivative of Φ_0. This result is true, even if the set of decision variables does not define the extremum. Therefore, these derivatives may be used to point the way towards the extremum. Of course, the usual caveats apply with respect to the possibility of a saddle point or local extremum.

RECEIVED July 12, 1983

Essergetic Functional Analysis for Process Design and Synthesis

ROBERT B. EVANS and PRASANNA V. KADABA

School of Mechanical Engineering, Georgia Institute of Technology, Atlanta, GA 30332

WALTER A. HENDRIX

Engineering Experiment Station, Georgia Institute of Technology, Atlanta, GA 30332

The concept of essergetic functional analysis is introduced as a tool for approaching a condition known as "thermoeconomic isolation" of the inter-dependent equipment components of a system or process. If an interdependent component is thermo-economically isolated, then that component may be suboptimized with respect to many new, underlying variables. The required essergy analysis proce-dures are illustrated by considering the synthesis and design of components of a large steam power plant.

The origin of essergetic functional analysis began mainly with work by Tribus (1) in 1956 in which he balanced the cost of lost work against the capital cost of the equipment required to prevent this lost work, resulting in the optimum design of this equipment. The concept of essergetic functional analysis is introduced as a tool for approaching a condition known as "thermoeconomic isola-tion" of the interdependent equipment components of a system or process. If an interdependent component is thermoeconomically isolated, then that component may be suboptimized with respect to many new, underlying variables which cannot be treated by any other existing optimization procedures (without unacceptable cost in computer programming manpower (7)). The required essergy analysis procedures are illustrated by considering a large steam power plant.

Many terms for potential work, such as thermodynamic "availability," "exergy," "exergetic," etc. are used in different senses by different authors, so that it is not possible to tell precisely what is meant by these terms without referring to an author's framework of definition (9,10,15,16,17,18,19,20,21,22).

0097–6156/83/0235–0239$06.50/0
© 1983 American Chemical Society

The writers' framework of definition is based upon the work of Gibbs (4) who in 1878 presented the following measure $\mathcal{E}$ of the potential work of any system:

$$\mathcal{E} = E - T_o S + P_o V - \sum_i \mu_{io} M_i \qquad \text{ESSERGY} \qquad (1)$$

where E is the system's energy including all kinetic and potential energy in addition to internal energy, S is the entropy, V is the volume, M_i is the mass (or moles) of species i, while T, P, and μ_i denote the absolute temperature, pressure, and Gibbs potential of species i respectively, with "o" denoting the reference state used to represent the environment of the system. Gibbs never named the quantity $\mathcal{E}$, but he did in effect usually refer to it as essential energy (i.e., energy in the form essential for power production) so that the term essergy (i.e., essential energy) is now used for $\mathcal{E}$ (3,6,7,8). The corresponding flow Ψ of essergy (excluding kinetic and potential energy) for any uniform mixture of masses (or moles {M_i}) is given by (2,5,6,7),

$$\Psi = H - T_o S - \sum_i \mu_{io} M_i \qquad \text{FLOW ESSERGY} \qquad (2)$$

where H is the enthalphy of the uniform mixture.

A Description of Essergetic Functional Analysis

As an example, let us consider a feedwater heater, such as illustrated by Component No. 6 in Figure No. 1. Let $\dot{Z}$ represent the annualized capital cost (say in dollars per year) of owning and operating the feedwater heater (including maintenance, overhead, etc., as well as interest). Also, let λ represent the unit cost of each type of lost work, while $T_o\dot{\Theta}$ represents the lost work, where $\dot{\Theta}$ represents the rate of entropy creation (or production) corresponding to each type of lost work in the feedwater heater (2,6,7). Then let λ_A, λ_B, and λ_H represent the unit costs of lost work $T_o\dot{\Theta}_A$, $T_o\dot{\Theta}_B$, and $T_o\dot{\Theta}_H$, due respectively to head loss (pressure drop) in the feedwater A, head loss in the condensing steam B, and heat transfer (temperature drop) from the condensing steam to the feedwater, denoted by H, so that the total annualized cost $\dot{\Gamma}$ attributable to the feedwater heater is,

$$\dot{\Gamma} = \dot{Z} + \lambda_A T_o \dot{\Theta}_A + \lambda_B T_o \dot{\Theta}_B + \lambda_H T_o \dot{\Theta}_H \qquad (3)$$

In order to determine the costs of essergy λ for each interconnecting stream between components, it was found necessary to assign a principal purpose of function to each component (or device, unit, etc.) which would be paid for by a principal product, measured by its essergy content. This scheme which we now call essergetic functional analysis (25), worked fairly well for direct sea water conversion (11,12,13,14,23,24), but it seemed to fail for a simple steam power plant cycle (which was

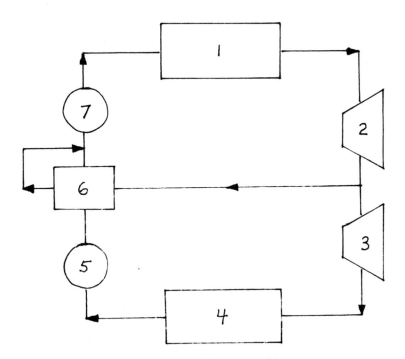

Figure 1. An abbreviated representation of a steam power plant cycle. The numbers 1,2,3,4,5,6, and 7 correspond respectively to the boiler, first turbine stage, second turbine stage, condenser, first feedwater pump, feedwater heater, and second feedwater pump. This first and second turbine stages may both lie inside a single-turbine (viewed as a single piece of equipment) with steam being tapped from a point near the middle of the turbine.

at that time considered in co-generation with sea water conversion)
because a unique function or purpose could not be found for the
main condenser.

Thus the concept of essergetic functional analysis was
abandoned for many years in favor of more typical thermoeconomic
decomposition techniques introduced by El-Sayed (3). El-Sayed
found the required unit costs of essergy λ to be equal to
Lagrange multipliers for essergy balances on the component in
question--such balances being taken as constraining equations for
minimizing the total cost of owning and operating a power system
with fixed product outputs (3,7). El-Sayed's decomposition
techniques have yielded many useful results (3,7,8).

Recent contributions by Smith (26) and Boteler (27) have
served to revive essergetic functional analysis as a viable
method for finding the unit essergy costs λ. Smith (26) showed
how to quantify the condenser's function of "getting rid of heat"
by having the condenser sell a form of "absolute essergy" known as
negentropy (28,29,30). Fig.2 shows results similar to Smith's for a
typical power plant (owned by the Wisconsin Electric Power Co.,
with thermodynamic data being taken from earlier publications by
Gaggioli (20)). Smith's results include the effects of turbine
capital costs (Gaggioli considered these to be sunk costs). The
unit cost of the essergy in the steam is seen to have the same
value at all feedwater heater tap points (plus or minus a few
percent--an "error" acceptable for basic design purposes). This
reflects the fact that the essergy content of the steam decreases
by a corresponding amount with lower temperature and pressure.
This near constancy of the unit cost of essergy vs. steam
temperature and pressure led naturally to Boteler's "Utilization
Function" (27), which appears in Requirements 2 and 3 of the
following paragraph. In view of these contributions by Smith
and Boteler, a description of essergetic functional analysis may
now be presented which is expected to be applicable to process
design and synthesis in general:

Essergetic Functional Analysis is a procedure for finding
unit essergy costs λ via the following three requirements:

1. Components (or devices, units, etc.) of the system being
    analyzed are to be defined in a manner such that each
    component has one and only one function (or purpose).
    Should a device have more than one function, further
    analysis must be carried out to meet this single
    function requirement--for example by subdividing
    the device into subcomponents each having only one
    function.

2. Each component's function (or purpose) is to be
    measured by an output flow of essergy $\dot{\varepsilon}$, such that
    for any given fixed value of $\dot{\varepsilon}$, the total annualized
    component cost $\dot{\Gamma}$ (given by an expression such as

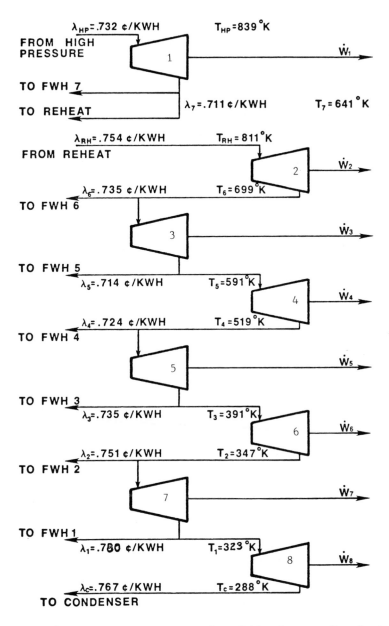

Figure 2. Useful energy costs λ_H of bleed steam for feed-water heaters.

Equation (3) above--or more generally by Equation
(13) of Reference 7--including any required
"utilization function costs" as described in
Requirement 3 below) is a minimum with respect to
all decision variables of that component--<u>including</u>
stream parameters such as $\dot{M}$, T, or P upon
which the essergy $\hat{\varepsilon}$ may depend (say via Equation
2, through the familiar dependence of
enthalpy and entropy upon temperature T and
pressure P--if $\dot{M}$, T, or P are not state variables,
in which case their value would result from decisions
in another component).

3. If necessary, utilization function costs are to be
    identified which will serve both to satisfy Require-
    ment 2 above and to yield a cost function $\dot{\Gamma}$ such
    that the marginal cost $\partial \dot{\Gamma}/\partial \hat{\varepsilon}$ (with all decision
    variables of that component held constant) is equal
    to the average unit cost $\dot{\Gamma}/\hat{\varepsilon}$ (with an arbitrary
    constant subtracted from $\dot{\Gamma}$) to within an accuracy
    bounded by the percentage degree of flatness of the
    minimum of $\dot{\Gamma}$ prescribed by Requirement 2.

It is to be understood that these three requirements are to be
met without significantly altering the optimum point of the
overall complex system. With utilization function costs
appropriately defined, this procedure is expected to apply to
complex systems in general--yielding unit essergy costs which
are approximately constant, as outlined in the remainder of this
section.

An example of a utilization function cost is the effect
upon the life of the high pressure turbine due to the creep rate
of the turbine blades as a function of temperature. Figure 3 is
a graph taken from work by Boteler (<u>27</u>) on the same power plant
considered by Smith (<u>26</u>), where the cost $\dot{\Gamma}/\hat{\varepsilon}$ for the boiler is
minimized with respect to the output steam temperature T. The
increasing component of $\dot{\Gamma}/\hat{\varepsilon}$ (with respect to T) represents a high
temperature utilization cost which the boiler pays to the high
pressure turbine as compensation for decreased blade life. It
is seen that the minimum of $\dot{\Gamma}/\hat{\varepsilon}$ occurs at $\lambda = 2.12\$/10^6$ BTU, at an
optimum steam temperature of 1040 °F, a value which agrees with
the value of $2.09 $/10^6$ BTU found by Smith (<u>26</u>) as shown in
Figure 2.

However, this result is based upon sparse creep-life data
for Timken 35-15 Stainless Steel (<u>31</u>), so that Figure 3 should be
regarded merely as a qualitative guide to such temperature
dependence. Considerably more material on creep-life-cost
relations must be gathered before reliable design values of the
optimum steam temperature can be obtained.

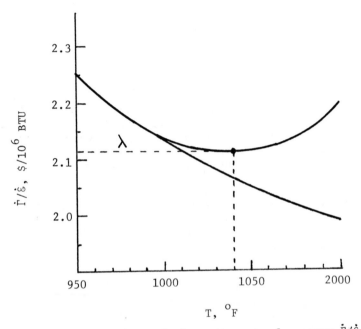

Figure 3. The minimum of the unit cost of essergy $\dot{\Gamma}/\dot{\varepsilon}$ with respect to the output steam temperature T for the power plant boiler.

But even as a qualitative guide, the result shows that the unit cost λ for the essergy of steam from the boiler is nearly constant at about \$2.1/10^6 BTU over the range from 1000F to 1070F, due to the relative flatness of the curve of $\hat{\Gamma}/\hat{\varepsilon}$ vs. T over this range.

Assuming that $\hat{\Gamma}/\hat{\varepsilon}$ complies with all three requirements of essergetic functional analysis, then λ will be nearly constant over a wide range of both temperature and pressure, provided of course that the surface is relatively flat in the region of the optimum (should this region not be sufficiently flat, then more detailed utilization functions should be introduced). Thus, in view of Requirement 3, the marginal cost λ of the essergy in the steam could be considered to be constant with respect to the magnitude of $\hat{\varepsilon}$, irrespective of whether a change in $\hat{\varepsilon}$ is caused by T,P, or $\dot{M}$.

Returning to Equation (3) for a feedwater heater, if all the unit essergy costs λ_A, λ_B, and λ_H were found to be absolutely constant in the manner described in the preceding paragraph, then the feedwater heater would be thermoeconomically isolated as discussed in Ref. 7 -- meaning that minimizing $\hat{\Gamma}$ thus defined would automatically coincide with optimizing the overall system as a whole -- without having to pay any attention whatsoever to variations in the rest of the system -- since such variations would be correctly represented by the constant values of λ_A, λ_B, and λ_H.

The significance of thermoeconomic isolation is discussed in the conclusions below. Let us now turn our attention to the treatment of thermal components such as a feedwater heater as represented by Equation (3), on the assumption that the unit costs λ_A, λ_B, λ_H will be constant as prescribed above over the range of significant variations.

Preliminary Second Law Analysis of the Main Thermal Components

Since our purpose is limited to introducing essergetic functional analysis, only simplified models will be considered. Let us first restrict our attention to closed feedwater heaters for which

$$\dot{Q} = \dot{M} \, c_p (T_2 - T_1) \tag{4}$$

where $\dot{M}$ is the feedwater rate.

The rate of entropy creation $\dot{\hat{\varepsilon}}_H$ due to heat transfer is ($\underline{14}$, $\underline{32},\underline{33}$),

$$\dot{\hat{\varepsilon}}_H = \dot{M} \, c_p (\ln \frac{T_2}{T_1} - \frac{T_2 - T_1}{T_c}) \tag{5}$$

where it is assumed that all heat leaves the condensing steam at the constant bulk temperature T_c.

The right side of Equation (5) is related to the heat transfer area A via ($\underline{33},\underline{34},\underline{35}$):

$$x \equiv \frac{UA}{\dot{M} \, c_p} \tag{6}$$

$$\frac{T_2 - T_1}{T_c - T_1} = 1 - e^{-x} \tag{7}$$

Here, x denotes the number of transfer units (often denoted by "NTU") and U is the overall heat transfer coefficient (or overall conductance) while the ratio $(T_2 - T_1)/(T_c - T_1)$ is the temperature "effectiveness." Substitution of Equation (7) into Equation (5) yields, defining $y \equiv T_1/T_2$,

$$\dot{\&}_H = \dot{M} c_p \frac{(y - 1)^2}{e^x - y} + \dot{M} c_p (y - 1 - \ln y) \tag{8}$$

In addition to the entropy creation $\dot{\&}_H$ due to heat transfer, there are entropy creations $\dot{\&}_A$ and $\dot{\&}_B$ due to viscous friction in the fluid in the tubing and on the shell side of the exchanger respectively. For example (32,33),

$$\dot{\&}_A = \frac{\dot{M}}{T_B} f \frac{L}{D} \frac{v^2}{2g_c} \tag{9}$$

where T_B is the log mean bulk temperature of the feedwater. The entropy creation rate $\dot{\&}_B$ is given by a similar expression, the detailed consideration of which is beyond the scope of this paper. The right side of Eqn. (9) will be recognized as being the product of the mass rate $\dot{M}$ times the head loss dissipation per unit mass divided by the bulk temperature T_B—it being noted that the average bulk velocity v is to be evaluated at this temperature.

The entropy creation $\dot{\&}_A$ due to head loss in the tubes is related to the heat transfer area A via the hydraulic diameter $D = 4A_cLN/A$. Since the mass rate $\dot{M}$ is given by $\dot{M} = \rho v A_c N \equiv \dot{m} A_c N$, Eqn. (9) yields

$$\dot{\&}_A = \frac{f \dot{m}^3}{T_B \rho^2 8g_c} A \tag{10}$$

And finally, one may use a typical empirical expression for the friction factor f (33) to obtain

$$\dot{\&}_A = \frac{0.046}{T_B \rho^2 2g_c} \left(\frac{d}{\mu}\right)^{-.2} \dot{m}^{2.8} A \tag{11}$$

Boilers may be treated in a similar manner, with $\dot{M}$ considered to be the mass rate of hot exhaust gas, as shown by Shen (36).

Let the unit costs of the useful energy (essergy) consumptions $T_o \dot{\&}_H$, $T_o \dot{\&}_A$, and $T_o \dot{\&}_B$ be denoted by constants λ_H, λ_A, and λ_B respectively; typical values lie in the range of 1 to 6¢/KW-hr. The origin of these constant values was considered above in the discussion of thermoeconomic isolation. In addition to the consumption costs $\lambda_H T_o \dot{\&}_H$, $\lambda_A T_o \dot{\&}_A$, and $\lambda_B T_o \dot{\&}_B$, the total cost of owning and operating the feedwater heater, condenser, or boiler must of course include its annualized capital cost $\dot{Z}$. Letting $\dot{K}$ include any other associated costs, the total cost $\dot{\Gamma}$ of owning and operating the feedwater heater, condenser, or boiler is

$$\dot{\Gamma} = \dot{Z} + \lambda_A T_o \dot{\&}_A + \lambda_B T_o \dot{\&}_A + \lambda_H T_o \dot{\&}_H + \dot{K} \tag{12}$$

The total cost $\dot{\Gamma}$ may be expressed in dimensionless form via dividing Eqn. (12) by $\lambda_H T_o \dot{M} c_p$--leading to dimensionless costs which are far less sensitive to the effects of external economic conditions than are the actual cost magnitudes. In order to simplify the resulting expression, let the cost per transfer unit γ_x be defined as follows,

$$\gamma_x \equiv \frac{\dot{Z} + \lambda_A T_o \dot{\&}_A + \lambda_B T_o \dot{\&}_B}{UA} \tag{13}$$

Utilization of this definition in (12) yields the following dimensionless expression for the total cost,

$$\frac{\dot{\Gamma}}{\lambda_H T_o \dot{M} c_p} = \frac{\gamma_x}{\lambda_H T_o} x + \frac{\dot{\&}_H}{\dot{M} c_p} + k \tag{14}$$

where $k \equiv \dot{K}/\lambda_H T_o \dot{M} c_p$.
The following definitions are also convenient:

$$\sigma_y \equiv y - 1 - \ln y \qquad\qquad \Gamma_D \equiv \dot{\Gamma}/\lambda_H T_o \dot{M} c_p \tag{15}$$

For lake-cooled condensers, σ_y must include the essergy wasted via the rejection of the warmed effluent cooling water, which from Eqn. (2) is given by $\dot{\&} = \dot{M} c_p T_o (\frac{1}{y} - 1 + \ln y)$ so that $\sigma_y = y - 1 - \ln y + (\frac{1}{y} - 1 + \ln y)$ which reduces to $\sigma_y = \frac{(y-1)^2}{y}$.

With these definitions, substitution of Eqn. (8) into Eqn.(14) gives,

$$\Gamma_D = \frac{\gamma_x}{\lambda_H T_o} x + \frac{(y-1)^2}{e^x - y} + \sigma_y + k \tag{16}$$

Under certain assumptions (discussed below), the cost per transfer unit γ_x is independent of the number of transfer units x, as shown following Eqn. (20) below. Figure 4 displays the minimum of this dimensionless total cost for a feedwater heater with constant parameters γ_x, y, λ_H, T_o, $\dot{M}$, c_p, and U. Similar curves result for condensers and boilers. The values used for the dimensionless parameters y and $\gamma_x/\lambda_H T_o$ are shown in the figure. From the definition $x \equiv UA/\dot{M} c_p$, it is seen that x is directly proportional to the heat transfer area A (since U, $\dot{M}$, and c_p are constant). Thus the optimum value of x (that is, the value of x which minimizes the total cost $\dot{\Gamma}$) corresponds to the optimum value of the heat transfer area A, which for these parameter values is 556 square meters. This optimum value x_{opt} may be expressed explicitly via setting $d(\dot{\Gamma}/\lambda_H T_o \dot{M} c_p)/dx = 0$, which yields the solution,

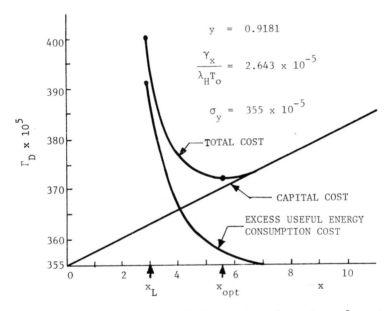

Figure 4.   Determination of the optimum heat transfer area for the feedwater heater.

$$x_{opt} = \ln \{y[1 + \frac{q}{2} (1 + \sqrt{1 + \frac{4}{q}})]\} \tag{17}$$

$$\text{where,} \quad q \equiv \frac{\lambda_H T_o (y - 1)^2}{\gamma_x y} \tag{18}$$

This result reflects the optimum trade-off between the excess use-ful energy consumption cost $(y - 1)^2/(e^x - y)$ and the capital cost $(\gamma_x/\lambda_H T_o)x$. The corresponding optimum savings in useful energy is given by $(y - 1)^2/(e^{x_L} - y) - (y - 1)^2/(e^{x_{opt}} - y)$, where x_L de-notes a typical low-capital-cost value of x such as the one shown in Figure 4.

The independence of γ_x and x may be easily demonstrated, for a linear cost model which neglects the two phase head loss and resistance, by substituting Eqn. (11) into Eqn. (13) to obtain (setting $\dot{\&}_B = 0$),

$$\gamma_x = \frac{\dot{z}_A + K_m \dot{m}^{2.8}}{U} \tag{19}$$

where for the linear model, $\dot{z}_A$ is a constant defined by,

$$\dot{z}_A \equiv \dot{Z}/A$$

$$\text{while,} \quad K_m \equiv \frac{0.046 \, \lambda_A T_o}{2g_c \, \rho^2 \, T_B} (\frac{d}{\mu})^{-.2} \tag{20}$$

Since the right side of Eqn. (19) is independent of the heat trans-fer area A, it follows that γ_x and x are independent of each other with respect to the variation shown in Figure 4--since U, $\dot{M}$, and c_p are constant in this variation. The inclusion of two-phase correc-tions does not significantly alter this result, as pointed out in the following section.

The optimum trade-off between the capital cost $\dot{z}_A/U$ and the useful energy consumption cost (via head loss) $K_m \dot{m}^{2.8}/U$ may, for this model neglecting the two-phase side, be found by expressing U in terms of $\dot{m}$ in Eqn. (19)--i.e., U is taken equal to the tube side heat transfer coefficient h, thus neglecting the resistances from the two-phase side and tubing wall -- these being reinstated in the corrections given in the following section. With this simplication, the overall conductance U may be expressed via Colburn's relation (33)

$$\frac{U}{\dot{m}c_p} = 0.023 (\frac{\mu c_p}{k})^{\frac{-2}{3}} (\frac{\dot{m} d}{\mu})^{-.2} \tag{21}$$

Thus one may write,

$$U \equiv [\, 0.023 \, (\frac{\mu c_p}{k})^{\frac{1}{3}} \, (\frac{d}{\mu})^{.8} \, \frac{k}{d} \,] \, \dot{m}^{0.8} \equiv K_h \dot{m}^{0.8} \quad (22)$$

Substitution into Eqn. (19) yields,

$$\gamma_x = \frac{\dot{z}_A}{K_h} \dot{m}^{-0.8} + \frac{K_m}{K_h} \dot{m}^2 \quad (23)$$

Now from Eqn. (16), it is seen that for any given value of x that the total cost $\hat{\Gamma}$ will be a minimum when γ_x is minimized. And since γ_x is independent of x as shown following Eqn. (20), it follows that γ_x as given by Eqn. (23) should be minimized with respect to the feedwater mass rate $\dot{m}$ in order to minimize the total cost $\hat{\Gamma}$. Setting the derivative $d\gamma_x/d\dot{m} = 0$ thereby yields the following expression for the optimum value $\dot{m}_{opt}$:

$$\dot{m}_{opt} = (\frac{.4\dot{z}_A}{K_m})^{\frac{1}{2.8}} \quad (24)$$

This optimum point is displayed in dimensionless form in Figure 5 where the minimum of $\gamma_x/\lambda_A T_0$ is shown with respect to the Reynolds number N_R via the following dimensionless form of Eqns. (23) and (24):

$$\frac{\gamma_x}{\lambda_A T_0} = \frac{\dot{z}_A (d/\mu)^{.8}}{\lambda_A T_0 K_h} N_R^{-.8} + \frac{K_m (d/\mu)^{-2}}{\lambda_A T_0 K_h} N_R^2 \quad (25)$$

$$N_{R,opt} = \frac{d}{\mu} (\frac{.4\dot{z}_A}{K_m})^{\frac{1}{2.8}} \quad (26)$$

The values used for the constant dimensionless parameters $\dot{z}_A(d/\mu)^{.8}/\lambda_A T_0 K_h$ and $K_m(d/\mu)^{-2}/\lambda_A T_0 K_h$ are shown in the figure. The resulting value for v_{opt} is 1.453 meters/sec. This result reflects the optimum trade-off between the excess useful energy consumption cost $[K_m(d/\mu)^{-2}/\lambda_A T_0 K_h](N_R^2 - 2.5 \times 10^7)$ and the capital cost $[\dot{z}_A(d/\mu)^{.8}/\lambda_A T_0 K_h]N_R^{-.8}$--it being noted that 2.5×10^7 is the square of the value 5000 for the Reynolds number--a value which marks a typical lower bound for fully developed turbulent flow in smooth pipes. The corresponding optimum savings in useful energy is given by $[K_m(d/\mu)^{-2}/\lambda_A T_0 K_h](N_{R,L}^2 - N_{R,opt}^2)$, where $N_{R,L}$ denotes a typical low-capital-cost value of N_R such as the one shown in Figure 5.

   The method used by Hendrix (32) and Smith (26) to find the unit essergy costs shown in Figure 2 is an extension of the technique employed by Gaggioli and Fehring (37). Hendrix (32) extended their technique by introducing turbine capital costs to find the unit costs λ for all of the useful energy flows in a Wisconsin Electric Power Co. Plant (37). Figure 2 shows his results for the steam flow through the eight turbines stages of this plant.

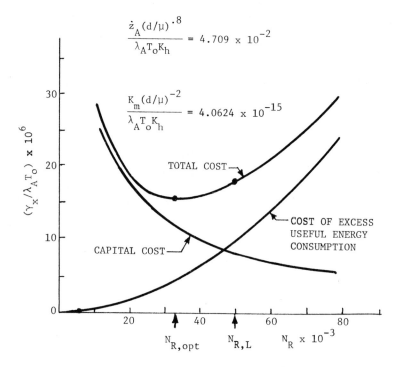

Figure 5. Determination of the optimum feedwater velocity
for the feedwater heater.

Since Feedwater Heater No. 5 consumes steam at $591^{\circ}K$, it is seen from Figure 2 that the unit cost λ_H for heat transfer essergy in Eqn. (12) will be $\lambda_H = \lambda_5 = 0.714¢/KW\text{-}hr$. And as demonstrated in Ref. 7, this feedwater heater will thus be thermoeconomically isolated, so that the unit costs λ_A, λ_B, and λ_H are each constant in Eqn. (12) with respect to all possible design or synthesis variation in that equation. Focusing our attention upon $\lambda_H = \lambda_5$ in Figure 2, it is seen that this unit cost does not vary greatly from the unit costs of essergy found by Hendrix for the steam going to each of the other feedwater heaters--the total variation lying between $0.711¢/KW\text{-}hr$ for FWH No. 7 and $0.780¢/KW\text{-}hr$ for FWH No. 1.

This suggests the possibility that for original design and synthesis purposes, unit costs such as $\lambda_H = \lambda_5$ may indeed be kept constant with respect to temperature variations of the steam extracted for components such as Feedwater Heater No. 5 via introducing primary turbine cost formulae which by definition would be insensitive to temperature and pressure--with any inaccuracies in these primary costs being offset through the use of utilization function costs (similar to the one introduced by Boteler, as discussed above) which would account for any anomalies in the effect of the steam temperature and pressure upon turbine cost and performance.

Through the use of such utilization function costs, the turbine costs are separated into primary turbine costs, which by definition are insensitive to the temperature and pressure of the steam, and secondary turbine costs which are automatically paid for by neighboring components via requiring their purchase of utilization functions. The primary turbine cost formulae then result in equal units costs λ_H of the bleed steam essergy for all of the feedwater heaters. Thus λ_H is then constant with respect to temperature, thereby meeting the condition of thermo-economic isolation of the feedwater heaters with respect to the turbines.

Further detailed consideration of the turbines and pumps is beyond the scope of this paper, other than to point out that the optimum number of turbine and pump stages is determined by finding the optimum number of feedwater heaters via the method outlined in a following section.

Internal Correction Factors

In addition to the secondary corrections for component inter-dependencies discussed at the end of the preceding section, introductory thermoeconomic equations such as Eqns. (13) through (26) require internal corrections (i.e., corrections internal to the component) before complex systems can be accurately syn-thesized and designed. These internal corrections may be incorporated into correction factors which leave the basic form of the introductory equations unchanged--thus preserving the insight gained from the introductory equations.

For example, the cost per transfer unit γ_x in Eqn. (13) might in general include dissipations $T_o\overset{.}{\&}_j$ other than just the head losses in the two streams so that in general,

$$\gamma_x \equiv \frac{\overset{.}{Z} + \sum_{j=1}^{m} \lambda_j T_o \overset{.}{\&}_j}{UA} \qquad (27)$$

The two-phase term $\sum_{j=2}^{m}\lambda_j T_o\overset{.}{\&}_j$ was neglected in the introductory expression for γ_x given by Eqn. (19). The numerator of Eqns. (19) and (23) may be corrected by defining an area cost factor f_A as follows,

$$f_A \equiv \frac{\overset{.}{Z} + \sum_{j=1}^{m} \lambda_j T_o \overset{.}{\&}_j}{1.4 \ \overset{.}{z}_A A} \qquad (28)$$

it being noted from Eqn. (24) that $K_m \overset{.}{m}^{2.8} = 0.4 \ \overset{.}{z}_A$ via Colburn's relation, a result which agrees with well-known results for similar heat exchangers (40,41).

Also, the overall thermal resistance $1/U$ must include all thermal resistances, so that the denominator of the introductory expression for γ_x given by Eqn. (23) must be corrected (since U as modeled by Eqn. (22) neglects the thermal resistances of the tubing wall and the two-phase side). The denominator of Eqn. (23) may be corrected by defining a thermal conductance factor f_U:

$$f_U \equiv \frac{K_h \ \overset{.}{m}_r^{0.8}}{U} \qquad (29)$$

where the subscript "r" denotes the reference value of $\overset{.}{m}$ given by Equation (24).

The factors f_A and f_U are not constant w.r.t. variation in $\overset{.}{m}$, as shown below, so that they are not convenient for differentiation. For that purpose, the following factors, introduced by H. Slack (38), are useful:

$$f_{Ar} \equiv \frac{\overset{.}{Z} + \sum_{j=2}^{m} \lambda_j T_o \overset{.}{\&}_j}{\overset{.}{z}_A A} \qquad (30)$$

$$f_m \equiv \frac{\overset{.}{m}}{\overset{.}{m}_r} \qquad (31)$$

$$f_{Ur} \equiv f_U + 1 - \frac{1}{f_m^{0.8}} \qquad (32)$$

Comparison of Eqns. (29), (31), and (32) shows that the factor f_{Ur} represents the following thermal resistance ratio,

$$f_{Ur} \equiv K_h \ \overset{.}{m}_r^{0.8} \left(\frac{1}{U} - \frac{1}{K_h \ \overset{.}{m}^{0.8}} + \frac{1}{K_h \ \overset{.}{m}_r^{0.8}} \right) \qquad (33)$$

while Eqn. (30) reduces to,

$$f_{Ar} = 1.4 \ f_A - 0.4 \ f_m^{2.8} \qquad (34)$$

Noting that,

$$\gamma_{xr} \equiv \frac{1.4 \ \dot{z}_A}{K_h \ \dot{m}_r^{0.8}} \tag{35}$$

one has for γ_x, in view of Eqns. (27) through (35),

$$\frac{1.4}{\gamma_{xr}} \gamma_x = f_{Ar} f_m^{-0.8} + 0.4 \ f_m^2 + (f_{Ur} - 1)(0.4 \ f_m^{2.8} + f_{Ar}) \tag{36}$$

For the case where f_{Ar} and f_{Ur} are not perfectly constant w.r.t. f_m, correction factors for their variation may be introduced in the simple manner of Appendix C of Chapter 1 of Ref. 24. Also, in view of Eqn. (9b), it will be observed that variation of $\dot{m}$ corresponds to variation of the number of tubes per pass N (at fixed $\dot{M}$ and A_c), so that γ_x would at first sight appear to be a non-differentiable step function of $\dot{m}$. However, γ_x is actually a differentiable function of $\dot{m}$, since as pointed out in Ref. 7, all quantities involved are actually statistical expectations or averages of the physical quantities represented, because of the uncertainties inherent in such thermodynamic and economic parameters (39). With this understanding the minimum of γ_x/γ_{xr} occurs at $d(\gamma_x/\gamma_{xr})/df_m = 0$, which yields

$$f_{m-opt} = \left(\frac{f_{Ar}}{1 + 1.4 \ (f_{Ur} - 1) \ f_{m-opt}^{0.8}} \right)^{\frac{1}{2.8}} \tag{37}$$

Equation (37) is expressed in a form for direct iteration--a form which converges very rapidly via the well-known Newton-Raphson technique. For example, with $f_{Ar} = 1.8$ and $f_{Ur} = 2$ one finds $f_{m-opt} = 0.9155$, which yields a statistical expected value of 570.4 for the optimum number of tubes per pass, so that either 570 or 571 tubes would be optimum. However, due to the flatness of the total cost curve in Figure 5, values of either 550 or 600 tubes per pass would make very little difference in the total cost, so that for the purpose of saving useful energy 600 tubes would be a more likely optimum design figure. The corresponding value of the overall conductance (for 600 tubes per pass) is 518.1 BTU/hr ft^{2}F, down from 1097 BTU/hr ft^{2}F for the case where the two-phase head loss and resistance were neglected. Such a drop in overall conductance reflects the value of f_{Ur} which was used ($f_{Ur} = 2$)--a value which reflects a large amount of fouling (it being recalled that for film-wise condensation, the two-phase convective coefficient is well-known to be several times higher than the convective coefficient for fully-developed turbulent flow of liquid water in smooth pipes).

Additional corrections are of course necessary. These include load factors, and cost corrections such as entrance and exit head losses and the manufacturing cost of the headers or water boxes. These last corrections must be included in order to find the optimum diameter of the tubing (via minimizing $f_{Aopt} f_{Uopt} \gamma_{xr}$ with respect to the tubing diameter). Also, the corrections for non-constancy of the factors f_{Ar} and f_{Ur} must not be overlooked. The mode of correction is outlined in Appendix C, Chapter 1 of Ref. 24, where sizable non-linearities are superimposed upon a linear base. In addition, the single pure evaporator model of a boiler must include a large correction for preheat as well as superheat--such corrections being applied by Boteler (27), and Hendrix (32).

And finally, the scope of the model itself must be enlarged to include such complexities as the exact pricing of co-generated steam for chemical processes such as paper making or distillation of alcohol. For land based plants, evaporation cooling towers should be included, the basic model for achieving the thermo-economic isolation of power plant cooling systems having been worked out by P.V. Kadaba and M.V. Smith (26) through the use of combined essergy-negentropy balances.

Optimum Number of Feedwater Heaters

In order to find the optimum number of feedwater heaters, the optimum value of $y \equiv T_1/T_2$ should first be found by minimizing $\dot{\Gamma}/\dot{Q}$ where $\dot{Q}$ and $\dot{\Gamma}$ are given by Eqns. (4,15, and 16). Letting T_1 be constant, one may achieve this minimum by minimizing γ_D, defined by,

$$\gamma_D = \frac{\dot{\Gamma} \; y}{\lambda_H T_o \dot{M} c_p (1-y)} = \frac{\Gamma_D y}{1-y} \tag{38}$$

In terms of γ_D, Eqn. (16) becomes,

$$\gamma_D = \frac{\gamma_x}{\lambda_H T_o} \left(\frac{xy}{1-y} \right) + \frac{y(1-y)}{e^x - y} + \frac{y}{1-y} (\sigma_y + k) \tag{39}$$

The parameter $(1-y)/xy$ is seen from Eqns. (4) and (6) to be equal to $\dot{Q}/UAT_1$, where $\dot{Q}/UA$ is often expressed as a corrected long-mean temperature difference in heat transfer texts (33,34,35). Considering y and $(1-y)/xy$ to be the key design variable (at constant T_1), the optimum value of $xy/(1-y)$ is found by multiplying x_{opt} in Eqn. (17) by $y/(1-y)$, while y_{opt} is found by minimizing γ_D at constant $xy/(1-y)$--for which case, the first term of this last equation is seen to be constant (since λ_H and T_o are constant inputs, while γ_x is at the constant minimum value determined by Eqn. (37). For large values of x_{opt}, variation of the second term may be neglected, so the base model for finding y_{opt} may be taken as,

$$\gamma_D = \frac{y}{1-y} (\sigma_y + k) + k_r \tag{40}$$

where k_r represents the first two terms of Eqn. (39), considered as constant, with corrections for smaller values of x_{opt} being incorporated into a correction factor obtained from differentiating the neglected second term of Eqn. (39), using $x = x_{opt}$ as given by Equation (17).

Equation (40) may be applied to lake-cooled power plant condensers as well as feedwater heaters. For the condensers, $\sigma_y = (y-1)^2/y$ so that Eqn. (40) reduces to,

$$\gamma_D = 1-y + \frac{k}{1-y} - k + k_r \tag{41}$$

Minimizing γ_D via $d\gamma_D/dy = 0$ gives the following basic result for lake-cooled condensers:

$$(1-y)_{opt} = \sqrt{k} \tag{42}$$

This result must of course be correted for small values of x_{opt} using the correction factor mentioned above. For feedwater heaters, a similar result is obtained by expanding the logarithm in Eqn. (15) to obtain a value of $\sigma_y = (y-1)^2/2y$ which needs another correction factor to conform to Eqn. (15) exactly. This merely introduces a factor of two into Eqn. (42) so that for feedwater heaters, the corresponding basic results is

$$(1-y)_{opt} = \sqrt{2k} \tag{43}$$

As indicated when $\dot{K} = k\lambda_H T_o \dot{M} c_p$ was first introduced just before Eqn. (12), $\dot{K}$ is constant with respect to decisions within each feedwater heater or condenser, but is a variable with respect to associated costs. Thus, for example, the dimensionless cost k includes the cost of cooling water inlet filters in the case of the condensers--while for feedwater heaters it includes the cost of the bleed steam duct and tap required for the addition of a new heater. It also includes the cost of end effects (such as headers and water boxes) which will be added automatically to k when the non-linearity correction factors mentioned above are included. When all these corrections are considered, the results of Eqns. (42) and (43) appear to be in good agreement with the values found in well-optimized plants (such as the Wisconsin Electric Plant from which the data for this work was taken), but it is beyond the scope of this paper to pursue this matter further.

Further optimization rests primarily in further minimizing the cost per transfer unit γ_x, a well-known criterion first pointed out by Tribus over 25 years ago (1,14,23). This criterion (often called minimizing the cost of "UA") is thus well-known by manufactures of heat transfer equipment. However, unless the heat exchange component is shown to be thermoeconomically isolated, this criterion has to be viewed with suspicion, and indeed it is currently used more as a "rule of thumb" than as a precise criterion.

But with the establishment of thermoeconomic isolation plus the use of the correction factors mentioned above, minimization of the cost per transfer unit γ_x becomes a precise criterion. It has already been shown above how the optimum value of the length and diameter of the tubes is obtained--as well as the optimum number of tubes per pass. Further synthesis of such heat exchangers results from minimizing $\gamma_{xr} f_{Aopt} f_{Uopt}$ with respect to material and geometric selection factors.

After all of the thermoeconomic synthesis-design equations (such as the expression for $\gamma_{xr} f_{Aopt} f_{Uopt}$) have been expanded into relevant material and geometric selection factors, each component is then synthesized via minimizing the cost w.r.t. these selection factors. Following that, the number of multiple components (such as feedwater heaters, pumps, and turbine stages) may be ascertained by considering the pumps and feedwater heaters to be "transmission systems" for mechanical and thermal exergy (3,24) respectively. Equation (43) is then applied repeatedly until the cost of the output thermal exergy is the same for the highest temperature feedwater heater as it is for the preheating of liquid water in the "economizer"--at which point the addition of another feedwater heater would produce no further savings.

This last step serves also to establish the optimum number of pumps and turbine stages, which completes the synthesis phase. The design phase is then carried out by re-optimizing each component via minimizing the unit cost of its essergy output--a process which increases the degree of constancy of these unit costs, thereby increasing the degree of thermoeconomic isolation of the corresponding components.

Conclusion

It has been shown how essergetic functional analysis may yield constant unit costs of output essergy from system components (devices, units, etc.--over the significant range of design variations), providing that the three requirements listed on page 3 can be met. If all the components of a system meet these requirements, then all essergy unit costs will be constants (over a significant range of design variations) throughout the system as a whole. Under this condition, all of the components are said to be thermoeconomically isolated--meaning that the optimum design point of each component (which by definition optimizes the system as a whole) may be found by referring only to the constant unit costs at which that component buys the various types of essergy from the rest of the system. This is, of course, a much stronger condition than is implied by the well-known existing techniques for thermoeconomic decomposition of systems and processes, such as dynamic programming, Lagrange Multiplier costing, etc.--all of which must account for variations in the rest of the system beyond what can be communicated via constant unit costs only.

It was then illustrated how a thermo-chemical process may be synthesized and designed if its components are thus thermoeconomically isolated. The illustration was carried out on components of a large steam power plant, using simplified primary equations with subsequent corrections--the primary equations serving to illustrate clearly the tradeoff between useful energy costs and investment in equipment.

It remains to be shown whether or not the three requirements of essergetic functional analysis are always consistent with proven thermoeconomic decomposition techniques such as El-Sayed's method of Lagrange multipliers. It could be that the proof of this consistency could only be obtained at the expense of new, stringent conditions upon the definition of the utilization functions needed to guarantee compliance with these three requirements.

If the problems implied in the preceding paragraph can be solved even to a good approximation, then system components could always be approximately isolated thermoeconomically. Being thus enabled to concentrate on a single component without having to account for the variations in the remainder of the system, one may investigate new variations (by permitting variations of previously constant quantities) which would otherwise lie out of reach of one's capabilities--whether or not enforced by computerization. In other words, if an interdependent component is thermoeconomically isolated, then that component may be suboptimized with respect to many new, underlying variables which cannot be treated by any other existing optimization procedure, without unacceptable cost in computer programming manpower.

Literature Cited

1. Tribus, M. et al. Thermodynamic and Economic Considerations in the Preparation of Fresh Water from Sea Water, UCLA Report No. 56-16, 1956.
2. Evans, R.B. "Thermodynamic Availability as a Resource and a Tool for System Optimization," (1958). University of CA, Dept. of Engineering, Los Angeles, Report No. 59-34, 1960.
3. El-Sayed, Y.M., and Evans, R.B., "Thermoeconomics and the Design of Heat Systems": Journal of Engineering for Power. Jan. 1970, pp. 27-35.
4. Gibbs, J.W. "On the Equilibrium of Heterogeneous Substances." 1878. The Collected Works, Yale University Press, vol. 1, p. 77, (1928).
5. Evans, R.B. A Contribution to the Theory of Thermo-Economics. Master's Thesis, University of CA, Dept. of Engineering, 1961.
6. Evans, R.B. "A Proof that Essergy is the Only Consistent Measure of Potential Work," Ph.D. Thesis, Dartmouth College, 1969.
7. Evans, R.B. "Thermoeconomic Isolation and Essergy Analysis." Energy: The International Journal, 5, 805, 1980.
8. El-Sayed, Y.M. and Tribus M. Strategic Use of Thermoeconomic Analysis for Process Improvement. This Volume.

9. Keenan, J.H. "Availability and Irreversibility in Thermo-
    dynamics." British Journal of Applied Physics. 2:183-193, 1951.
10. Szargut, J. "Grenzen fuer die Anwendungsmoeglichkeiten des
    Exergiebegriffs. Brennstoff-Waerme-Kraft. v. 19, no. 6,
    June 1967, pp. 309-13.
11. Tribus, M., and Evans, R.B. "Economic and Thermodynamic Aspects
    of Sea Water Conversion." Proceedings: Conference on Water
    Research at the University of California, May 1960.
12. Tribus, M., and Evans, R.B. "Thermo-Economic Considerations
    in the Preparation of Fresh-Water From Sea-Water." Dechema
    Monographien, NR. 781-834 BAND 47. Verlag Chemie, GMBH,
    Weinheim/Bergstrasse, 1962.
13. Tribus, M. and Evans, R.B. "Optimum-Energy Technique for Deter-
    mining Costs of Saline-Water Conversion." J. Am. Water Works
    A., 1962.
14. Tribus, M. and Evans, R.B. "The Thermo-Economics of Sea-Water
    Conversion." UCLA, Engr. Dept., Report No. 62-53, 1963.
15. Rant, Z. "Exergie, ein neues Wort fur, Technische Arbeits-
    fahigkeit." Forsch, Ing-Wes., v.22, no.1, pp,36-37, 1956.
16. Penner, S., Editor "2nd Law Analysis of Energy Devices and
    Processes," Energy--The Int. J., 5, 665-1011, 1980.
17. Gouy, G.J. Physique (8), 501 (1889).
18. Stodola, A. Zeitschrift d. VDI 32, 1086 (1898).
19. Wepfer, W.J. "Applications of Available Energy Accounting,"
    pp 161-186, Thermodynamics: 2nd Law Analysis. R.A. Gaggioli,
    Editor, ACS Symp. Series 122, Am. Chem, Soc., Wash.D.C., 1980.
20. Gaggioli, R.A., Editor. Thermodynamics: 2nd Law Analysis. ACS
    Symp. Series 122, Am. Chem. Soc., Wash. D.C., 1980.
21. Gibbs, J.W. "Elementary Principles in Statistical Mechanics"
    (1901). The Collected Works, v.II, Yale U. Press, 1948.
22. Jaynes, E. "Information Theory and Statistical Mechanics",
    Phys. Rev., 106, pp.620-630, 1957.
23. Evans, R.B. and Tribus, M. "Thermo-Economics of Saline Water
    Conversion." I&EC Proc. D. and D., 4, 2, 195-206, 1965.
24. Evans, R.B. "Basic Relationships Among Entropy, Exergy, Energy,
    and Availability" (1963). App. A, Chap. 1, Princ. of Desalina-
    tion, ed. by K. Spiegler, Acad. Press, NY, 2nd.ed. (1980).
25. Dickerson, S. and Robertshaw, J. Planning and Design, Lexing-
    ton Books, 1957.
26. Smith, M. Effects of Condenser Design upon Boiler Feedwater
    Essergy Costs in Power Plants. M.S. Thes., Ga. Tech, 1981.
27. Boteler, K. Essergy Anal. of Boilers. M.S. Thes., Ga. Tech,
    1982.
28. Brillouin, L. Science and Information Theory. 2nd ed. Academic
    Press, Inc., New York, 1962, (1st ed., 1956).
29. Voigt, H. "Evaluation of Energy Processes Through Entropy and
    Exergy." RM-78-60, Int. Inst. App. Sys. Anal., Austria, 1978.
30. Evans, R. and Smith, M. "The Uniqueness of Reference States
    for Essergetic Analysis." 2nd W. Cong. Ch. Engr., Montreal,
    Canada, 1981.

31. Clauss, F. "An Examination of High Temperature Stress Rupture Correlating Parameters." Proce. ASTM, 60, 905, 1960.
32. Hendrix, W.A. "Essergy Optimization of Regenerative Feedwater Heaters," M.S. Thesis, Ga. Inst. of Technology, June 1978.
33. Giedt, W. Princ. of Engr. Heat Transfer. Van Nostrand,NY,1957.
34. Jacob, M. Heat Transfer, v. 2, Wiley, 1956.
35. Kays, W. and London, A. Compact Heat Exchangers. 2nd Ed., McGraw-Hill, 1964.
36. Shen, L.S. "Essergy Analysis of Basic Vapor Compression Refrigeration Systems," M.S. Thesis, GA Tech, 1979.
37. Fehring, T. and Gaggioli, R., "Economics of Feedwater Heater Replacement." Trans. ASME, July 1977, pp.482-488.
38. Slack, Henry. "Corrections for Shell-Side Head Loss Resistance." Special Project, GA Tech, M.E., 1980.
39. Evans, R.B. "A New Approach for Deciding Upon Constraints in the Maximum Entropy Formalism," The Maximum Entropy Formalism. M.I.T. Press, Cmbridge, MA, 1978, pp.169-203.
40. Bejan, A. "General Criterion for Rating Heat-Exchanger Performance, Int. J. Heat and Mass Transfer, 21, 5, 655-658, 1978.
41. Jenssen, S.K. "Heat Exchanger Optimization." Chemical Engr. Progress, v.65, no.7, pp.59-66 July 1969.

RECEIVED July 7, 1983

Thermoeconomic Optimization
of a Rankine Cycle Cogeneration System

ROBERT M. GARCEAU[1] and WILLIAM J. WEPFER

School of Mechanical Engineering, Georgia Institute of Technology, Atlanta, GA 30332

This paper presents the application of thermo-
economics to the optimization of a Rankine cycle
cogeneration system which supplies process heat
in the form of hot water. The method described
in this paper is well-suited for application to
thermal systems. This technique does not require
that available-energy (exergy) be used explicitly
in the optimization although the effects of
available-energy dissipations are evaluated via
the shadow and marginal prices. The design space
for this cogeneration system consists of five
independent variables and three parameters that
reflect "market" conditions; fuel cost,
electricity cost, and the required hot water
supply temperature. Results (suboptimizations)
are presented for several given sets of market
conditions.

Thermal systems can be completely described using balance
equations for mass, energy, and entropy in conjunction with
thermophysical property relations and/or equations of state,
equipment performance characteristics, thermokinetic or rate
equations, and boundary/initial conditions. With the thermal
system adequately described, it can be optimized by any current
technique. Although the approach presented in this paper is not
explicit in Second Law terms, it never-the-less will yield the
optimal design and with the appropriate transformations, will
yield any desired Second Law quantity.

The variables which are used to describe the system usually
are not all independent since there exist many equations of constraint. It may be possible to substitute the constraint
equations into the objective function, leaving only independent

[1]Current address: Aerospace Systems Division, Harris Corporation, Box 37,
Melbourne, FL 32901

0097–6156/83/0235–0263$07.25/0

variables to be optimized. Unfortunately, it is not always
mathematically desirable to eliminate all the dependent variables
from the objective function. One alternative is the use of
Lagrangian multipliers.

The application of Lagrange's Method to large scale thermal
systems is well-known and wide-spread. Brosilow and Lasdon (1-2)
and Gembicki (3) applied this technique to complex chemical processing plants. El-Sayed and Evans (4) developed a method whereby
a complex thermal system is decomposed into its component parts,
each component buying and selling available-energy with other
components. This approach requires that the problem be cast in
terms of available-energy coordinates, that is, the constraint
equations representing the internal economic transactions (supply
and demand equations) must be explicit in available-energy flows.
The benefit of such a transformation is that the Lagrange multipliers represent prices that describe the internal sales and
purchases of available-energy. In turn, these prices can be used
to show the economic trade-offs between capital investment and
available-energy destruction for each component of the system.

In order to obtain the available-energy balance equations
(the intercomponent available-energy sales and purchases), it is
necessary to incorporate all the relevant thermodynamic constraints. In general no guidelines exist concerning the selection
of the form for these constraint equations. Furthermore, the
restriction that all state variables (temperatures, pressures,
mole fractions, mass flows ...) be transformed to available-energy
functions is difficult and quite tedious. As a rule, this transformation from state variables to available-energy flows results
in a set of highly non-linear algebraic equations which must be
solved in order to obtain the set of optimum decision variables.

The technique to be described in this paper follows that of
Tribus and El-Sayed (5) in that no transformation is made to
available-energy coordinates. Rather the problem is optimized
using thermodynamic and equipment coordinates. In this way any
desired Second Law quantities can be obtained upon conclusion of
the optimization using an appropriate transformation.

Description of the Optimization Technique

This method requires the objective function to be expressed in
terms of the dependent (or state) variables, $\{x_i\}$, and independent
(or decision) variables, $\{y_k\}$.

$$\phi_o = \phi_o \ (\{x_i\}, \{y_k\}) \qquad (1)$$

The equations of constraint are divided into two groups. One
set of constraints, referred to as substitution constraints, are
used to eliminate selective dependent variables from the objective
function, ϕ_o, and from the set of Lagrange constraints, ϕ_j. The
other set, called Lagrange constraints are used directly in the

optimization scheme, each having an associated Lagrange multiplier, λ_j. The constraints are defined by equations of the form

$$\phi_j = \bar{\phi}_j - x_j \tag{2}$$

where ϕ_j is the equation of constraint to be used in Lagrange's Method (a Lagrange constraint). The constraint must be equal to zero at the optimum. $\bar{\phi}_j$ is the defining equation that describes the relation between the associated state variable, $\{x_j\}$, and the other state variables, $\{x_i\}_{i \neq j}$, and decision variables $\{y_k\}$.

With these definitions, the Lagrange constraints are then expressed as functions of the state and decision variables

$$\phi_j = \phi_j \; (\{x_j\}, \{y_k\}) \qquad j = 1,n \tag{3}$$

Using this approach, the problem is formulated as

Extremize $\qquad \phi_0 = \phi_0 \; (\{x_i\}, \{y_k\}) \tag{4}$

Constrained by $\quad \phi_j = \phi_j \; (\{x_i\}, \{y_k\}) \qquad j = 1,n \tag{5}$

With the selection of the objective function and Lagrange constraints, the Lagrangian, L, is defined by

$$L = \phi_0 + \sum_j \lambda_j \, \phi_j \tag{6}$$

In accordance with the Method of Lagrange, to guarantee that L is independent of the set of state variables, $\{x_i\}$ (i.e. L is unconstrained or in other words, that the optimum of L is equal to the constrained optimum of ϕ_0), we must have

$$\frac{\partial L}{\partial x_i} = 0 \qquad\qquad i = 1,n \tag{7}$$

or

$$\frac{\partial \phi_0}{\partial x_i} + \sum_{j=1}^{n} \lambda_j \frac{\partial \phi_j}{\partial x_i} = 0 \qquad\qquad i = 1,n \tag{8}$$

This yields a set of n equations that are linear in the unknown Lagrangian multipliers, $\{\lambda_j\}$. Equation 8 may be put in matrix form and rearranged to yield

$$\left[\frac{\partial \phi_j}{\partial x_i}\right] \; [\lambda_j] = - \left[\frac{\partial \phi_0}{\partial x_i}\right] \tag{9}$$

or
$$[\lambda_j] = - \left[\frac{\partial\phi_o}{\partial x_i}\right]\left[\frac{\partial\phi_j}{\partial x_i}\right]^{-1} \tag{10}$$

The Lagrangian multipliers may be interpreted as prices which reveal the economic value associated with a differential change in the corresponding state variables, and are referred to as shadow prices (5).

When the objective function and Lagrange constraints are well-behaved analytic functions, the optimum design is defined by

$$\left[\frac{\partial L}{\partial y_k}\right] = \left[0\right] \tag{11}$$

or rewriting

$$\left[\frac{\partial\phi_o}{\partial y_k}\right] + [\lambda_i]\left[\frac{\partial\phi_i}{\partial y_k}\right] = \left[0\right] \tag{12}$$

The optimum design is therefore located by solving the resulting simultaneous equations to obtain each optimum decision variable, y_k.

When optimizing real world problems it is often convenient to define a marginal price, θ_k, associated with each decision variable, y_k. The set of marginal prices are defined

$$\theta_k = \frac{\partial L}{\partial y_k} \tag{13}$$

$$= \frac{\partial\phi_o}{\partial y_k} + \sum_{i=1}^{n} \lambda_i \frac{\partial\phi_i}{\partial y_k} \qquad \text{for} \quad k = 1, m \tag{14}$$

or in matrix form

$$[\theta_k] = \left[\frac{\partial L}{\partial y_k}\right] = \left[\frac{\partial\phi_o}{\partial y_k}\right] + [\lambda_i]\left[\frac{\partial\phi_i}{\partial y_k}\right] \tag{15}$$

The physical interpretation of these marginal prices is that they are prices associated with a differential change in the corresponding decision variables, $\{y_k\}$, and may be used to indicate the potential benefit of changing a decision variable.

Equation 10 represents n linear equations in the unknown Lagrangian multipliers and Equation 12 represents m simultaneous equations with the decision variables as unknowns. For many practical applications of this method, the n+m equations corresponding to Equations 10 and 12 are highly non-linear and quite unwieldy to solve. Consequently, a numerical solution may be necessary. One such procedure is to obtain numerical values for the shadow prices, $\{\lambda_i\}_\ell$, from Equation 10 for a particular set of design variables, $\{y_k\}_\ell$. These shadow prices are then introduced to Equation 15 and the marginal prices, $\{\theta_k\}_\ell$, evaluated. The marginal prices represent the derivatives of the objective function, ϕ_o, at that particular design point, ℓ. It is then possible to use the marginal prices to point in the direction toward the optimum design. A new generation of design variables, $\{y_k\}_{\ell+1}$ is determined using

$$y_{k,\ell+1} = y_{k,\ell} \pm \Delta y \qquad (16)$$

where Δy_k is some predetermined iteration increment.

When the marginal price, $\theta_{k,\ell}$, is negative the derivative of the objective function is negative. If we wish to minimize the objective function, this indicates that the decision variable needs to be increased in order to approach the optimum. Similarly, if the marginal price, $\theta_{k,\ell}$, is positive, the decision variable needs to be decreased in order to approach the optimum.

In many cases it is informative to have Second Law based prices associated with a change in thermodynamic value streams corresponding to θ_k or λ_i. These Second Law based prices may be obtained by using the chain rule

$$\lambda_s = \lambda_i \frac{\partial x_i}{\partial A_s} = \lambda_i \frac{1}{\left(\dfrac{\partial A_s}{\partial x_i}\right)} \qquad (17)$$

$$\theta_s = \theta_k \frac{\partial y_k}{\partial A_s} = \theta_k \frac{1}{\left(\dfrac{\partial A_s}{\partial y_k}\right)} \qquad (18)$$

where λ_s and θ_s are prices associated with changes in an available-energy stream, A_s (which has y_k and x_i as variables), through changes in y_k and x_i. A non-dimensional marginal price, $\bar{\theta}_k$, is often desired (5). This is accomplished using Equation 19

$$\bar{\theta}_k = \theta_k (y_k/\phi_o) \qquad (19)$$

System Model

Cogeneration has been utilized by industry for many years but recent trends in purchased energy costs and the enactment of PURPA

have significantly influenced the economics to a more favorable
position. This method of producing low temperature heat has an
advantage over the more conventional system of burning fuel for
low temperature applications in that a cogeneration system has a
much lower available-energy consumption than the conventional
system (6-9). In other words, cogeneration systems can provide
low temperature heat at higher Second Law efficiencies than the
alternative of producing low temperature steam from a low pressure
boiler or furnace.

The optimization must focus on how much shaft work (or
electricity) the cogeneration system should produce, given a
specific product heat requirement, for present and future market
conditions. The optimization must simultaneously locate the
optimal capital investments, performance specifications, and
operating characteristics for the system.

The working fluid for the Rankine cycle is steam, the
properties of which are usually available in tabular or graphical
form. This suggests that a completely analytical solution is not
desirable and that a numerical technique should be employed.

The cogeneration system studied in this paper consists of
four major components, boiler, turbine/generator, condenser and
boiler feed pump (Figure 1). The system produces two products,
electricity and base load hot water. The hot water produced is
the exiting cooling water from the condenser. This means that
the steam must condense at a relatively high temperature and
pressure since it is the condensing steam which supplies the
required energy to the hot water.

The hot water is supplied to process at a high temperature,
TB, and is returned at a lower temperature, TC. The supply and
return temperatures are specified by the process and are therefore
treated in the optimization as constants.

The approach selected to optimize the system is that of
suboptimizing the system at various fixed electrical outputs.
This will yield an optimal design for each specified electrical
output. These suboptimizations will not be functions of the price
of electricity and will depend only on the relative costs of fuel,
equipment, and of capital. Once the value of electricity is
known, the overall optimum design can be selected from these
suboptimizations.

The problem must be set up such that the objective function
and the Lagrange constraint equations are functions of the state
and decision variables (Equations 4 and 5). A major deviation
from the procedure outlined by Tribus and El-Sayed (5) is in the
selection of the Lagrange constraint equations and state variables. The added complexity of having steam as the working fluid
(compared to an ideal gas in the gas turbine optimization performed by Tribus and El-Sayed) makes it impractical to select
state variables that correspond to available-energy flows. Consequently, this requirement was relaxed entirely. This gives the
designer the opportunity to use any variable as a state variable,

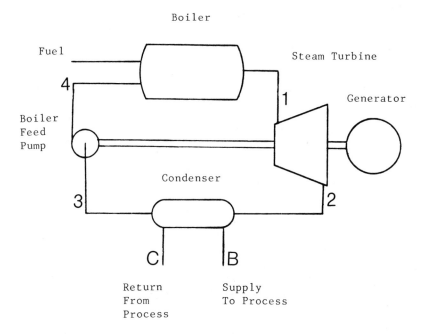

Figure 1.   Rankine cycle cogeneration system.

thus allowing considerably more freedom in the selection of state
and decision variables which in turn simplifies the objective
function, Lagrange constraint equations, and subsequent
derivatives.

Objective Function. The objective of the optimization is to
minimize the total cost of owning and operating the system (at
fixed product output). The objective function must be expressed
in terms of the system's costs (fuel, equipment, and capital) by
using an amortized capital investment, Z, for each component and
the total fuel cost, CFUEL, written per operating hour. The
objective function may be written as

$$\underset{[\{x_i\}\{y_k\}]}{\text{Minimize}} \quad \phi_o = \text{CFUEL} + \text{ZA} + \text{ZB} + \text{ZC} + \text{ZD} \qquad (20)$$

Cost Estimation. The capital costing equations used in the
cogeneration problem have been designed to yield approximate
capital and maintenance expenditures and to reflect the consequence of changing the system's variables on these costs. The
form of these equations expresses equipment costs in terms of
stream and performance variables. In all cases a capital recovery
factor is used to account for the cost of capital (i = 15%) and
estimated useful life (n = 40 years).

The approach taken to develop these costing equations was to
single out the most important parameters that influence cost, and
use them to yield a base cost, designating them with a prime
(i.e. Z'). This base cost is then adjusted by multiplication
factors so as to incorporate the influence of other factors. The
form of these equations has been suggested in the literature
(10-12) and by experienced engineers, then curve fit to available
data. However, extreme care must be exercised when applying any
of these equations in the field.

The interested reader should refer to Reference (13) for a
detailed description of costing equations for the boiler, turbine,
condenser, and pump. The costing equations for this system are
listed in Table I.

Fixed Charges. The costing equations previously discussed
determine the cost associated with each component of the system.
The total system cost is composed of the sum of the component
costs (ZTOT) plus any other charges attributable to the system.
These other charges, called fixed charges, include such costs as
the piping between components, foundation charges, building
charges, operating personnel charges, etc. Fixed charges are
estimated at 1.5 times the sum of the component costs. These
costs are considered constant for a specified heat and work output
requirement. Because the Lagrange optimization scheme is at fixed
product, these costs need not be incorporated into this part of
the optimization. However, these charges must be considered in

Table I. Amortized Costing Equations for the Cogeneration System

BOILER	ZA	= ZA(STM, P1, T1, AN, AR)
	ZA	= X11 * FAP * FAM * FAT * FAN * FAR
where	X11	= CRF * C11
	FAP	= exp(B11 * P1)
	FAM	= exp(B12 * alog STM)
	FAT	= 1.0 + C12 * exp[(T1-T1S)/B13]
	FAN	= 1.0 + [(1.0-ANS)/(1.0-AN)]**B14
	FAR	= 1.0 + [(1.0-ARS)/(1.0-AR)]**B15
TURBINE &	ZB	= ZB(STM, P1, P2, T1, T2, BN)
GENERATOR	ZB	= X21 * FBW * FBT * FBN
where	T1R	= T1 + 460
	T2R	= T2 + 460
	X21	= CRF * C21
	FB1	= B22 * BN * STM
	F2T	= CPS * (T1R-T2R-T2R*alog(T1R/T2R))
	F2P	= R * T2R * alog (P1/P2)
	FBW	= exp{B21 * alog [FB1 * (F2T+F2P)]}
	FBT	= 1.0 + (C22 * exp [(T1-T1S)/B23])
	FBN	= 1.0 + [1.0-BNS)/(1.0-BN)]**B24
CONDENSER	ZC	= ZC(CA, P3, P2, PB, PC, T2, TB)
	ZC	= X31 * FCA1 * FCR * FCPW * FCP * FCB
for		$100 \leqslant CA \leqslant 3000 \ ft^2$
	ZC	= X31 * FCA2 * FCR * FCPW * FCP * FCB
for		CA > 3000
where	X31	= CRF
	FCA1	= CA * C31 * exp(B31 * alog CA)
	FCA2	= CA * C36
	FCR	= [P3 * ((1/CR)-1.0)/C35]**B32
	FCPW	= [(PC-PB)/C35]**B33
	FCP	= C32 + C33 * P2 + C34 * (P2**2)
	FCB	= exp(B34/(T2-TB-5))
PUMP	ZD	= ZD(STM, P4, P3, DN)
	ZD	= X41 * FD1 * FDN
where	X41	= CRF * C41
	Y2	= B42 * STM * V34 * (P4-P3)/DN
	FD1	= exp[B41 * alog Y2]
	FDN	= 1.0 + [(1.0-DNS)/(1.0-DN)]**B43
FUEL	CFUEL	= CF * HF

order to select the optimum work/heat ratio from the set of suboptimizations (each at constant product).

Constraint Equations. The selection of fixed parameters and
decision variables is given in Table II. The reason for this
selection will become apparent later.

Table II. Fixed Parameters and Decision Variables for the
                      Cogeneration System

$\{y_f\}$ FIXED PARAMETERS

TB	Condenser hot water outlet temperature
TC	Condenser hot water inlet temperature
PB	Condenser hot water outlet pressure
PC	Condenser hot water inlet pressure
HWM	Required hot water mass flow rate
WA	Net turbine shaft work output
CR	Condenser shell-side pressure loss coefficient
X2	Turbine exit quality
U	Condenser overall heat transfer coefficient
CPW	Specific heat of water at constant pressure
CN	Condenser First Law efficiency

$\{y_k\}$ DECISION VARIABLES

AN	Boiler efficiency
AR	Boiler pressure drop coefficient
P1	Turbine inlet pressure
P2	Condenser inlet pressure
DN	Pump isentropic efficiency

    The equations of constraint link the cost estimate through
the system's thermodynamic performance to fuel costs. The thermodynamic analysis must relate the variables used to describe the
system's performance to those used in the cost estimate. In this
problem, costing equations are used which are generally in terms
of stream and performance variables. Thus the thermodynamic
analysis need only be in terms of these variables. Sixteen
equations of constraint have been developed from a thermodynamic
analysis of the cycle, and are given in Table III.
    Some constraints, such as ϕ_2 use stream variables to describe
the thermodynamic state of the working fluid. Numerical values
for steam properties are generated using a computer subroutine
called STEAM (14). Thus not all constraint equations are in
analytical form.
    The problem has now been reduced to five independent
variables. The form of the constraint equations has been arranged
such that each state variable can be obtained explicitly and the

Table III. Equations of Constraint for the Cogeneration System

i	State Variable x_i	Thermodynamic Constraints Defining Relation ϕ_i'
1	H2	f(P2,X2)
2	T2	f(P2, quality)
3	P3	P2 * CR
4	H3	f(P3, SAT. LIQ.)
5	P4	P1/AR
6	H4	H3 + CF1 * V34 * (P4-P3)/DN
7	STM	HWM * CPW * (TB-TC)/{CN*(H2-H3)}
8	WP	STM * (H4-H3)
9	H1	(WA+WP)/STM + H2
10	T1	f(H1, P1)
11	H2S	f(P1,P2,T1)
12	BN	(WA+WP)/(STM * (H1-H2S))
13	HF	STM * (H1-H4)/AN
14	T3	f(P3, SAT. LIQ.)
15	TM	{(T3-TC) - (T2-TB)}/alog{(T3-TC)/(T2-TB)}
16	CA	HWM * CPW * (TB-TC)/(TM*U)

Such that Lagrange Constraints are of the form

$$\phi_i = \phi_i' - x_i \qquad i = 1,16$$

order selected such that the resulting matrix is diagonalized. The next step is to obtain equations for the solution of the shadow and marginal prices. This requires the evaluation of various derivatives of the constraint equation matrix. However, because not all the constraints are in algebraic form (those constraints that are functions of steam table properties) numerical derivatives must be evaluated. One other note, there are two condenser costing equations. This means that two separate derivatives must be taken and the derivative corresponding to whichever costing equation is valid for that value of condenser area, is the one that should be used.

System Solution. The shadow prices, $\{\lambda_j\}$, are evaluated using Equation 10. Table IV lists the algebraic solution for the shadow prices (made very easy by choice of diagonalized constraint equations). The marginal prices $\{\theta_k\}$ are evaluated with Equation 15. Again, the choice of fixed parameters and decision variables makes it possible to solve Equation 15 explicitly for each marginal price. Table V lists these expressions.

Table IV. Shadow Price Equations for the Cogeneration System

SP(16) for	=	(ZC/CA)*(B31+1) $100 \leqslant CA \leqslant 3000 \ ft^2$
SP(16) for	=	(ZC/CA) $CA > 3000 \ ft^2$
SP(15)	=	-SP(16)*CA/TM
SP(14)	=	SP(15)*{(alog DUM1)-1+1/DUM1}/{alog(DUM1)}**2
SP(13)	=	CF
SP(12)	=	X21*FBT*FBW*[FBN*B21/BN+(FBN-1)*B24/(1-BN)]
SP(11)	=	SP(12)*BN/(H1-H2S)
**SP(10)	=	SP(11)*DER1+X11*FAP*FAN*FAR *FAM*(FAT-1)/B13+X21*FBN *FBW*[(FBT-1)/B23+FBT*B21 *CPS*(1-T2R/T1R)/(F2T+F2P)]
**SP(9)	=	SP(10)*DER2-SP(12)*BN/(H1-H2S)+SP(13)*STM/AN
SP(8)	=	[SP(9)+SP(12)/(H1-H2S)]/STM
SP(7)	=	[SP(8)*WP-SP(9)*(WP+WA)/STM -SP(12)*BN+SP(13)*HF+ZA*B12+ZB*B21+ZD*B41]/STM
SP(6)	=	SP(8)*STM-SP(13)*STM/AN
SP(5)	=	SP(6)*CF1*V34/DN+ZD*B41/(P4-P3)
SP(4)	=	SP(6)+SP(7)*STM/(H2-H3)-SP(8)*STM
**SP(3)	=	SP(4)*DER3-SP(6)*CF1*V34/DN+SP(14) *DER4+ZC*B32/P3-ZD*B41/(P4-P3)
***SP(2)	=	SP(15)*[-alog(DUM1)+DUM1-1]/[alog(DUM1)]**2+ZB*B21 *[-CPS*alog(T1R/T2R)+R*alog(P1/P2)] /(F2T+F2P)-ZC*B34/(T2-TB-5)**2
SP(1)	=	-SP(7)*STM/(H2-H3)+SP(9)

where

    **DER# = numerical derivative number #.
    ***DUM1 = (T3-TC)/(T2-TB)

Table V. Marginal Price Equations for the Cogeneration System

---

PM(1)   = -SP(13)*HF/AN+X11*FAP*FAM*FAT*FAR*(FAN-1)*B14/(1-AN)

PM(2)   = -SP(6)*(H4-H3)/DN+X41*FD1*[-FDN*B41/DN+(FDN-1)
          *B43/(1-DN)]

*PM(3) = SP(5)/AR+SP(10)*DER5+5P(11)*DER6+ZA*B11+ZB*B21*
          R*T2R/[P1*(F2T+F2P1)]

**PM(4) = DUM2+DUM3*FCA1
  for   100 ≤ CA ≤ 3000 ft^2

**PM(4) = DUM2+DUM3*FCA2
  for   CA > 3000 ft^2

PM(5)   = -SP(5)*P1/AR**2+X11*FAP*FAM*FAT*FAN*
          (FAR-1)*B15/(1-AR)

---

*DER# = numerical derivative number #.

**DUM2 = SP(1)*DER7+SP(2)*DER8+SP(3)*CR+SP(11)*DER9
          -ZB*B21*R*T2R/(P2(F2T+F2P))

DUM3 = X31*FCP*FCR*FCPW*FCB*(C33+2*C34*P2)

The solution procedure requires the designer to select a feasible set of decision variables $\{y_k\}$ for the first iteration. Once this initial set of five decision variables has been chosen, the entire design (for that iteration) is fixed and the set of state variables $\{x_i\}$, and cost estimates are determined. With values assigned to all the state and decision variables, the set of shadow prices $\{SP(i)\}$, is evaluated, and in turn, the set of marginal prices, $\{PM(i)\}$, is determined. These marginal prices are then used to direct the iteration as described by Equation 16.

This procedure is repeated until the marginal prices are small. One such measure of smallness for the marginal price vector is

$$\varepsilon = \Sigma(w_i \ \bar{\theta}_i)^2 \tag{21}$$

where the w_i are weighting factors.

Results and Conclusions

The computational procedure used to optimize this Rankine cycle cogeneration system operates in one of two modes. When a particular set of decision variables is far from the optimal set, there will be a large difference in the total operating cost between two successive iterations. This difference can be used as a measure of the distance from the optimum. When the difference is large, the entire set of decision variables is changed based on the input set of marginal prices. When the difference is smaller than some pre-determined value, only one decision variable is changed and a new set of state variables, shadow prices, and marginal prices are evaluated. Using this new set of marginal prices, another decision variable is changed and a new set of marginal prices computed. This is automatically repeated until all the decision variables have been changed. Each time a complete set of new decision variables is generated, the program displays the parameters necessary to evaluate the system design.

This procedure can be used to generate a variety of data. It is possible to parametrically vary any or all of the fixed decision variables. The parameters that were varied (Table VI) included fuel cost (CF), work/heat ratio (W/Q), and hot water temperature (TB). The strategem was to monitor the change in the optimal design by changing fuel cost and work output for several hot water requirements. In this manner, given the economic conditions and hot water requirements, the optimal amount of shaft work can be selected from these suboptimizations.

Table VI. Parametrically Varied Fixed Variables for the
Cogeneration System

Parameter Varied		Range Varied	Increment
CF	Fuel Cost	2–4 $/$10^6$ Btu	1.0 $/$10^6$ Btu
W/Q	Work/Heat Ratio	0.175 – 0.400	0.025
TB	Required Hot Water Temperature	225 – 300°F	25°F

Results of Suboptimizations. The trends in stream and performance variables associated with increasing fuel costs are that of increasing the system's performance. As the cost of fuel rises the boiler efficiency, pump isentropic efficiency, and turbine isentropic efficiency all increase. The condenser's thermodynamic performance increases because the steam's condensing temperature (or pressure as can be seen in Figure 4) decreases, approaching the required hot water temperature. This general increase in performance allows the steam mass flow rate, boiler pressure, and pump work all to decrease. As an example of this increase in the system's performance, Figures 2 and 3 clearly show the increase in boiler efficiency and pump isentropic efficiency associated with the increasing fuel costs for different values of the work/heat ratio.

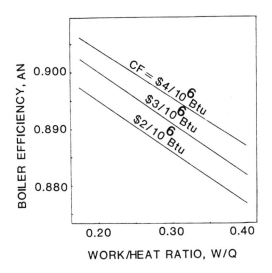

Figure 2.   Optimum boiler efficiency as a function of work/heat ratio for various fuel prices.

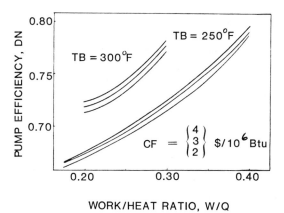

Figure 3.   Optimum pump efficiency versus work/heat ratio for various fuel prices for hot water supply temperatures of 250°F and 300°F.

This increase in system performance is not free. The incremental economic benefit of decreased fuel consumption is balanced by the incremental increase in the required capital investment.

The one decision variable that remains relatively unaffected by the cost of fuel is the boiler pressure drop coefficient, AR. This decision variable seems to decrease slightly or remain stable as fuel costs rise. This seems to indicate that the effect of the boiler pressure drop, AR, is dominated by the benefit which can be gained by altering other system parameters.

The magnitude of the change in some system parameters associated with increasing fuel costs is small. From a practical standpoint, these changes in the optimal parameters may even be of little engineering significance. For example, when the required hot water temperature, TB, equals 250°F and the work/heat ratio is 0.175, the pump efficiency, DN, goes from 0.665 to 0.670 as fuel costs go from two to four dollars per million Btu, however, over the same range, the condensing temperature, T2, changes from 262.2°F to 260.5°F which may be significant in the design of the condenser.

More significant trends occur as the work/heat ratio increases. Varying this parameter is extremely important in order to locate the optimal work output given a specific heat requirement and fuel cost. Trends in the system parameters and costs associated with changing work/heat ratio are plotted in Figures 2-8.

The effect of the work/heat ratio, W/Q, on the optimal boiler efficiency, BN, is that of decreasing BN as W/Q increases. This is an interesting trend because its justification may not be immediately apparent. For low work/heat ratios the condenser (and hence its products) play a dominating role in the design of the system. The condenser has a specific energy requirement which the boiler must supply. The energy supplied to the stream (hence the First Law efficiency) is very closely tied to the condenser. The shaft work produced by the turbine is dependent not only on the amount of energy supplied by the boiler but also on the temperature and pressure at which it is supplied. In other words, the shaft work is dependent on the available-energy input to the turbine. Thus for larger shaft work outputs, one needs to increase boiler pressure and to a lesser extent temperature, that is an increase in the boiler Second Law efficiency. However, the boiler First Law efficiency is predominately a function of temperature (assuming that the superheated steam behaves ideally). Therefore, for larger values of the work/heat ratio, the turbine dominates and requires a large available-energy input from the boiler. This may be accomplished by an increase in the boiler exit pressure and temperature (the Second Law efficiency) which is not fully reflected by boiler First Law efficiency. On the other hand, when the condenser dominates it requires a larger heat input to the working fluid and thus a maximum First Law efficiency regardless of the boiler outlet pressure.

Figure 3 shows the isentropic pump efficiency plotted versus the work/heat ratio. There is an upward trend for pump efficiency as the work/heat ratio rises. This is because the increased pressure demanded by the turbine requires additional pump work. The pump's increased importance in the system therefore calls for improved performance (efficiency). This same trend is shown in Figure 3 for several different fuel costs and for two different required hot water temperatures.

Two ways of increasing the amount of shaft work produced (at fixed Q) are (1) to decrease the turbine outlet pressure, P2, and/or (2) to increase the turbine inlet pressure, P1. These two effects are clearly demonstrated in Figures 4 and 5. Figure 4 shows how the optimum turbine outlet pressure decreases as the amount of work produced increases. These curves drop steeply until pressures are reached which correspond to condensing temperatures near 255°F (TB+5°F). The condenser cost factor, FCB, prevents the temperature from going below this mark. The optimal values for the turbine inlet and outlet pressures are determined with the optimization balancing the incremental pump, boiler, and turbine costs resulting from the increasing of P1 against the incremental condenser and turbine costs associated with lowering P2.

The boiler pressure drop coefficient increases as the work/heat ratio increases. This is clearly demonstrated in Figure 6. As the shaft-work becomes more important in the system, improved system design tends in the direction of decreasing the pump work by lessening the pressure drop in the boiler.

The system's Second Law efficiency rises as the work/heat ratio increases (Figure 7). This is partially due to improved performance of the turbine, pump, and condenser and the higher temperature steam from the boiler. This considerably decreases the available-energy destruction due to heat transfer in the boiler. Thus, the turbine can take advantage of this for the production of shaft work.

Altering TB (holding TC and HWM constant) changes the amount of heat, Q, which must be supplied to the hot water. When comparing the thermodynamic stream and performance variables for various TB, it is then important that a comparison be made for the same value of the work/heat ratio. It must be kept in mind that the scale of the systems being compared is different.

Increasing the required hot water temperature, TB, implies the turbine outlet pressure (temperature) must also increase. This means that for the same work/heat ratio, the turbine inlet pressure must also be higher as is clearly demonstrated in Figure 5. Other trends associated with increasing TB include the decreasing of the maximum temperature difference in the condenser, an increase in pump efficiency (Figure 3), and an increase in the system's Second Law efficiency.

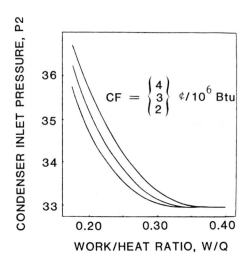

Figure 4.   Optimum condenser inlet pressure as a function
of work/heat ratio for various fuel prices.

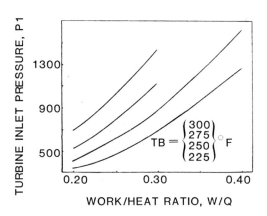

Figure 5.   Optimum turbine inlet pressure as a function of
work/heat ratio for various hot water supply temperatures
at a fixed fuel price of $3 per million Btu.

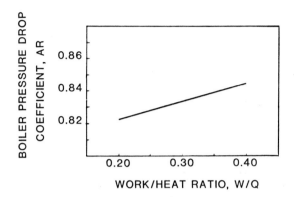

Figure 6.   Optimum boiler pressure drop coefficient as a function of work/heat ratio for a hot water supply temperature of 250°F at a fuel price of $2 per million Btu.

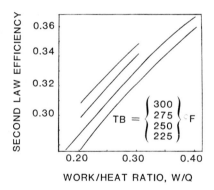

Figure 7.   System second law efficiency as a function of work/heat ratio for various hot water supply temperatures at a fuel price of $3 per million Btu.

Comparison with Alternative and the Selection of Overall Optimum.
The alternative to a cogeneration system is typically taken as a
low pressure boiler or furnace, and purchasing electricity from
the utility. It is when the economy of the optimally designed
cogeneration system is compared to the alternative, that a
system's true potential can be shown.

The amortized capital cost attributable to the low pressure
furnace is estimated using the relation

$$Z_f = CRF * 153.964 * 10^{0.89476 \, Log \, HP} \qquad (22)$$

where the boiler horsepower is given by

$$HP = 33500 * Q$$

Fixed charges for the low pressure furnace, FC_f, are estimated in the same manner as the cogeneration system, at one and a
half times the equipment cost. The fuel cost of producing hot
water, $CFUEL_f$, is estimated using the unit cost of fuel, CF, the
heat input to the water, Q, and the estimated furnace efficiency

$$CFUEL_f = \frac{CF * Q}{\eta_f} \qquad (23)$$

where the furnace efficiency, η_f, is taken to be 0.80. The total
cost of producing the hot water by the low pressure furnace,
$CTOT_f$, is estimated by

$$CTOT_f = CFUEL_f + FC_f + Z_f \qquad (24)$$

The hot water (steam) produced from a cogeneration system is
charged with $CTOT_f$, the cost of producing hot water from a low
pressure boiler. The balance of the cost of the cogeneration
system is then attributable to the shaft work (or electricity)
produced. The cost per unit of electricity produced, CE, can be
calculated as the difference of the total cost of the cogeneration
system and the cost allocated to the hot water, all divided by the
amount of electricity produced,

$$CE = \frac{ZTOT + FC - ZTOT_f}{E} \qquad (25)$$

where $E = W * \eta_g$ and the generator efficiency, η_g, is assumed to
be 95%.

In order to select the electricity production that maximizes
the profit returned from its sale, the market price of electricity, MPE, must be known. The net revenue generated by the sale of

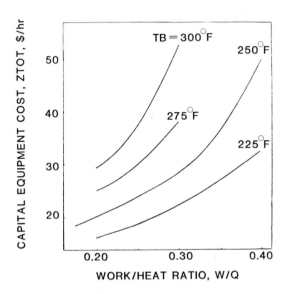

Figure 8.   Equipment cost as a function of work/heat ratio for various hot water supply temperatures.

the cogenerated electricity, NRG, is then expressed as

$$NRG = (MPE-CE) * E \qquad (26)$$

For a particular hot water requirement the optimal work output will correspond to the point where net revenue generated is a maximum. Figures 9-11 illustrate net revenue curves as a function of the work/heat ratio for a required hot water temperature of 250°F and for various market conditions.

Examination of these curves shows an increase in the optimal amount of electricity as the market price of electricity increases. Also, as fuel costs rise, the optimal amount of electricity production decreases.

Closure

The optimization method presented in this paper is uniquely suited for application to thermal systems. Its matrix notation and adaptability to numerical methods are easily programmed for computer solution. This method requires no prior knowledge about the economic value of the interconnecting streams nor about the commodity of value being transferred.

Second Law quantities such as available-energy and negentropy need not be introduced explicitly into the optimization although their effects are evaluated (indirectly) by the shadow and marginal prices. Furthermore, any required Second Law based prices can be obtained by using Equations 17 and 18 or by way of a thermoeconomic accounting technique (5,15).

In the application of this method to a Rankine cycle cogeneration system, generalized costing equations for the major components have been developed. Also, the utility of the method was extended by relaxing the rule that each state variable (and hence each Lagrange constraint) must correspond to an available-energy flow. The applicability was further extended by the introduction of numerical techniques necessary for the purpose of evaluating partial derivatives of steam table data.

The design space for the cogeneration system described in this paper consists of five independent variables and three key parameters that reflect the "market" conditions; fuel cost, electricity cost, and the required hot water temperature. Then, for each set of market conditions the system was suboptimized. Each suboptimization reflects the optimum thermodynamic system configuration for a given set of market conditions. The net result of this approach is the reduction from an eight-dimension to a three-dimension design space. This makes the task of choosing the optimal design substantially easier.

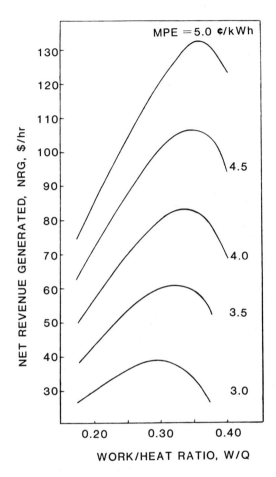

Figure 9.   Net revenue generated as a function of work/
heat ratio for various market prices of electricity, a
hot water supply temperature of 250°F, and a fuel cost of
$2 per million Btu.

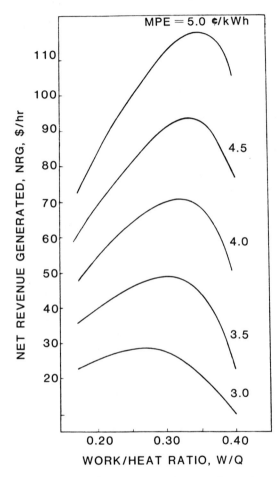

Figure 10.   Net revenue generated as a function of work/
heat ratio for various market prices of electricity, a hot
water supply temperature of 250°F, and a fuel cost of $3
per million Btu.

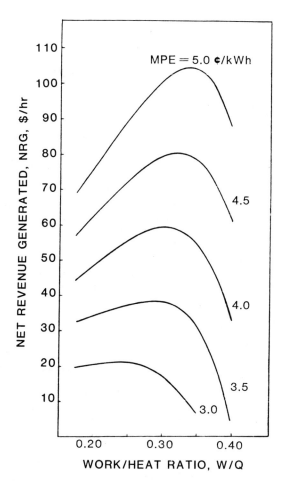

Figure 11. Net revenue generated as a function of work/ heat ratio for various market prices of electricity, a hot water supply temperature of 250°F, and a fuel cost of $4 per million Btu.

Acknowledgments

The authors would like to acknowledge the financial support
provided by the School of Mechanical Engineering, Georgia
Institute of Technology and the Celanese Corporation.

Literature Cited

1. Brosilow, C. B.; Lasdon, L., "An Optimization Technique for
    Recycle Processes" paper presented at the IChE-AIChE joint
    meeting, London, England, 1964.
2. Lasdon, L., Systems Research Center Report 50-C-64-19, Case
    Institute of Technology, Cleveland, 1964.
3. Gembicki, S. Ph.D. Thesis, Dartmouth College, Hanover, NH,
    1968.
4. Evans, R. B.; El-Sayed, Y. M., Trans. ASME, Journal of
    Engineering Power, 1970, 92, 27-35.
5. El-Sayed, Y. M.; Tribus, M., "The Strategic Use of Thermo-
    economic Analysis for Process Improvement," paper presented
    at the AIChE Meeting, Detroit, August, 1981.
6. Gaggioli, R. A.; et al., Proceedings American Power
    Conference, 1975, 37, 671-679.
7. Petit, P. J.; Gaggioli, R. A., in "Thermodynamics: Second Law
    Analysis;" Gaggioli, R. A., Ed.; ACS SYMPOSIUM SERIES No. 122,
    American Chemical Society: Washington, D.C., 1980; Chap. 2.
8. Gaggioli, R. A.; Wepfer, W. J., "Available-Energy Accounting--
    A Cogeneration Case Study," paper presented at the 85th AIChE
    Meeting, Philadelphia, June, 1978.
9. Wepfer, W. J. and Gaggioli, R. A., International Journal of
    Mechanical Engineering Education, 1981, 9, 283-295.
10. Clark, F. D. and Lorenzoni, A. B., "Applied Cost Engineering,"
    Marcel Dekker, Inc.: New York, 1978.
11. Humphreys, K. K. and Katell, S., "Basic Cost Enginering,"
    Marcel Dekker, Inc.: New York, 1981.
12. Peters, M. S. and Timmerhaus, K. D., "Plant Design and
    Economics for Chemical Engineers," 2nd Ed., McGraw-Hill:
    New York, 1968.
13. Garceau, R. M., M. S. Thesis, Georgia Institute of Technology,
    Atlanta, Georgia, August, 1982.
14. "Steam," Computer Subroutine, Georgia Institute of Technology,
    Atlanta, Georgia, based on ASME Steam Tables, 3rd ed., ASME,
    New York, 1977.
15. Wepfer, W. J.; Crutcher, B. G., Proc. American Power
    Conference, 1981, 43, 1070-1082.

RECEIVED July 7, 1983

Application of Second Law Based Design Optimization to Mass Transfer Processes

BRYAN B. MOORE[1] and WILLIAM J. WEPFER

School of Mechanical Engineering, Georgia Institute of Technology, Atlanta, GA 30332

This paper provides a framework for the application of Second Law based design methodology to separation systems. A relationship is derived for the available-energy destruction in a binary separation column as a function of the reflux ratio and the feed and product mass fractions. This derivation is limited to separations in which the entropy production is predominately due to mass transfers. This relationship is used in an application to a simple binary system to balance the trade-offs between inefficiency (fuel costs) and capital investment. The Second Law optimization yields results identical to those obtained from a traditional direct-search technique.

In general thermoeconomic optimization requires the derivation of expressions for entropy production, via nonequilibrium thermodynamics, due to each independent extensive property transport. In order to obtain such expressions it is necessary to apply thermodynamic property relations for multicomponent systems in conjunction with material and energy balances, heat, mass, and momentum transport equations.

Once the necessary expressions for the entropy productions are developed, the thermodynamic variables must be transformed into the relevant process design variables. These various equations can then be coupled with capital cost expressions to allow system optimization by any current technique (Lagrange multipliers, surrogatic worth trade-off, ...).

Specifically this paper describes an expression for the entropy production due to the mass fluxes in binary mass transfer systems with application to continuous differential contactors.

[1]Current address: Fibers Operations, Celanese Corporation, Box 32414, Charlotte, NC 28232

0097–6156/83/0235–00289$06.00/0

A stagnant film model is used for two-phase boundaries (1-2), which in effect, isolates the mass transfer process to a thin region at the interface stagnant film. Once the expressions for entropy production in terms of pressure, temperature, and composition are available a transformation is made to process variables such as reflux ratio, column height, packing or tray geometry, column diameter and column efficiency. Results of this design optimization model are compared with the results obtained via traditional methods.

Derivation of the Entropy Production Equation

Consider a two-phase, binary mass transfer system as shown in Figure 1. It is assumed that a stagnant film region exists such that 1) the vapor and liquid are in equilibrium, 2) there is a net transfer of component A from liquid to vapor and a corresponding net transfer of component B from vapor to liquid, and 3) the bulk flow of vapor and liquid in the z-direction within the stagnant film is negligible. As the derivation proceeds, the constant molal overflow assumption will be invoked (2). In addition, a form of the Gibbs equation will be used to eliminate the time derivatives. Finally, heat transfer induced entropy productions within the column will be neglected. The resulting expression for steady-state entropy production arising from mass transfer in a binary system will be cast in terms of the process variables.

Thermodynamic Governing Equations. Derivation of the expression for entropy production arising from mass transfer requires application of the fundamental balance equations. Potential and kinetic energy effects as well as momentum effects are neglected. With these assumptions the governing equations are given as follows:

Mass

$$\frac{\partial \rho_i}{\partial t} + \nabla \cdot \dot{\underline{n}}_i = 0 \tag{1}$$

Energy

$$\frac{\partial (\rho u)}{\partial t} + \nabla \cdot \underline{u} = 0 \tag{2}$$

Entropy

$$\frac{\partial (\rho s)}{\partial t} + \nabla \cdot \underline{s} = \dot{s}_p \tag{3}$$

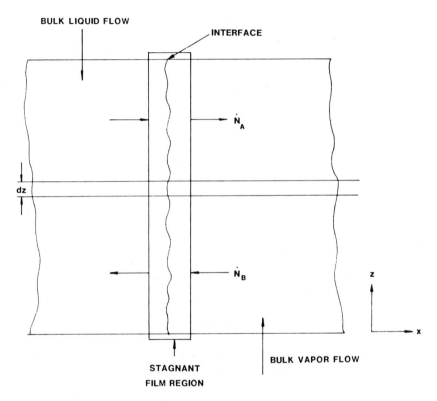

Figure 1. Two-phase stagnant-film, binary mass transfer system.

Entropy Current Density $(\underline{3}-\underline{4})$

$$\underline{\dot{s}} = \frac{1}{T}\underline{\dot{u}} + \frac{\mu_i}{T}\underline{\dot{n}}_i \tag{4}$$

Time Derivative of the Gibbs Equation $(\underline{3}-\underline{4})$

$$\frac{\partial(\rho u)}{\partial t} = T\frac{\partial(\rho s)}{\partial t} + \sum \mu_i \frac{\partial \rho_i}{\partial t} \tag{5}$$

Taking the divergence of Equation 4 and solving for $T(\nabla \cdot \underline{\dot{s}})$ yields

$$T(\nabla \cdot \underline{\dot{s}}) = \nabla \cdot \underline{\dot{u}} - \underline{\dot{s}} \cdot \nabla T - \sum \mu_i (\nabla \cdot \underline{\dot{n}}_i) - \sum \underline{\dot{n}}_i \cdot \mu_i \tag{6}$$

Substitution of Equation 6 into the entropy balance yields

$$T\frac{\partial(\rho s)}{\partial t} = \sum \mu_i (\nabla \cdot \underline{\dot{n}}_i) + \sum \underline{\dot{n}}_i \cdot \nabla \mu_i + \underline{\dot{s}} \cdot \nabla T - \nabla \cdot \underline{\dot{u}} + T\dot{s}_p \tag{7}$$

Equation 7 can be further simplified by substitution of the mass and energy balance equations

$$T\frac{\partial(\rho s)}{\partial t} = -\sum \mu_i \frac{\partial \rho_i}{\partial t} + \frac{\partial(\rho u)}{\partial t} + \underline{\dot{s}} \cdot \nabla T + \sum \underline{\dot{n}}_i \cdot \nabla \mu_i + T\dot{s}_p \tag{8}$$

However, a form of the Gibbs Equation, Equation 5, can be factored out of Equation 8 to give

$$T\dot{s}_p = -\underline{\dot{s}} \cdot \nabla T - \sum \underline{\dot{n}}_i \cdot \nabla \mu_i \tag{9}$$

Equation 9 relates the basic extensive properties, the respective driving potentials, and the entropy production in the diffusion process. The first term, $\underline{s} \cdot \nabla T$, arises from heat transfer effects while the second term is due to mass transfer. For processes wherein the entropy production due to column heat transfer is small relative to the mass-transfer, the $s \cdot \nabla T$ term is negligible and Equation 9 simplifies to

$$T\dot{s}_p = -\sum \underline{\dot{n}}_i \cdot \nabla \mu_i \tag{10}$$

Employing the geometry of Figure 1, Equation 10 is reduced to a one-dimensional form

$$T\dot{s}_p = -\dot{n}_A \frac{d\mu_A}{dx} - \dot{n}_B \frac{d\mu_B}{dx} \tag{11}$$

To obtain the total entropy production in the tower, Equation 11 must be integrated over the total diffusion volume

$$\int_V T\dot{s}_p dV = \int_V \left[-n_A \frac{d\mu_A}{dx} - n_B \frac{d\mu_B}{dx} \right] dV \qquad (12)$$

The volume differential, dV, can be expressed in terms of cartesian coordinates dxdydz. The mass flux normal to the interface is a function of the concentration difference between the bulk fluid and the interface (at which the liquid and vapor are in equilibrium) which is a function of tower height, z, in the column. Integration of Equation 12 over dx yields

$$-T\dot{S}_p = \int\int \left[\dot{n}_A(z)\Delta\mu_A - \dot{n}_B(z)\Delta\mu_B \right] dydz \qquad (13)$$

where $\Delta\mu_A = \mu_A - \mu_{AE}$ and $\Delta\mu_B = \mu_B - \mu_{BE}$. The actual limits of integration in the y direction are arbitrary because the integrand is a function of z only. Integration of Equation 13 over an arbitrary distance δy gives

$$-T\dot{S}_p = \int \left[\Delta\mu_A \dot{n}_A(z)\delta y - \Delta\mu_B \dot{n}_B(z)\delta y \right] dz \qquad (14)$$

The total mass transfer is equal to the integration of the mass flux over the interfacial area

$$\dot{N}_A = \int \dot{n}_A(z) dydz \qquad (15)$$

Since the mass flux is independent of y, taking the derivative of Equation 15 gives

$$\frac{d\dot{N}_A}{dz} = n(z)\delta y \qquad (16)$$

Substitution of Equation 16 into 14 yields

$$-T\dot{S}_p = \int \Delta\mu_A d\dot{N}_A - \int \Delta\mu_B d\dot{N}_B \qquad (17)$$

Assuming constant molal overflow (2), the mass flow rate of component A must be related to the mass flow rate of component B

$$d\dot{N}_A = d\dot{N}_B \tag{18}$$

Equation 17 can now be simplified to

$$-T\dot{S}_p = \int \left[\Delta\mu_A - \Delta\mu_B \right] d\dot{N}_A \tag{19}$$

The differences in chemical potential, at constant pressure and temperature may be expressed as (5)

$$\Delta\mu_A = RT \ \ell n \frac{y_A}{y_{AE}} \tag{20}$$

$$\Delta\mu_B = RT \ \ell n \frac{y_B}{y_{BE}} \tag{21}$$

Equation 19 can now be rearranged to yield

$$\dot{S}_p = R \ \int \ell n \left[\frac{y_E(1-y_A)}{y_A(1-y_{AE})} \right] d\dot{N}_A \tag{22}$$

where the identities $y_A + y_B = 1$ and $x_A + x_B = 1$ have been used to simplify the result.

A mass balance on component A of the bulk vapor flow gives (2)

$$d\dot{N}_A = \bar{V} dy_A \tag{23}$$

where $\bar{V}$ is the bulk vapor molar flow rate and dy_A is the change of component A of the bulk vapor. Substitution of Equation 23 into 22 produces the needed equation for the total entropy production due to mass transfer in a binary system in which heat transfer entropy productions within the column are small

$$\dot{S}_p = \bar{V}R \int_{y_{A,in}}^{y_{A,out}} \ell n \left[\frac{y_{AE}(1-y_A)}{y_A(1-y_{AE})} \right] dy_A \tag{24}$$

The parameters controlling the rate of entropy production in the tower are now obvious; the vapor flow rate $\bar{V}$ (a function of the reflux ratio), the inlet and outlet mole (or mass) fractions, and the relationship between y_A and y_{AE} (a function of the reflux ratio and the relative volatility).

Two important points need to be mentioned. First, as the
mole fraction, y_A, approaches the equilibrium mole fraction, y_{AE},
the integrand approaches zero. Thus the point of minimum entropy
production coincides with that of minimum reflux. Second,
$y_{AE} > y_A$ guarantees that the argument of the logarithm cannot be
less than unity, which means that $\dot{S}_p > 0$. Finally Equation 24 is
only applicable to cases wherein the diffusion processes represent
the dominant mechanism for entropy production.

Application to Simple Tower Design

Consider a simple distillation system (Figure 2) designed to
process 700 lbm/hr (318 kgmol/hr) of feed (6). The unit is to
operate continuously for 8500 hrs/yr at a total pressure of one
atmosphere. The feed contains 45 mol % benzene and 55% toluene.
The feed is saturated at its boiling temperature of 201°F(94°C).
The objective of this separation process is to divide the feed
into two product streams; an overhead product consisting of 92%
benzene and a bottoms containing 95% toluene. Table I lists
additional information about this system. The distillation tower
is a continuous contactor having a packing material cost of
$38/ft^3. The purchase costs for tower shell, condenser, and

Table I. System Design Information

Relative volatility, α	2.5
Molal heat capacity of the liquid mixture, C	40.0 Btu/lbmol°F
Molal heat of vaporization of mixture, h	13700 Btu/lbmol
Heat transfer coefficient of the reboiler, h_r	80 Btu/hr ft^2 °F
Heat transfer coefficient of the condenser, h_c	100 Btu/hr ft^2 °F
Initial ΔT of condenser cooling water	50 °F
Inlet cooling water temperature	90 °F
Inlet state of steam	saturated vapor at 60 psia
Heat of condensation of steam	915.5 Btu/lbm
Boiling temperature of feedstock	201 °F
Overhead product purity	92% benzene
Bottoms product purity	95% tolvene
Feed composition	45% benzene
Feed flow rate	700 lbmol/hr

reboiler were obtained from manufacturer's data based on weight
for the shell and area for the heat exchangers. The sum of

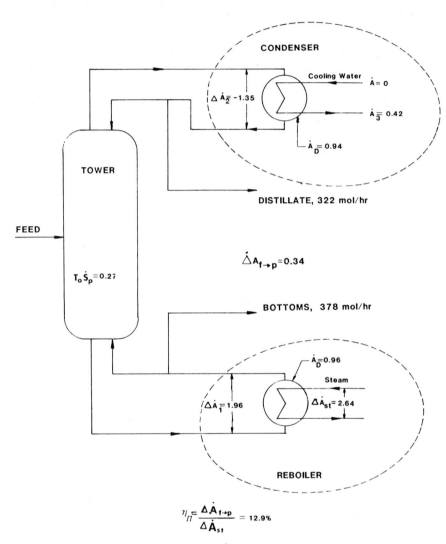

Figure 2. Simple distillation system for the separation of a 45%-55% benzene-toluene feed into 92% benzene distillate and a 95% toluene bottoms product. Available-energy flows and destructions are given in 10^6 Btu/hr.

piping, insulation, and instrumentation etc. are taken to be 60% of the cost for the installed equipment. The steam supplied to the reboiler is saturated at 292°F and is priced at $0.75/10^3$ lbm, and the cooling water is priced at $0.045/10^3$ gal.

Available-Energy Analysis. Using standard techniques (7-8) an available-energy analysis was performed on the system operating at a reflux ratio of 1.2, and shows the overall Second-Law efficiency to be 12.9% (9). The relevant equations for this analysis are given in Table II. The reboiler has a Second-Law efficiency equal to $\Delta\dot{A}_1/\Delta\dot{A}_{st}$ or 74.2%. The tower efficiency, defined as $\Delta\dot{A}_{f\to p}/(\Delta\dot{A}_1 + \Delta\dot{A}_2)$ is 55.7%. Note that it is of little use to

Table II. Available-Energy Balance Equations

$$\Delta\dot{A}_{st} = \dot{Q}_{rblr} \left[1 - \frac{T_0 s_{fg}}{h_{fg}} \right]$$

$$\dot{A}_{D,rblr} = T_0 \dot{S}_{p,rblr} = \frac{T_0 \dot{Q}_{rblr} \left[T_{st} - T_{pr} \right]}{T_{st} T_{pr}}$$

$$\Delta\dot{A}_1 = \Delta\dot{A}_{st} - \dot{A}_{D,rblr}$$

$$\dot{A}_{D,cond} = T_0 \dot{S}_{p,cond} = \dot{m}_{cw} c_{cw} T_0 \left[\ell n \frac{T_3}{T_0} - \frac{T_3 - T_0}{T_{cond}} \right]$$

$$\dot{A}_3 = \dot{m}_{cw} c_{cw} T_0 \left[\frac{T_{out}}{T_{in}} - 1 - \ell n \frac{T_{out}}{T_{in}} \right]$$

$$\Delta\dot{A}_2 = -\dot{A}_3 + \dot{A}_{D,cond}$$

$$\dot{A}_{D,col} = T_0 \dot{S}_{p,cond} = \bar{V}RT_0 \int_{y_{A,in}}^{y_{A,out}} \ell n \left[\frac{y_{AE}(1-y_A)}{y_A(1-y_{AE})} \right] dy_A$$

$$\dot{A}_{f\to p} = \Delta\dot{A}_1 + \Delta\dot{A}_2 - \dot{A}_{D,col}$$

evaluate a condenser efficiency without consideration of the system as a whole inasmuch as the role of the condenser is to rid the system of low-temperature heat (entropy). The tower available-energy destruction was computed using Equation 24.

Calculations showed that the thermal contribution to
$\Delta A_{feed \rightarrow product}$ is only 8% of the total. It is also interesting
to note that the value of $\Delta A_{feed \rightarrow product}$ computed above compares
closely to the calculated value for the isothermal minimum work
of separation, $0.36(10^6)$ Btu/hr, (2). Therefore this column
qualifies as one in which the mass transfer process dominates over
the heat transfer processes.

Thermoeconomic Governing Equations. The objective of design
optimization is the selection and/or specification of system
hardware which minimizes the total expenditure for capital, fuel,
and other costs (10-11). The use of an explicit Second Law design
strategy enables a large complex system to be split into much less
complex parts. Each subsystem can then be optimized individually.
The primary advantage of such an approach is the additional
insight gained by dealing with much simpler systems.

The basis of any thermoeconomic analysis is the application
of a money balance to the system or subsystem of interest

$$\dot{\$}_{product} = \sum \dot{\$}_{fuel} + \sum \dot{\$}_{capital}, \text{ etc.} \tag{25}$$

The cost of the product of an energy conversion system must equal
the sum of fuel expenses and capital (and labor....) charges. It
is often convenient to express the product cost in terms of the
average unit cost of product, λ_p, and the total amount of product,
P_p, or $\dot{\$}_{product} = \lambda_p P_p$. Similarly the fuel costs may be expressed
as $\dot{\$}_{fuel} = \lambda_f P_f$. Equation 25 may now be written as

$$\lambda_p = \frac{\sum \lambda_f P_f}{P_p} + \frac{\sum \dot{\$}_{capital}}{P_p} \tag{26}$$

The first term in this equation reflects fuel costs and alternatively could be expressed in terms of the system inefficiencies
and the system utility costs. The second term is indicative of
the capital investment. However, the system inefficiencies and
the capital investment are functions of the design variables.
Thus minimization of the unit cost of product involves a function
that is dependent on only the utility or fuel costs and the
design variables.

The strategy to be employed in this paper requires the
application of a money balance to each system component. That is,
each system component is viewed as an energy converter which
processes fuel from one form to another and sends its product
along to the next system component for further processing. Then,
beginning with a "working design" each system component is
optimized for minimum unit product cost. These suboptimizations
are done successively for each component in an iterative fashion
until the design converges. This procedure requires the use of

available-energy as the measure of thermodynamic value for fuel
inputs and product outputs (10).

In the system of interest the reboiler transfers some of the
available-energy in steam to the bottoms stream. The column uses
the increase in available-energy of the bottoms to transform feed
into products. The condenser processes the reflux for subsequent
re-use in the column. In the reboiler and column available-energy
is supplied in one form and converted into a product. However,
the condenser exists for the sole purpose of decreasing the
available-energy of the reflux. Or, in other terms, the condenser
serves to eliminate entropy from the system.

This however complicates the optimization procedure. The
problem can be alleviated by using an entropy penalty function $\dot{P}_s$,
which serves to properly apportion the total condenser costs to
the reboiler and column. The entropy penalty function, which
Evans describes as "negentropy" (12-13), can also be viewed as the
commodity or product which the condenser sells to the tower and
reboiler. The entropy penalty function leaving the condenser is
given by (9, 13-14)

$$\dot{P}_{s,cond} = \dot{Q}_{cond} - T_0 \dot{S}_{p,cond} - \dot{A}_{net\ exit,\ cond} \tag{27}$$

The entropy penalty function attributable to the reboiler is given
by (9)

$$\dot{P}_{s,reboiler} = \frac{T_0}{T_1} \dot{Q}_{rblr} \tag{28}$$

and the entropy penalty function attributable to the column is
found by difference

$$\dot{P}_{s,col} = \dot{P}_{s,cond} - \dot{P}_{s,rblr} \tag{29}$$

Thus by requiring the reboiler and column to purchase $\dot{P}_{s,rblr}$ and
$\dot{P}_{s,col}$ from the condenser, the condenser costs are properly
included in the optimization procedure. However, the money
balances will need to include the money flows associated with the
entropy penalty function transactions.

Essentially the column is viewed as a system through which
the circulating process stream is converted to product streams.
The circulating process stream passes through the reboiler and
condenser extracting and depositing available-energy. From this
perspective it is only reasonable to price the process stream at a
constant cost λ, which is analogous to the extraction method (15).
This constant price λ, obtained from the reboiler money balance,
is used to cost the inefficiencies in both the tower and the
condenser.

A money balance on the reboiler yields

$$\lambda \Delta \dot{A}_1 = \lambda_{st} \Delta \dot{A}_{st} + \dot{Z}_{rblr} + \lambda_s \dot{P}_{s,rblr} \qquad (30)$$

where λ represents the cost of the circulating process stream, λ_{st} is the unit cost of steam, $\dot{Z}_{rblr}$ is the capital cost of the reboiler, and λ_s is the unit cost of the entropy penalty function. Similarly a money balance on the condenser gives

$$\lambda_s \dot{P}_{s,cond} = \lambda \Delta \dot{A}_2 + \lambda_{cw} \dot{m}_{cw} + \dot{Z}_{cond} \qquad (31)$$

However with the aid of an available-energy balance (see Table II) Equation 31 can be rearranged to yield

$$\lambda_s \dot{P}_{s,cond} = \lambda T_0 \dot{S}_{p,cond} + \lambda \dot{A}_{cond,exit}$$

$$+ \lambda_{cw} \dot{m}_{cw} + \dot{Z}_{cond} \qquad (32)$$

where λ represents the cost of the circulating process stream flowing between the column and condenser and $T_0 \dot{S}_{p,cond}$ is the available-energy destruction in the condenser.

Equations 31 and 32 can be solved for λ and λ_s

$$\lambda = \frac{\lambda_{st} \Delta \dot{A}_{st} + \dot{Z}_{rblr} + (\lambda_{cw} \dot{m}_{cw} + \dot{Z}_{cond}) \dfrac{\dot{P}_{s,rblr}}{\dot{P}_{s,cond}}}{\Delta \dot{A}_1 - \left[T_0 \dot{S}_{p,cond} + \dot{A}_{cond,exit} \right] \dfrac{\dot{P}_{s,rblr}}{\dot{P}_{s,cond}}} \qquad (33)$$

and

$$\lambda_s = \frac{\lambda_{cw} \dot{m}_{cw} + \dot{Z}_{cond} + \lambda T_0 \dot{S}_{p,cond} + \lambda \dot{A}_{cond,exit}}{\dot{P}_{s,cond}} \qquad (34)$$

A money balance on the tower yields

$$\lambda_p = \frac{\lambda (\Delta \dot{A}_1 + \Delta \dot{A}_2) + \lambda_s \dot{P}_{s,col} + \dot{Z}_{col}}{\Delta \dot{A}_{f \to p}} \qquad (35)$$

Equations 33-35 are the basic thermoeconomic governing equations for the Second Law based optimization. It should be noted that although column entropy productions due to heat transfer are neglected, the analysis nevertheless includes the fact that the column "buys" thermal available-energy from the reboiler in the thermoeconomic governing equations.

Optimization Procedure. Given a set of design variables (a working design), capital cost equations as functions of the design variables (9), the unit costs of utilities, and Equations 33-35, the Second-Law based optimization may be performed.

The first step in this procedure is to perform an analysis of a working design of the system in order to obtain values for the unit cost of the process stream, λ, and the entropy cost penalty function, λ_s. In turn unit costs are used to optimize the column.

The first subsystem to be optimized is the column. Given a set of specified product purities and column pressure the reflux ratio remains as the only column variable, that is, λ_p is a function of reflux ratio. Using values for λ and λ_s obtained from the analysis of the working design, λ_p is computed from Equation 35 for several values of reflux ratio (Table III). The optimal reflux ratio is obtained via a search of these values.

The next component to be optimized is the reboiler. The reboiler area is fixed once the values for steam temperature, T_{st}, and process stream temperature, T_1, are fixed (assuming a constant heat transfer conductance). The process stream temperature is fixed by the column pressure and the product purities. Thus only T_{st} remains as a variable. The unit cost of the process stream, λ, is then optimized via Equation 33 with respect to the steam temperature (Table III).

The last component to be optimized is the condenser. The condenser area is fixed once the process stream temperature, T_2, and the exit cooling water temperature, T_3, are fixed. However, the process stream temperature is fixed by the column pressure and the product purities. Thus the condenser is optimized by minimizing the unit cost of the entropy penalty function, λ_s, with respect to T_3.

The result of these procedures is a new set of design variables as well as a new set of unit product costs, λ_p, λ, and λ_s. The above procedure is then repeated until the design variables and unit costs converge. It is the author's experience that convergence is usually attained in two or three iterations for simple systems.

Results and Conclusions

The principle results of applying this Second Law based design methodology to the previously described separation system are contained in Table III. In addition, the results of applying a traditional method - that of a direct search through the design

SECOND LAW ANALYSIS OF PROCESSES

Table III. Results of Second Law Optimization

Reflux Ratio	Column Height (Diameter) (ft)	Column Available Energy Consumption (10^5 Btu/hr)	Column Cost (10^3\$/yr)	Reboiler Cost (10^3\$/yr)	Condenser Cost (10^3\$/yr)	Steam Cost (10^3\$/yr)	Cooling Water Cost (10^3\$/yr)	Second-Law Based Unit Product Cost (\$/10^6 Btu)	Unit Product Cost Obtained from a Direct Search Method (\$/10^6 Btu)
1.14	37(7.43)	2.54	23.7	5.93	3.51	66.5	8.66	37.64	37.71
1.16	32(7.47)	2.59	21.4	5.97	3.53	67.1	8.74	37.13	37.19
1.18	30(7.50)	2.64	20.2	6.00	3.55	67.7	8.82	37.00	37.03
1.20	28.5(7.54)	2.70	19.4	6.03	3.56	68.3	8.90	37.00	37.00
1.25	25.7(7.63)	2.84	18.2	6.11	3.61	69.8	9.11	37.28	37.22
1.30	23.9(7.72)	2.98	17.4	6.19	3.66	71.4	9.31	37.63	37.61
1.40	21.4(7.89)	3.26	16.5	6.35	3.75	74.5	9.71	38.85	38.60
1.60	18.7(8.23)	3.82	15.8	6.66	3.93	80.6	10.52	41.43	40.94
1.80	17.2(8.55)	4.38	15.6	6.92	4.11	86.7	11.33	44.19	43.43
2.00	16.1(8.86)	4.95	15.5	7.18	4.30	92.9	12.14	47.03	46.0

space – are also presented in Table III. As can be seen, both methods yield an optimum reflux ratio equal to 1.20. More detailed results are given in (9).

The results presented in Table III show that as the capital investment in the tower (Z_{tower}) increases at larger tower heights, the available-energy destruction decreases. Thus the optimal design reflects the classical trade-off between capital investment and fuel cost. It is important to note that heat exchanger design plays a major role in separation systems (16). At the optimum reflux ratio, the reboiler and condenser represent 21% and 12% of the total capital investment. Similarly, at optimum design, the fuel costs, steam and cooling water, represent 64% and 8% of total system costs.

The fundamental advantage of the Second Law methodology is the fact that the optimization of a separation system may be simplified by decomposing the system into its individual components and individually suboptimizing each one in order to achieve a global optimum.

The number of variables in this simple system is not large enough to preclude traditional methods. The three optimizing variables are the steam condensing temperature, the reflux ratio, and the cooling water exit temperature. The product purities were assumed fixed by market demands and the column pressure was fixed at one atmosphere. In the separation of multicomponent mixtures (more than two components) there are multiple distillation columns and many different configurations are possible. Because of the complexity of multicomponent separation systems traditional optimization techniques are very tedious and therefore heuristic methods (sometimes involving many rules-of-thumb) are employed to choose the best configuration. The application of the Second Law design methodology promises to reduce the number of variables involved as well as to provide the designer with greater insight because one is working with each system component on an individual basis.

Closure

This paper has provided a framework for further application of Second Law based design methodology to separation systems. It has done so by providing a relationship that gives the available-energy destruction for a binary separation as a function of the process variables for the case in which the entropy production is primarily due to mass transfer effects. The Second Law methodology has been described and applied to a simple binary separation system. The method yields results identical to those obtained from a traditional direct search technique, and accurately indicates the respective trade-offs between fuel costs and capital investment.

Legend of Symbols

A	Available energy
$\dot{m}$	Mass flow rate
$\dot{n}_i$	Mass flux
p	Pressure
R	Gas Constant
S	Entropy per unit mass
$\dot{s}$	Entropy flux
S_p	Entropy production
t	Time
T	Temperature
u	Internal energy per mass
$\dot{u}$	Internal energy flux
V	Volume
$\bar{V}$	Bulk vapor molar flow rate
x,y,z	Cartesian coordinates
Δ	Difference, output minus input
δ	Thickness
λ	Unit cost
λ_{cw}	Unit cost of cooling water
λ_f	Unit cost of feed
λ_p	Unit cost of product
λ_s	Unit cost of entropy penalty function
μ_i	Chemical potential of component i
ρ_i	Mass density of component i
$\$$	Money flow

Legend of Subscrips

A	Component A
AE	Component A in the equilibrium state with B
B	Component B
BE	Component B in the equilibrium state with A
col	Column
cond	Condenser

cw   Cooling water

E   Equilibrium

f   Feed

i   Component i

P   Product, production

pr   Process

rblr   Reboiler

S   Entropy penalty function

0   Dead state

Acknowledgments

The authors would like to express their appreciation to the Engineering Foundation and the Gulf Oil Company for their support of this work.

Literature Cited

1. Sherwood, T. K., Pigford, R. L., Wilkie, C. R., "Mass Transfer," McGraw-Hill, New York, 1975.
2. King, C. J., "Separation Processes," 2nd ed., McGraw-Hill, New York, 1980.
3. Callen, H. B., "Thermodynamics," J. Wiley, New York, 1960.
4. DeGroot, S. R., "Thermodynamics of Irreversible Processes," North Holland Publ. Co., 1951.
5. Denbigh, K., "The Principles of Chemical Equilibrium," 3rd ed., Cambridge University Press, 1971.
6. Peters, M. S., Timmerhaus, K. D., "Plant Design and Economics for Chemical Engineers," 3rd ed., McGraw-Hill, New York, 1980.
7. Gaggioli, R. A., et. al., Proc. American Power Conf., 1975, 37, 671-679.
8. Gaggioli, R. A., Petit, P. J., Chemtech, 1977, 7, 496-506.
9. Moore, B. B., M. S. Thesis, Georgia Institute of Technology, August, 1982.
10. Reistad, G. M., Gaggioli, R. A., in "Thermodynamics: Second Law Analysis;" Gaggioli, R. A., Ed.; ACS SYMPOSIUM SERIES No. 122, American Chemical Society: Washington, D.C., 1980, pp. 143-160.
11. Wepfer, W. J., in "Thermodynamics: Second Law Analysis;" Gaggioli, R. A., Ed.; ACS SYMPOSIUM SERIES No. 122, American Chemical Society: Washington, D.C. 1980; pp. 161-186.
12. Evans, R. B., Energy-The International Journal, 1980, 5, 8-9, 805-822.
13. Smith, M. S., M. S. Thesis, Georgia Institute of Technology, March 1981.

14.   Evans, R. B.; et. al., "Essergetic Functional Analysis for
      Process Design and Synthesis," Paper No. 28c, A.I.Ch.E.
      Annual Meeting, Detroit, MI, August 1982.
15.   Wepfer, W. J., Crutcher, B. G., Proc. Amer. Power Conf.,
      1981, 43, 1070-1082.
16.   Benedict, M., Gyftopoulos, E. P., in "Thermodynamics: Second
      Law Analysis;" Gaggioli, R. A., Ed.: ACS SYMPOSIUM SERIES
      No. 122, American Chemical Society: Washington, D.C., 1980;
      pp. 195-203.

RECEIVED July 12, 1983

Multiobjective Optimal Synthesis

L. T. FAN and J. H. SHIEH

Department of Chemical Engineering, Kansas State University, Manhattan, KS 66506

Conventionally, a food and/or chemical process
system has been designed optimally by considering
only one objective of economic efficiency (profit
or cost), although it is often difficult or even
impossible to express variables and parameters
associated with such a system in a monetary unit.
For example, the available energy is invaluable to
a nation where the energy resources are seriously
depleted, because the available energy once lost
in a process cannot be recovered by any means; yet
the cost of energy resources can be unreasonably
low because of the artificial or manipulated world
market condition. Furthermore, the excessive loss
of available energy may result in severe
pollution, which in turn, may lead to destruction
of the environment or of human life. Again it is
difficult to assign monetary values to such
destruction. It is therefore, natural that the
concept of multiobjective analysis should be
introduced in synthesizing a chemical or biochemical process system. This paper discusses the
basic concepts and terminologies of a multiobjective problem and reviews methods for solving
it. The methods are illustrated with an example
of the milk evaporation process.

A chemical or biochemical process system has been designed
conventionally by considering only one objective function of
economic efficiency (profit or cost). However, many of the
objectives are often difficult or even impossible to express in
a common monetary unit. It is, thus, natural that the concept
of multiobjective optimization be introduced in synthesizing a
chemical or biochemical process system.

0097-6156/83/0235-0307$07.25/0

A multiobjective optimization problem can be mathematically stated as:

Optimize (minimize or maximize)

$$\underline{J} = \begin{bmatrix} \overline{f_1(\underline{x})} \\ f_2(\underline{x}) \\ \cdot \\ \cdot \\ \cdot \\ \cdot \\ \cdot \\ \cdot \\ \cdot \end{bmatrix} = [f_i(\underline{x})], \; i = 1, \, 2, \, \ldots \quad (1)$$

subject to

$$\underline{x} \; \varepsilon \; X$$

$$g_j \leq 0, \; j = 1, \, 2, \, \ldots$$

where the underline denotes the vector, X indicates the design variables space, and g_i's represent inequality constraints characterizing physical and practical limitations and equality constraints which are the governing mathematical models of the process system and its subsystems; i = 1 corresponds to the case of a single objective function.

There are basically two approaches for solving a multi-objective optimization problem. The first is to attempt to find the optimal or preferred solution directly, and the second is to generate the so-called non-inferior solutions (set) and then locate the preferred solution among them. The latter approach is widely adopted and will be used in this work.

A variety of methods are available for solving a multi-objective function problem. These methods include the weighting method (1, 2), ε-constraint method (2,3), analytical approach (4,5), goal programming (6-8), surrogate worth trade-off method (9), and others (10,11). These approaches take into account interactions among different objectives (requirements) and resolve conflicts among them in decision making.

In this work, the ε-constraint method (2) is employed for generating a trade-off curve (or non-inferior set), and the surrogate worth trade-off method (9) for selecting the preferred decision. A milk evaporation process is optimized for illustration.

Generation of the Non-Inferior Set

The ε-constraint method used here is based on the Kuhn-Tucker condition for non-inferior decision (2). The first equation of the Kuhn-Tucker condition can be rewritten as

$$
\left(\frac{\partial f_1}{\partial \underline{x}}\right)^T w_1 + \sum_{i=2}^{n} \left(\frac{\partial f_i}{\partial \underline{x}}\right)^T w_i + \left(\frac{\partial \underline{g}}{\partial \underline{x}}\right)^T \underline{\lambda} = 0 \tag{2}
$$

Since only relative values of the weights are significant, we can assume that w_1 is 1 without loss of generality. Thus, Equation 2 becomes

$$
\left(\frac{\partial f_1}{\partial \underline{x}}\right) + \left(\frac{\partial \underline{\tilde{f}}}{\partial \underline{x}}\right)^T \underline{\tilde{w}} + \left(\frac{\partial \underline{g}}{\partial \underline{x}}\right)^T \underline{\lambda} = 0 \tag{3}
$$

where

$$
\underline{\tilde{f}} = [f_2(\underline{x}), f_3(\underline{x}), \ldots, f_n(\underline{x})]^T
$$
$$
\underline{\tilde{w}} = [w_2, w_3, \ldots, w_n]^T
$$

This equation allows us to interpret $\underline{w}$ in the second term as a Lagrangian multiplier vector. This interpretation implies that a non-inferior decision satisfying the above equation can be obtained by solving the optimization problem:

        Minimize

$$
J = f_1(\underline{x}) \tag{4}
$$

subject to

$$
\left. \begin{array}{l} \underline{\tilde{f}}(\underline{x}) \le \underline{e} \\[2mm] \underline{g}(\underline{x}) \le 0 \quad (\text{or } \underline{x} \in X) \end{array} \right\} \tag{5}
$$

where $\underline{e}$ is an $(n-1)$-dimensional constant vector. $\underline{e}$ is varied parametrically to yield the set of non-inferior decisions. Note that each e_i must not be smaller than a certain value in order to render the feasible decision set defined by constraint (5) to be non-empty. To identify the minimum value of e_i, the following auxiliary problem with a single objective must be solved.

$$
J_i = f_i(x)
$$
subject to
$$
\underline{x} \in X
$$

where other objectives $f_j(\underline{x})$, $j \neq i$, are entirely neglected.
The optimal value of J_i is the minimum value of e_i. Although
the ε-constraint method is somewhat intricate compared with the
weighting methods, it is widely used because of its
applicability to non-convex problems.

Selection of the Preferred Decision

The algorithm of the surrogate worth function method ($\underline{9},\underline{12}$)
consists of two parts. One is the generation of the non-
inferior set which forms the trade-off surface in the objective
space. The other is the search for the preferred decision in
the non-inferior set. The feature of this method is that the
preferred decision is located by the use of the surrogate worth
function introduced by Haimes and Hall ($\underline{9}$). The second part is
used here.

The so-called trade-off surface (curve) represents the non-
inferior solution obtained in the preceding section, and the
trade-off ratio between the i-th and j-th objectives is defined
as

$$T_{ij} = - \frac{\partial f_i}{\partial f_j}$$

The surrogate worth function, W_{ij}, estimates the desirability of
the trade between a decrease of T_{ij} units in the i-th objective
and an increase of one unit in the j-th objective; the other
objectives remain at their current values. Thus, W_{ij} is a
function of the trade-off ratio, T_{ij}, and the non-inferior
objective, $\underline{f}(\underline{x})$. Haimes and Hall[9] have defined W_{ij} in such
a way that

$$W_{ij} > 0 \qquad\qquad\qquad\qquad\qquad (6)$$

when the trade is desirable, i.e., T_{ij} units of $f_i(\underline{x})$ are
preferred over one unit of $f_j(x)$ for a given $\underline{f}(\underline{x})$;

$$W_{ij} = 0 \qquad\qquad\qquad\qquad\qquad (7)$$

when the trade is even; and

$$W_{ij} < 0 \qquad\qquad\qquad\qquad\qquad (8)$$

when the trade is undesirable. The larger the absolute value of
W_{ij}, the greater the desirability or undesirability of the
trade. The numerical value of W_{ij} will depend on the decision
maker's response to the question: Is it desirable to reduce
$f_i(\underline{x})$ by T_{ij} units when $f_j(\underline{x})$ is increased by one unit and other

objectives are maintained at their current levels? It should be relatively simple to answer this question, since the attained levels of all objectives are known (12).

The surrogate worth function, W_{ij}, can be more easily understood in terms of the marginal rate of substitution, M_{ij}. M_{ij} is defined as the slope of the indifference curve, $v(\underline{f})$, i.e., M_{ij} units of $f_j(\underline{x})$ is equivalent to one unit of $f_i(\underline{x})$ according to the decision maker's preference. If the difference, $(T_{ij} - M_{ij})$, is positive for a non-inferior decision, an increase of one unit in $f_j(x)$ will result in a further decrease in $f_i(\underline{x})$ than that required to maintain the decision at the same value of $v(\underline{f})$. Thus, such change is desirable, and W_{ij} must be positive at the decision. Consequently, W_{ij} is essentially identical to the difference, $(T_{ij} - M_{ij})$, as illustrated in Figure 1 (13). Obviously,

$$T_{ij} - M_{ij} = 0, \quad j = 1, 2, \ldots, i=1, i+1, i-1, \ldots, n \qquad (9)$$

at the preferred decision, i.e.,

$$W_{ij} = 0, \quad j = 1, 2, \ldots, i-1, i+1, \ldots, n \qquad (10)$$

The computational scheme for this method is as follows:
1. Determine the relationship between the trade-off ratio, T_{ij}, and one of the objective functions, f_i.
2. Select the marginal rate of substitution, M_{ij}.
3. Obtain the preferred solution by equating the trade-off ratio, T_{ij}, and the marginal rate of substitution, M_{ij}.

Example

The procedure for solving a multiobjective optimization problem is demonstrated here by applying it to the optimal design of a milk evaporation process shown in Figure 2. From the standpoint of energy conservation, we wish to minimize the dissipation of available energy. However, an effort to reduce the dissipation of available energy tends to increase the heat transfer area or size of the evaporator and vice versa. Because of this con- flicting relationship, they are considered to be two objective functions of the system to be minimized (14). [Those interested in the thermodynamic basis for evaluating the dissipation of available energy are referred to numerous articles and treatises available on the subject (14-25; also see Appendix A).]

According to the ε-constraint method, this problem can be formulated as (see Appendix B):

Minimize

$$J_2 = f_2 = A \qquad (11)$$

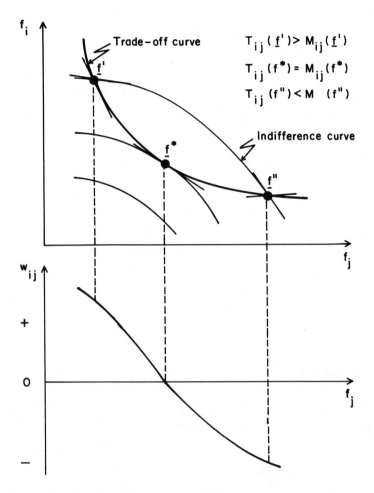

Figure 1. Trade-off curve, indifference curve and surrogate worth function.

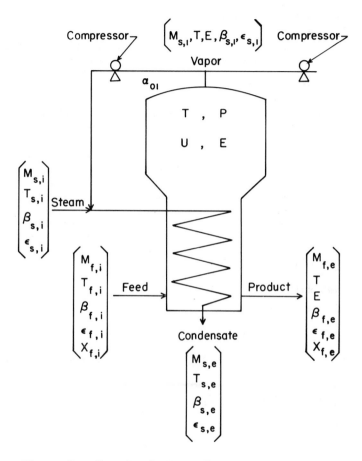

Figure 2.   Sketch of the milk evaporation process.

subject to $\quad\quad$ $\underline{g} \geq 0$ $\hfill$ (12)

$\quad\quad\quad\quad\quad\quad$ $f_1 \leq e$ $\hfill$ (13)

where A represents the contact area of the evaporator, and Equation 12 the entire set of equality and inequality constraints (see Appendix B). The trade-off curve is obtained by solving this optimization problem by parametrically changing e. Note that e cannot be less than a certain value which is the minimum value of f_1.

This problem contains 31 variables and 29 equality constraints (or governing equations) including the objective function. This gives rise to 2 variables as independent (or decision) variables. For a practical reason, the saturation pressure for steam, P, and the fraction of steam generated in the evaporator, which is reused for heating, α_{01}, are selected as the independent variables. A random search technique (26) is adopted to locate the optimal point for each given e. The results are tabulated in Table I, and the trade-off curve is plotted in Figure 3. The relationship between these two objectives is obtained by the least square method as

$$f_2 = 9.0336 - 43.0112 \times 10^{-2} f_1$$

$$+ 7.0598 \times 10^{-4} f_1^2 + 192.7518 \times 10^{-6} f_1^3, \quad f_1 > 22 \quad (14)$$

$$f_2 = 4.6244 - 20.9404 \times 10^{-2} f_1$$

$$- 0.95197 \times 10^{-4} f_1^2 + 178.6031 \times 10^{-6} f_1^3, \quad 7.15 < f_1 \leq 22 \quad (15)$$

$$f_2 = 3.41565 - 1.33327 \times 10^{-2} f_1$$

$$+ 0.511598 \times 10^{-4} f_1^2 - 345.9425 \times 10^{-6} f_1^3, \quad 1.15 < f_1 \leq 7.15 \quad (16)$$

$$f_2 = 19.97505 - 2392.055 \times 10^{-2} f_1$$

$$+ 10548.842 \times 10^{-4} f_1^2 + 6389034.56 \times 10^{-6} f_1^3, \quad 0 < f_1 < 1.15 \quad (17)$$

To select the preferred solution according to the computational scheme presented, we first determine the relationship between the trade-off ratio, T_2, and each of the f_1's of the four regions represented by Equations 14 through 17, respectively, as

Table I. Heat Transfer Area, Available Energy Dissipation and Independent Variables

Heat Transfer Area, A m^2	Available Energy Dissipation, $(T_0 \sigma)$ kcal	Pressure in the Evaporator, P_1 atm	Fraction of the Generated Steam Reused, α_{01}
1.7416	30.3734	0.0475	0.4411
1.7441	27.4108	0.0483	0.8224
1.7587	25.4023	0.0470	0.9829
1.7614	25.1468	0.0470	1.0
1.7930	23.8349	0.0470	0.9518
1.9197	21.3418	0.0470	0.8163
2.1315	14.3121	0.3822	0.0474
2.9062	9.0298	0.047	1.0
3.1873	7.1516	0.7289	0.439
3.3949	1.4807	0.047	0.5353
3.4185	1.0485	0.949	1.0
16.8646	0.1314	6.0	1.0

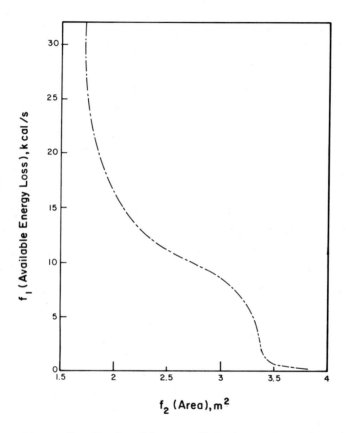

Figure 3.   Trade-off curve for the milk evaporation
process.

Table II.   Preferred Decision of the Milk Evaporation System

f_1 = 9.029829 kcal/s

f_2 = 2.9062 m^2

$T_{s,i}$ = 450 K

$T_{s,e}$ = 450 K

$M_{s,i}$ = 0.4084 kg/s

$M_{s,1}$ = 1.4175 kg/s

$M_{f,i}$ = 1.88996 kg/s

$M_{f,e}$ = 0.4725 kg/s

α_{01} = 1.0

T = 305.16 K

P = 0.047 atm

$$T_{21} = - \frac{\partial f_2}{\partial f_1}$$

$$= 0.430112 - 14.1196 \times 10^{-4} f_1 - 578.2554 \times 10^{-6} f_1^2,$$

$$f_1 > 22 \tag{18}$$

$$T_{21} = - \frac{\partial f_2}{\partial f_1}$$

$$= 0.209404 + 1.90394 \times 10^{-4} f_1 - 535.8093 \times 10^{-6} f_1^2,$$

$$7.15 < f_1 \leq 22 \tag{19}$$

$$T_{21} = - \frac{\partial f_2}{\partial f_1}$$

$$= 0.0133327 - 1.023196 \times 10^{-4} f_1 + 1037.8275 \times 10^{-6} f_1^2,$$

$$1.15 < f_1 \leq 7.15 \tag{20}$$

and

$$T_{21} = - \frac{\partial f_2}{\partial f_1}$$

$$= 23.92055 - 21097.684 \times 10^{-4} f_1 - 19167103.68 \times 10^{-6} f_1^2,$$

$$0 < f_1 \leq 1.15, \tag{21}$$

Second, for simplicity and illustration, the marginal rate of substitution, M_{21}, is selected as

$$M_{21} = 0.167435$$

Third, from Equation 9, i.e.,

$$T_{21} - M_{21} = 0,$$

we have, from Equations 18 through 21, respectively,

$$0.430112 - 14.1196 \times 10^{-4} f_1 - 578.2554 \times 10^{-6} f_1^2 = 0.167435 \tag{22}$$

$$0.209404 + 1.90394 \times 10^{-4} f_1 - 535.8093 \times 10^{-6} f_1^2 = 0.167435 \tag{23}$$

$$0.0133327 - 1.023196 \times 10^{-4} f_1 + 1037.8275 \times 10^{-6} f_1^2 = 0.167435 \tag{24}$$

and

$$23.92055 - 21097.684 \times 10^{-4} f_1 - 19167103.68 \times 10^{-6} f_1^2$$

$$= 0.167435 \tag{25}$$

Equation 22 gives

$$f_1 = 20.127391 \text{ kcal/s} \tag{26}$$

and

$$f_1 = -22.5692 \text{ kcal/s} \tag{27}$$

Since these two values of f_1 are not in the feasible region defined by Equation 14, they should be discarded. Equation 23 gives rise to

$$f_1 = 9.0298 \text{ kcal/s} \tag{28}$$

and

$$f_1 = -8.674458 \text{ kcal/s} \tag{29}$$

The second of these two values is negative and thus, is in the infeasible region; it should be discarded. The first of these two values is in the feasible region defined by Equation 15. From Equation 24, we obtain

$$f_1 = 12.2348 \text{ kcal/s} \tag{30}$$

and

$$f_1 = -12.1363 \text{ kcal/s} \tag{31}$$

Since both values are not in the feasible region defined by Equation 16, they should be discarded. Similarly, from Equation 25, we have

$$f_1 = 2.42955 \text{ kcal/s} \tag{32}$$

$$f_1 = -2.5396 \text{ kcal/s} \tag{33}$$

Again both values are in the infeasible region and should be discarded. Substituting Equation 28 into Equation 15 yields

$$f_2 = 2.9062 \text{ m}^2 \tag{34}$$

This is the optimal solution for the second objective function, corresponding to Equation 28, which is the optimal solution for the first objective function. The values of the objective functions and design variables for the preferred solution are tabulated in Table II. The resultant optimal configuration of

the system is shown in Figure 4. The energy and exergy of each
stream in the optimal design are summarized in Table III.

Discussion and Conclusion

As illustrated in Figure 1, the surrogate worth trade-off method
provides a means of locating the preferred (or optimal) solution
by determining the point of tangency between the trade-off curve
(function) and the so-called indifference curve (function).
This is also demonstrated numerically in the example presented
in the preceding section. An indifference surface (or curve) is
defined as a locus of different conditions in the objective
space, any two of which cannot be distinguished by the
preference criterion of the decision maker. An indifference
curve or surface can be expressed in terms of the value
function, $v(f)$, as

$$\{\underline{f} \mid v(\underline{f}) = \text{constant}\} \tag{35}$$

A different value of the constant gives rise to a separate
indifference surface. These indifference surfaces do not
intersect each other, and, therefore, every point in the
objective space lies on one and only one indifference surface.
The trade-off surface is tangent to one of the indifference
surfaces at the preferred point. As mentioned earlier, the
marginal rate of substitution of f_i for f_j, M_{ij}, is expressed as

$$M_{ij} = - \frac{\left(\frac{\partial v}{\partial f_j}\right)}{\left(\frac{\partial v}{\partial f_i}\right)} = - \frac{\partial f_i}{\partial f_j} \tag{36}$$

Since the trade-off and indifference surfaces are tangent to
each other at the preferred point, $\underline{f}^*$, we have

$$T_{ij}(\underline{f}^*) = M_{ij}(\underline{f}^*) \tag{37}$$

Note that the indifference surfaces are obtained without knowing
the function form of the value function, $v(\underline{f})$. They are
generally determined by directly comparing many sampled points
in the objective space based on the decision maker's preference.
   Figure 3 indicates that in the example, the trade-off curve
(or the non-inferior solutions) is not completely concave in
shape and that the feasible region is not exactly convex; however, the two objectives, capital investment, f_2, in terms of
the heat transfer area and available energy, f_1, are always in
conflict with each other in the region under consideration. A
reduction in the heat transfer area will always give rise to an
increase in the dissipation of available energy.
   The resultant optimal configuration, Figure 2; in other
words, the generated steam is totally reused.

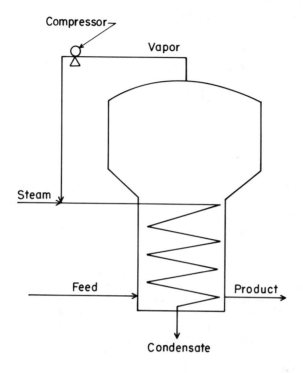

Figure 4.   Optimal configuration of the milk evaporation.

Table III. Energy and Exergy of Each Steam in the Optimal Design	
$\beta_{f,i}$ = 578.7506 kcal/kg,	$\varepsilon_{f,i}$ = 589.3042 kcal/kg
$\beta_{s,i}$ = 636.5436 kcal/kg,	$\varepsilon_{s,i}$ = 120.6537 kcal/kg
$\beta_{s,1}$ = 609.0213 kcal/kg,	$\varepsilon_{s,1}$ = 116.0266 kcal/kg
$\beta_{f,e}$ = 2360.9779 kcal/kg,	$\varepsilon_{f,e}$ = 2362.6292 kcal/kg
$\beta_{s,e}$ = 151.85 kcal/kg,	$\varepsilon_{s,e}$ = 29.1163 kcal/kg

The traditional single-objective (or scalar) optimization problem, corresponding to the multiobjective optimization problem under consideration, can be stated as:

Minimize

or

$$J_2' = \omega_1 f_1(\underline{x}) + \omega_2 f_2(\underline{x})$$

$$J_2 = f_1(\underline{x}) + \omega_{12} f_2(\underline{x}) \tag{38}$$

subject to

$$\underline{x} \in X$$

$$g_i \leq 0, \quad j = 1, 2, \ldots$$

where ω_1 and ω_2 are the weighting factors, and ω_{12} is the ratio of ω_2 to ω_1. Notice that ω_{12} can be viewed as a Lagrange multiplier.

For simplicity, suppose that ω_1 and ω_2 are such that

$$\omega_{12} = 5.972$$

in the scalar optimization problem corresponding to the example considered in the preceding section. Minimization of J_2 over $\underline{x}$ can be accomplished by any classical optimization technique. For convenience, the adaptive random search technique (26) is used here. The optimal solution obtained is located at

$$\left. \begin{array}{l} x_1 = P_1 = 0.0475 \\ \\ x_2 = \alpha_{01} = 1.0 \end{array} \right\} \tag{39}$$

where

and

$$\left. \begin{array}{l} J_2 = 26.48 \\ \\ f_1 = 9.1 \text{ kcal/s} \\ \\ f_2 = 2.91 \text{ m}^2 \end{array} \right\} \tag{40}$$

Notice that the optimal f_1 and f_2 are essentially identical to those obtained in the multiobjective optimizaton problem. In fact, it is known (9) that, for the system with a convex feasible region,

$$M_{12} = T_{12} = \omega_{12} \tag{41}$$

The value of ω_{12} selected here for illustration is equal to T_{12} which is the inverse of T_{21} or 0.167435. The feasible region of the system under consideration is not exactly convex; however, the relationship given in Equation 41 is apparently satisfied.

As can be seen from the values of f_1 and f_2 in Equation 40 and Figure 4, the optimal scalar objective function is located on the trade-off curve. In other words, as long as Equation 41 is satisfied, the search for the scalar objective function, J_2, need be carried out only along the trade-off curve. Thus, we see that under the optimal condition the appropriate cost of the dissipation of available energy, ω_1, and that of the evaporator, ω_2 should be such that this ratio, ω_{12}, be equal to M_{12}, or reversibly, the value of M_{12} be equal to ω_{12}.

Nomenclature

A   contact area of the evaporator, m^2

BR1   available energy to the evaporation system through compressor 1, kcal

BR2   available energy to the evaporation system through compressor 2, kcal

E   boiling point elevation, K

e   allowable level vector for the objective function vector, $\underline{f}$

$\underline{f}$ objective function vector

$\underset{\sim}{\underline{f}}$ objective function vector defined as $[f_2, f_3, \ldots, f_n]^T$

f_i scalar function associated with the i-th subsystem

$\underline{g}$ constraint function vector

g_i vector function associated with the i-th subsystem

J   scalar objective function

$\underline{J}$ vector objective function

J_j j-th objective function

$M_{f,i}$ mass flow rate of the aqueous milk solution at the inlet, kg/s

$M_{f,e}$ mass flow rate of the aqueous milk solution at the exit, kg/s

$M_{s,1}$ mass flow rate of the vapor, kg/s

Continued on next page

$M_{i,j}$ marginal rate of substitution of f_i for f_j

$M_{s,i}$ mass flow rate of steam at the inlet, kg/s

$M_{s,e}$ mass flow rate of condensate at the exit, kg/s

P pressure in the evaporator, atm

T temperature of the aqueous milk solution and vapor at the exit, K

$T_{f,i}$ temperature of the aqueous milk solution at the inlet, K

$T_{i,j}$ trade-off ratio between the i-th and j-th objectives

$T_{s,i}$ temperature of steam at the inlet, K

$T_{s,e}$ temperature of steam at the exit, K

U heat transfer coefficient, kcal/s, m^2, K

$\underset{\sim}{w}$ weighting coefficient vector defined as $[w_2, w_3, \ldots, w_n]^T$

w_i i-th component of the weighting coefficient vector, $\underline{w}$

w_{ij} surrogate worth function associated with the i-th and j-th objectives

$\underline{x}$ decision vector

$x_{f,i}$ concentration of the milk in the aqueous solution at the inlet in mole fraction

$x_{f,e}$ concentration of the milk in the aqueous solution at the exit in mole fraction

Greek Symbols

β specific enthalpy measured in reference to the dead state, kcal/kg

$\varepsilon_{s,1}$ specific exergy for the steam generated in the evaporator, kcal/kg

ε specific exergy, kcal/kg

λ Lagrangian multiplier

$\underline{\lambda}$ Lagrangian multiplier vector

α_{21} structural parameter from unit 1 to unit 2

α_{11} structural parameter from unit 1 to unit 1

α_{01} structural parameter from unit 1 to unit 0

Literature Cited

1. Kuhn, H.W.; Tucker, A.W., "Nonlinear Programming"; Proc. of the Second Berkeley Symposium on Mathematics and Probability: Univ. of California Press, Berkeley, Ca, 1951; pp. 481-692.
2. Cohon, J.L.; Marks, D.H., "A Review and Evaluation of Multi-objective Programming Techniques"; Water Resources Research 1975, 11, 521.
3. Huang, S.C., "Note on the Mean-Square Strategy for Vector Valued Objective Functions"; Journal of Optimization Theory Applications 1972, 9, 364.
4. Zadeh, L.A., "Optimality and Non-Scalar Valued Performance Criteria"; IEEE Transactions 1963, AC8, 59.
5. Reid, R.W.; Vemure, V., "On the Non-Inferior Index Approach to Large Scale Multi-criteria Systems"; Journal of the Franklin Institute 1971, 291, 4, 241.
6. Charnes, A; Cooper, W.W., "Management Models and Industrial Application of Linear Programming"; John Wiley: NY, 961.
7. Major, D.C., "Benefit-Cost Ratios for Projects in Multiple Objectives Investment Programs"; Water Resources Research 1969, 5, 1174.
8. Lee, S.M., "Goal Programming for Decision Analysis"; Auervach: Philadelphia, 1972.
9. Haimes, Y.Y.; Hall, W.A., "Multiobjectives in Water Resources Systems Analysis: The Surrogate Worth Trade-Off Method"; Water Resources Research 1974, 10, 615.
10. Cohon, J.L.; Marks, D.H., "Multiobjective Screening Models and Water Resource Investment"; Water Resources Research 1973, 9, 521.
11. Roy, B., "Problems and Methods with Multiple Objective Functions"; Mathematical Programming 1971, 1, 50.
12. Haimes, Y.Y., et al., "Multiobjective Optimization in Water Resources Systems: The Surrogate Worth Trade-Off Method"; Elsevier Scientific Publishing: Amsterdam, 1975.
13. Nakayama, H.; Sawaragi, Y., "Decision Making with Multiple Objectives and its Applications (in Japanese)"; Systems and Control 1976, 20, 511-520.
14. Fan, L.T.; Shieh, J.H. "Thermodynamically Based Analysis and Synthesis of Chemical Process Systems"; Energy 1980, 5, 955.

15. Denbigh, K.C., "The Second-Law Efficiency of Chemical Processes"; Chem. Engr. Sci. 1956, 6, 1.

16. Bruges, E.A., "Available Energy and the Second Law Analysis"; Academic Press: London, 1959.

17. Gaggioli, R.A., "Thermodynamics and the Non-Equilibrium System"; Ph.D. Dissertation: University of Wisconsin, 1961.

18. Baehr, von H.D.; Schmidt, E.F., "Definition und Berechnung von Breenstaffexergien"; BWK 1963, 15, 375.

19. Szargut, J; Petela, R., "Egzergia"; Wydawnictwa Naukowo-Techniczne: Warsawa, Poland, 1965 (in Polish).

20. Evans, R.B.; Tribus, M., "Thermo-Economics of Saline Water Conversion"; I & EC Process Design and Development 1965, 4, 195.

21. Reistad, G., "Availability: Concepts and Application"; Ph.D. Dissertation: University of Wisconsin, 1970.

22. Riekert, L., "The Efficiency of Energy-Utilization in Chemical Processes"; Chem. Engr. Sci. 1974, 29, 1613.

23. Gaggioli, R.A.; Petit, P.J., "Use the Second Law First"; Chemtech 1977, August, 496.

24. Rodriguez, L. in "Calculation of Available Energy Quantities "; Gaggioli, R.A, Ed.: ACS SYMPOSIUM SERIES NO. 122, American Chemical Society: Washington, D.C., 1980; p. 39.

25. Wepfer, W.J.; Gaggioli, R.A. in "Reference Datumo for Available Energy"; Gaggioli, R.A., Ed.: ACS SYMPOSIUM SERIES NO. 122, American Chemical Society: Washington, D.C., 1980; p. 77.

26. Chen, H.T.; Fan, L.T., "Multiple Minima in a Fluidized Reactor-Heater System"; AICLE J. 1976, 22, 680.

Appendix A: Thermodynamic Background

In carrying out the thermodynamic first- and second-law analyses of a process system, the calorific thermodynamic functions of material species involved in the process are necessary. Conventionally the enthalpy, the entropy and the Gibbs free energy are evaluated on the basis of a reference state where all are known elements are in their pure states, the temperature is 25°C (298.15 K) and the pressure is 1 atm.   An adoption of such a reference state often yields negative energy and Gibbs free energy, and these, in turn, render the analysis of process systems difficult, if not impossible.   To circumvent such a difficulty a great deal of effort has been spent for determining the thermodynamically meaningful reference state that tends to yield a positive calorific function.   One of the frequently employed reference states is the so-called dead state where the reference material species (datum level materials) are essentially the products of complete combustion, the reference concentrations (datum level concentrations) are the environmental concentrations of these products, the reference temperature (datum level temperature) is the environmental temperature, and the reference pressure (datum level pressure) is the environmental pressure (14,18,19,22,23,25).   This reference state has been adopted in this work for evaluating the energy and available energy contents for carrying out the thermodynamic first- and second-law analyses.

The specific enthalpy relative to the dead state, β, is defined as (14, 19,24)

$$\beta \equiv h - h_0 \tag{A-1}$$

where h is the specific enthalpy at any state and h_0 that at the dead state.   By adding the enthalpy at the standard state, h^0, into and subtracting it from Equation A-1, we have

$$\beta = (h - h^0) + (h^0 - h_0) \tag{A-2}$$

Since

$$h - h^0 = {}_{T^0}\!\int^T c_p dT + {}_{P^0}\!\int^P [v - T\left(\frac{\partial v}{\partial T}\right)_P]dP, \tag{A-3}$$

Equation A-2 becomes

$$\beta = (h^0 - h_0) + {}_{T^0}\!\int^T c_p dT + {}_{P^0}\!\int^P [v - T\left(\frac{\partial v}{\partial T}\right)_P]dP \tag{A-4}$$

By defining (14)

$$\beta^0 \equiv h^0 - h_0 \tag{A-5}$$

$$\beta_T \equiv {}_{T^0}\!\int^T c_p dT \tag{A-6}$$

$$\beta_P \equiv \int_{P^0}^{P} [v - T(\tfrac{\partial v}{\partial T})_P] dP, \tag{A-7}$$

Equation A-3 can also be expressed as

$$\beta = \beta^0 + \beta_T + \beta_P \tag{A-8}$$

Here, β^0 is termed the specific chemical enthalpy, β_T the specific thermal enthalpy and β_P the specific pressure enthalpy. The combination of the specific thermal enthalpy, β_T, and the specific pressure enthalpy, β_P, may be named the specific physical enthalpy. When the material species is one of the components in a solution, Equations A-1 through A-7 are valid, provided that the specific quantities are changed to the partial molar quantities. Note that superscript 0 refers to the standard state, and subscript 0 refers to the dead state; c_P is the specific heat, and v is the specific volume.

The specific exergy, ε, is defined as ([14],[16],[18-23])

$$\varepsilon \equiv (h - h_0) - T_0(s - s_0)$$
$$\equiv (h^0 - h_0) - T_0(s^0 - s_0) + (h - h^0) - T_0(s - s^0) \tag{A-9}$$

Substituting Equation A-3 and

$$s - s^0 = \int_{T^0}^{T} \frac{c_P}{T} dT - \int_{P^0}^{P} (\tfrac{\partial v}{\partial T})_P dP \tag{A-10}$$

into Equation A-8 gives rise to

$$\varepsilon = (h^0 - h_0) - T_0(s^0 - s_0) + \int_{T^0}^{T} c_P(1 - \frac{T_0}{T}) dt$$
$$+ \int_{P^0}^{P} [v - (T - T_0)(\tfrac{\partial v}{\partial T})_P] dP \tag{A-11}$$

where s is the specific entropy at any state, s_0 the specific entropy at the dead state, and T_0 the datum level temperature. By defining ([14],[19],[22],[23])

$$\varepsilon^0 \equiv (h^0 - h_0) - T_0(s^0 - s_0) \tag{A-12}$$

$$\varepsilon_T \equiv \int_{T^0}^{T} c_P(1 - \frac{T_0}{T}) dP \tag{A-13}$$

and

$$\varepsilon_P \equiv \int_{P^0}^{P} [v - (T - T_0)(\tfrac{\partial v}{\partial T})_P] dP, \tag{A-14}$$

Equation A-11 reduces to

$$\varepsilon = \varepsilon^0 + \varepsilon_T + \varepsilon_p \tag{A-15}$$

where ε^0 is termed the specific chemical exergy, ε_T the specific thermal exergy, and ε_p the specific pressure exergy. Again, the combination of the specific thermal exergy, ε_T, and the specific pressure exergy, ε_p, may be named the specific physical exergy (14,19,22,23).

Equations A-1 through A-15 are the working formulas for evaluating the energy and available energy contents of a material species which is involved in a chemical process.

Appendix B: List of the Performance Equations for Determining the Non-Inferior Set

The objective function to be minimized is

$$J_2 = f_2 = A \tag{B-1}$$

This is Equation 13 in the text. The equality constraints included in Equation 12 in the text are

$$M_{f,i}\, x_{f,i} - M_{f,e}\, x_{f,e} = 0 \tag{B-2}$$

$$M_{f,i} - M_{f,e} - M_{s,1} = 0 \tag{B-3}$$

$$M_{f,i}\beta_{f,i} + (M_{s,i} + M_{s,1}\alpha_{01})\beta_{s,i} - M_{f,e}\beta_{f,e} - M_{s,1}\beta_{s,1}$$
$$- (M_{s,i} + M_{s,1}\alpha_{01})\beta_{s,e} = 0 \tag{B-4}$$

$$\alpha_{01} + \alpha_{11} + \alpha_{21} = 0$$

$$T_0\sigma = M_{f,i}\varepsilon_{f,i} + M_{s,i}\varepsilon_{s,i} + BR_1 + BR_2$$
$$- (M_{s,i} + M_{s,1}\alpha_{01})\varepsilon_{s,e} - M_{f,e}\varepsilon_{f,e} - M_{s,1}\alpha_{21}\varepsilon_{s,2} \tag{B-5}$$

$$U = 0.0474 - 0.0339\, x_{f,e} \tag{B-6}$$

$$A = \frac{(\beta_{s,i} - \beta_{s,e})(M_{s,i} + M_{s,1}\alpha_{01})}{U(T_{s,i} - T - E)} \tag{B-7}$$

$$\beta_{s,i} = 55.5556(6.9756 + 0.013023T_{s,i} + 1.557 \times 10^{-7}\ T_{s,i}^2$$
$$- 1.547 \times 10^{-8}\ T_{s,i}^3) \tag{B-8}$$

$$\beta_{s,e} = T_{s,e} - 298.15 \tag{B-9}$$

$$\beta_{s,1} = 55.5556(6.9756 + 0.013023T + 1.557 \times 10^{-7}\ T^2$$
$$- 1.547 \times 10^{-8}\ T^3) \tag{B-10}$$

$$\beta_{f,i} = (T_{f,i} - 298.15) + x_{f,i}(5978.918 - 0.642\ T_{f,i}) \tag{B-11}$$

$$\beta_{f,e} = (T - 298.15) + x_{f,e}(5978.918 - 0.642\ T) \tag{B-12}$$

$$\varepsilon_{s,i} = 55.5556\{-0.153674 - 2.29605\ \ell n\left(\frac{T_{s,i}}{298.15}\right)$$
$$+ 0.007564T_{s,i} - 1.387475 \times 10^{-7}T_{s,i}^2$$
$$+ 1.38838 \times 10^{-9}T_{s,i}^3 - 5.648159 \times 10^{-12}T_{s,i}^4$$
$$+ 0.59242\ \ell n[\frac{P_{s,i}}{(P_{s,i})_{sat.}}]\} \tag{B-13}$$

$$\varepsilon_{s,e} = 55.5556(0.56207 - 0.003433T_{s,e}$$
$$+ 8.296 \times 10^{-8}T_{s,e}^2) \tag{B-14}$$

$$\varepsilon_{f,i} = 5900.506\ x_{f,i} + (1 - 0.742\ x_{f,i})(T_{f,i} - 298.15$$
$$- 298.15\ \ell n\ \frac{T_{f,i}}{298.15}) + 2.296\ [x_{f,i}\ell nx_{f,i}$$
$$+ (1 - x_{f,i})\ell n(1 - x_{f,i})] \tag{B-15}$$

$$\varepsilon_{f,e} = 5900.506\ x_{f,e} + (1 - 0.742\ x_{f,e})(T - 298.15$$
$$- 298.15\ \ell n\ \frac{T}{298.15}) + 2.296[x_{f,e}\ell n\ x_{f,e}$$
$$+ (1 - x_{f,e})\ \ell n(1 - x_{f,e})] \tag{B-16}$$

$$\varepsilon_{s,1} = 55.5556[-0.153674-2.29605 \ \ell n \ \frac{T}{298.15} + 0.007564T$$

$$- 1.387475 \times 10^{-7}T^2 + 1.38838 \times 10^{-9}T^3$$

$$- 5.648159 \times 10^{-12}T^4 + 0.59242 \ \ell n(\frac{P}{P_{sat}})] \qquad \text{(B-17)}$$

$$\varepsilon_{s,2} = 55.5556[-0.153674 - 2.29605 \ \ell n \ \frac{T_{s,2}}{298.15}$$

$$+ 0.0075646 \ T_{s,2} - 1.387475 \times 10^{-7}T^2_{s,2}$$

$$+ 1.38838 \times 10^{-9}T^3_{s,2} - 5.648159 \times 10^{-12}T^4_{s,2}$$

$$+ 0.59242 \ \ell n(\frac{P}{P_{sat}})] \qquad \text{(B-18)}$$

$$BR_1 = \frac{\varepsilon_{s,i} - \varepsilon_{s,1}}{0.75} \qquad \text{(B-19)}$$

$$BR_2 = \frac{\varepsilon_{s,2} - \varepsilon_{s,1}}{0.75} \qquad \text{(B-20)}$$

$$T_{s,e} = T_{s,i} \qquad \text{(B-21)}$$

$$T = 297.8095 + 226.6964P - 45.4813P^2_1 - 836.3919P^3_1 ,$$

$$0.047 \leq P \leq 0.4 \qquad \text{(B-22)}$$

$$T = 349.1699 + 25.8431P - 1.16936P^2_1 - 0.13278P^3_1 ,$$

$$0.4 < P \leq 6.2 \qquad \text{(B-23)}$$

$$M_{f,i} = 1.88996 \ kg/s \qquad \text{(B-24)}$$

$$x_{f,i} = 0.1 \qquad \text{(B-25)}$$

$$x_{f,e} = 0.4 \qquad \text{(B-26)}$$

$$T_{s,i} = 450 \ K \qquad \text{(B-27)}$$

$$E = 0 \ K \qquad \text{(B-28)}$$

$$T_{f,i} = 298.15 \text{ K} \tag{B-29}$$

$$\alpha_{11} = 0 \tag{B-30}$$

The inequality constraints included in Equation 12 in the text are

$$A > 0 \tag{B-31}$$

$$T_{s,i} > T \tag{B-32}$$

$$T_0 \sigma > 0 \tag{B-33}$$

The inequality constraint of Equation 13 in the text is

$$J_1 = f_1 = (T_0 \sigma) \leq e \tag{B-34}$$

where $(T_0 \sigma)$ is the available energy dissipation evaluated from Equation B-5.

RECEIVED July 12, 1983

Multiobjective Analysis for Energy and Resource Conservation in an Evaporation System

H. NISHITANI and E. KUNUGITA

Department of Chemical Engineering, Osaka University, Toyonaka, Osaka, 560, Japan

Efficient use of both energy and resource in an evaporation system was studied based on multi-objective
analysis. The exergy consumption and the total investment cost were used to measure energy and resource conservation, respectively. The trade-off
curve between the two objectives shows the change
in the optimal solution as the unit cost of exergy
is changed.

A process system is composed of various pieces of equipment and
is operated by many different types of energy sources. The size
of each piece of equipment and the amount of each source of
energy should be as small as possible. Usually, energy and resource conservation are achieved based on the economic cost of
the commodities. Although relative economics plays an important
role in all decisions concerning the system, the physical units
of measure will enable the engineers to investigate the energy
and resource conservation from the point of view of technology.
They will then be able to understand the problems more easily
and will therefore be in a better position to improve the efficiency of the system.

In this paper the problem of energy and resource conservation
was considered at the process design stage. Since energy conservation can be achieved with additional equipment, there exists
a trade-off between the two objectives. Recently the cost of
energy has been changing rapidly in comparison with the cost of
other materials. In other words, reflection of the value of
energy in price is less reliable than reflection of the value of
equipment materials in price. Therefore, it is beneficial to
discover the trade-off between the two objectives, which shows
the optimal design under various energy conditions.

0097-6156/83/0235-0333$06.00/0

Energy and Resource Conservation in the Process System

Energy may be supplied by means of steam, electricity, etc.
A quantity of energy can be assigned a value only when certain
conditions are known. Since the enthalpy does not pertain to the
quality of energy, the available energy (exergy) should be used
to measure the value of a commodity for operating and sustaining
a process system (1).

A second law analysis based on exergy has been conducted with
respect to a set of data for the various flows of vapor and
liquid contained in the specified system (1-3). The analysis
shows the locations of the major inefficiencies, and hence the
pieces of equipment or steps in the process system which could
be improved. However, this analysis gives no hint as to the
nature of the changes that might be made. When any change which
can affect energy conservation is assumed, whether in the system
structure or in the state of the various flows, a minimization
problem is formulated by introducing parameters which describe
the changes (4). The objective function is calculated from the
dissipation of the exergy in the system as follows:

$$f_1 = \Sigma\{\text{exergy input to the system}\}-\Sigma\{\text{usable exergy output from the system}\} \qquad (1)$$

What ever is discharged from any outlet flow into the environment
is regarded as lost and is not included in usable outputs. The
optimal solution for f_1 gives indications with respect to energy
conservation.

On the other hand, efficient use of resource also should be
discussed using physical units of measures as well as usage of
energy. However the equipment materials such as basic metals are
less substitutable than energy because each metal has inherent
values. Since there is no common physical unit of measure to
evaluate the material usage, the investment cost for each piece
of equipment is used as a substitute in this paper. Consequently,
the total investment cost should be minimized from the standpoint
of resource conservation. The investment cost of each piece of
equipment is correlated with the size of the equipment based on
the logarithmic relationship.

$$f_2 = \sum_J \{c_J(z_J)^{b_J}\} \qquad (2)$$

where z_J is the variable which shows the size of equipment J;
b_J and c_J are constant.

When a process system is designed, both the total dissipation
of exergy and the total investment cost are considered as the
objective functions to be minimized. Consideration of two criteria naturally gives rise to a two-objective optimization problem.

$$\min\ (\underline{f}(\underline{y},\underline{z})) \qquad (3)$$

subject to $\quad \underline{h}(\underline{y},\underline{z}) = \underline{0}, \ \underline{g}(\underline{y},\underline{z}) \leq \underline{0}$

where
$$\underline{f} = (f_1, f_2)^t$$
$$\underline{y} = (y_1, y_2, \ldots, y_m)^t$$
$$\underline{z} = (z_1, z_2, \ldots, z_n)^t$$
$$\underline{h} = (h_1, h_2, \ldots, h_n)^t$$
$$\underline{g} = (g_1, g_2, \ldots, g_r)^t$$

The equality constraints composed of the mass and heat balances and the performance equations in each subsystem, thermodynamic properties of the flows, and specifications for design are represented by the functions $\underline{h}$ which are in the form of n equations with m+n variables. These equations are easily arranged in the order of precedence based on structural analysis. The number of independent variables (parameters), $\underline{y}$, corresponds to the degrees of freedom in the system. When the value of the parameters is given, n equations are solved with respect to n variables, $\underline{z}$. Thereupon, the **inequality** constraints, if any, are checked and the objective functions are calculated. Therefore, the problem is rewritten simply as follows:

$$\min \ (f_1(\underline{y}), f_2(\underline{y}))$$
subject to
$$\underline{g}(\underline{y}) \leq 0 \qquad\qquad (4)$$

Two-Objective Analysis

The non-inferior solution set for the two-objective problem can be solved by various methods (5,6). In this paper the ε-constraint method is used by taking into account its applicability to the non-convex problems.

$$\min \ f_2(\underline{y})$$
subject to
$$f_1(\underline{y}) \leq \varepsilon, \quad \underline{y} \in Y \qquad\qquad (5)$$

where ε is a one dimensional constant vector and Y is the feasible region defined by the inequality constraints in R^m.

$$Y \equiv [\underline{y} | \underline{g}(\underline{y}) \leq 0] \qquad\qquad (6)$$

With the assumption that the inequality constraints are given as upper and lower limits with respect to each parameter, the following is obtained.

$$Y = \{\underline{y} | y_{i \ min} \leq y_i \leq y_{i \ max} \ (i=1,2,\ldots,m)\} \qquad\qquad (7)$$

Under these conditions the Kuhn-Tucker conditions for the ε-constraint problem can be represented by using the new variables,

S_i's, which are the negative values of the ratio of the partial derivative of f_2 with respect to y_i and that of f_1 with respect to y_i.

$S_i \geq \mu$ for i where $y_i = y_{i\ max}$ & $D_i > 0$
$(y_i = y_{i\ min}$ & $D_i < 0)$

$S_i = \mu$ for i where $y_{i\ min} < y_i < y_{i\ max}$ 　　　　　　 (8)

$S_i \leq \mu$ for i where $y_i = y_{i\ min}$ & $D_i > 0$
$(y_i = y_{i\ max}$ & $D_i < 0)$

where,

$$S_i \equiv -(\partial f_2/\partial y_i)/D_i$$
$$D_i \equiv \partial f_1/\partial y_i \neq 0$$
$$(i = 1, 2, \ldots, m) \qquad (9)$$

and where μ is the multiplier for the ε-constraint, $(f_1 - \varepsilon)$, in the Lagrangian function.

The conditions are interpreted to indicate that the sensitivities are equal to μ for the variables of which the optimal values are between the lower and upper limits, but the sensitivities are smaller (greater) than μ for the variables of which the optimal values are at the lower (upper) limits in the case of the positive D_i.

Trade-Off Between Two Objectives. By increasing ε a family of ε-constraint problems is solved successively using the max-sensitive method ($\underline{7}$). Using the usual ε-constraint method the problems are solved separately for the various values of ε. However with this method the optimal solution for the new value of ε is pursued into the neighbourhood of the optimal solution for the last value by examining the sensitivities for all parameters. Incidentally, the Lagrange multiplier for the ε-constraint is obtained from among the sensitivities for the parameters.

The optimal solution line for the family of the ε-constraint problems is a one dimensional manifold in the parameter space. It defines a trade-off curve in the objective function space.

$$G(f_1, f_2) = 0 \qquad (10)$$

The trade-off ratio between f_2 and f_1, T_{21}, is defined as the direction ratio of the normal to the curve.

$$T_{21} \equiv (\partial G/\partial f_1)/(\partial G/\partial f_2) = -\partial f_2/\partial f_1 \qquad (11)$$

When the ε-constraint is active, the following relationship is satisfied based on the characteristics of the Lagrange multiplier.

$$\partial f_2/\partial f_1 = \partial f_2/\partial \varepsilon = -\mu \tag{12}$$

Therefore, the trade-off ratio is equal to μ, i.e.

$$T_{21} = \mu \tag{13}$$

As shown above, the trade-off ratio is obtained with the trade-off curve using this method.

Illustrative Example: Energy and Resource Conservation for a Milk Concentration Process

An evaporation system to concentrate a solution is composed of three subsystems as shown in Figure 1, including:
(a) an evaporator to vaporize water from the process feed,
(b) a multi-jet condenser to condense the vapor and produce a vacuum,
(c) a heat exchanger to recover exhaust heat energy from the steam condensate.
In this system, steam, cooling water, and steam condensate operate and sustain the three subsystems. The state variables of each flow, that is, temperature, pressure, and composition are shown in Table I together with the flow rate.

The system equations for each subsystem. The system equations for the evaporator are derived from the overall material balance, the material balance for the non-volatile product, the enthalpy balance, and the heat transfer rate.

$$m_f = m_\ell + m_v \tag{14}$$

$$m_f x_f = m_\ell x_\ell \tag{15}$$

$$m_s \beta_s + m_f \beta_i = m_s \beta_d + m_\ell \beta_\ell + m_v \beta_v \tag{16}$$

$$m_s(\beta_s - \beta_{\dot{d}}) = A_e U_e(x_\ell)(T_s - T_\ell) \tag{17}$$

where,

$$T_\ell = T_1 + E$$

The system equations for the multi-jet condenser are composed of the enthalpy balance and the equation which approximates the performance data in the Chemical Engineering Handbook (8).

$$m_v \beta_1 + m_c \beta_c = (m_v + m_c)\beta_x \tag{18}$$

$$\ell n(m_c/m_v) = a_0 + a_1 Z + a_2 Z^2 \tag{19}$$

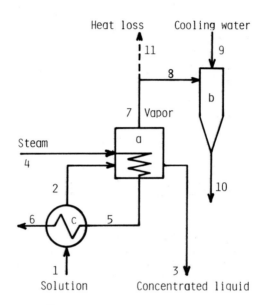

Figure 1.   Flowsheet of an evaporator system.

Table II.   All Design Conditions and Data for a Milk
Concentration Process

T_f = 298.15 K, m_f = 2.0 kg/sec, x_f = 0.1, x_ℓ = 0.4,

c_p (milk) = 1.0-0.642x kcal/kg/K,

λ = 776.77-0.64T kcal/kg,

U = 0.00136(350-250x) kcal/m2/K/sec, F=0.0 K,

P_1 = 0.0013 exp (20.39-5126/T_1) atm,

P_c = 3 atm, T_c = 298.15 K,

P_0 = 1 atm, T_0 = 298.15 K,

a_0 = 17.684, a_1 = -4.817, a_2 = 0.3829, a_3 = 0.05539,

b_1 = 0.67, b_2 = 0.6, b_3 = 0.67,

c_1 = 1015, c_2 = 84, c_3 = 1015

where, $Z = \ln(P_1) - a_3(T_c - T_0)$

The system equations for the counter-current heat exchanger are derived from the enthalpy balance and the heat transfer rate.

$$m_s\beta_d + m_f\beta_f = m_s\beta_e + m_f\beta_i \tag{20}$$

$$m_s(\beta_d - \beta_e) = A_h U_h(x_f)\Delta T \tag{21}$$

where,

$\Delta T = ((T_s - T_i) - (T_e - T_f))/\ln((T_s - T_i)/(T_e - T_f)), R \neq 1$

$\Delta T = T_s - T_f, R = 1,$

$R = m_f c_p/m_s$

Thermodynamic Properties of Vapor and Liquid. The pressure in the evaporator, P_1, is related to the saturated temperature of the vapor, T_1, based on the following equation.

$$P_1 = \exp(e_1 - e_2/T_1) \tag{22}$$

where e_1 and e_2 are constant.
The specific enthalpy and the specific exergy are calculated from the following simplified formulas.
Water, saturated steam, and superheated steam :

$$\beta = (T - T_0) + \lambda + 0.45E \tag{23}$$

$$\varepsilon = (T - T_0 - T_0\ln(T/T_0)) + \lambda(1 - T_0/T) + 0.45(E - T_0\ln((T+E)/T)) \tag{24}$$

where λ is the latent heat of the vaporation and E is the boiling point rise.
Solution:

$$\beta = \beta^0 x + T_0 \int^{T+E} c_p dT \tag{25}$$

$$\varepsilon = \varepsilon^0 x + T_0 \int^{T+E} c_p(1 - T_0/T) dT \tag{26}$$

where β^0 and ε^0 are the specific chemical enthalpy and the specific chemical exergy, respectively; x is the weight fraction of solids in the solution.
Pressure exergy of cooling water: Since cooling water used in the multi-jet condenser is compressed to P_c by a pump, its pressure exergy must be evaluated.

$$\varepsilon = (P_c - P_0)/\rho \tag{27}$$

Two Objectives. The objective function, f_1, is calculated by performing an exergy balance of the system shown in Figure 1.

$$f_1 = m_f \varepsilon_f - m_\ell \varepsilon_\ell + m_s \varepsilon_s + m_c \varepsilon_c - \delta[m_s \varepsilon_e + (m_v + m_c)\varepsilon_x] \tag{28}$$

Since drains and used cooling water are discharged into the environment, they are regarded as lost. Therefore, δ takes 0 in Equation (28).

The total investment cost for the system is sum of the costs of the evaporator, the multi-jet condenser, and the heat exchanger (9).

$$f_2 = c_1(A_e)^{b1} + c_2(m_c)^{b2} + c_3(A_h)^{b3} \tag{29}$$

where A_e and A_h are the heat transfer areas of the evaporator and the heat exchanger, respectively; m_c is the flow rate of cooling water.

Parameters. In this problem, the steam temperature, T_s, the saturated temperature of vapor, T_1, which is the normal boiling point in the evaporator, and the preheated temperature of the solution, T_i, are selected as the parameters among various other possibilities. T_1 corresponds to the extent of a vacuum using the multi-jet condenser and T_i is related to the presence or absence of the heat exchanger. The two-objective optimization problem is stated as follows:
To determine the values for T_s, T_1, and T_i, which minimize both f_1 and f_2, when the feed rate, m_f, the weight fraction of solids in the feed, x_f, and the weight fraction of solids in the concentrated liquid, x_ℓ, are specified.

Results. In this example, 2.0kg/sec of milk having 10% milk solids and at a temperature of 298.15 K is to be concentrated into a solution containing 40% milk solids. All design conditions and necessary data are summarized in Table II.

The feasible domain of the parameters is shown as a box in Figure 2. Point 1 is the solution for minimizing the annual exergy consumption. From the viewpoint of energy conservation a combination of 1) the lowest temperature of the steam, 2) an intermediate value of the saturated temperature of the vapor, i.e., intermediate vacuum, and 3) the maximum heat recovery from the steam condensate, is recommended. This solution implies that energy of lower quality should be used and graded usage of heat energy should be considered. These are the general indications from the second law analysis. On the other hand, point 6 shows the solution for minimizing the annual investment cost. The trajectory of the non-inferior solution for the two-objective problem is illustrated by the line 1-2-3-4-5-6 in Figure 2.

Table I.　Flow Conditions for an Evaporator System

stream No.	Flow Rate kg/sec	State Variables Composition wt.%	State Variables Temperature K	Pressure atm	Thermodynamic Properties Enthalpy kcal/kg	Thermodynamic Properties Exergy kcal/kg
1	m_f	x_f	T_f	P_0	β_f	ε_f
2	m_f	x_f	T_i	P_0	β_i	ε_i
3	m_ℓ	x_ℓ	T_ℓ	P_0	β_ℓ	ε_ℓ
4	m_s	–	T_s	P_s	β_s	ε_s
5	m_s	–	T_s	P_0	β_d	ε_d
6	m_s	–	T_e	P_0	β_e	ε_e
7	m_v	–	T_v	P_v	β_v	ε_v
8	m_v	–	T_1	P_1	β_1	ε_1
9	m_c	–	T_c	P_c	β_c	ε_c
10	$m_c + m_v$	–	T_x	P_0	β_x	ε_x
11	–	–	T_ℓ	–	β_w	ε_w

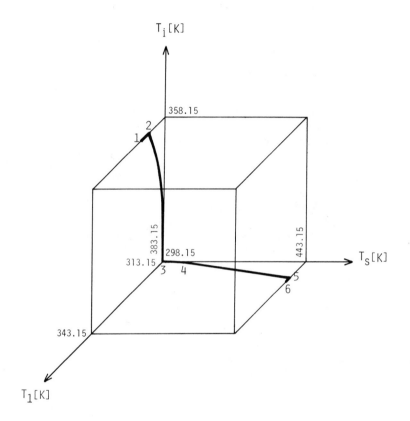

Figure 2.   Non-inferior solution set.

The solution moves along the boundary of the feasible domain.
The trade-off curve in the objective function space is non-
convex as shown in Figure 3. Figure 4 shows the profiles of the
sensitivities, S_j 's ($j=T_s,T_1,T_i$), along the non-inferior solution curve. The sensitivity profile drawn in bold strokes
shows the changes in the Lagrange multiplier. It is equal to the
trade-off ratio along the non-inferior solution curve based on
Equation (13). It is not continuous at point 3.

Overall Evaluation.   From the point of view of economic optimization the cost of energy and resource are but one of the important factors. The operating cost is estimated based on the
exergy consumption, f_1, multiplied by the unit cost of exergy
(10). The unit cost of exergy is a macroscopic index which
shows the energy situation at the time. Since the annual investment cost has been used to measure resource conservation in this
paper, the annual total cost is represented as follows,

$$\phi_1 = \alpha f_1 + f_2 \tag{30}$$

where α is the unit cost of exergy.
The trade-off curve shows the change in the optimal solution
as the unit cost of exergy is changed. Once a unit cost of exergy is specified, an optimal solution can be determined directly using the profile of the Lagrange multiplier in Figure 4. The
solution is the intersection of the curve drawn in bold strokes
and the horizontal line which corresponds to the unit cost of
exergy. There are two intersections in the case of a recent
exergy cost of 28 dollars per million kcal-exergy. This fact
arises from the non-convexity of the trade-off curve. The optimal solution is the left-hand intersection and is located near
point 2.
Another viewpoint in addition to economics can be introduced
for the overall evaluation. The cumulative exergy-consumption,
which comprises all stages of the production process from the
raw materials to the final product, may serve the purpose of
selecting the proper technological process from the viewpoint
of saving unrenewable natural resouces (11). This may be called
an ecological evaluation. With the assumption that the exergy
consumption required to make all the equipment is directly proportional to the investment cost. Then, the total cumulative
exergy consumption is represented very approximately as follows,

$$\phi_2 = f_1 + (\omega f_2)/\alpha \tag{31}$$

where ω is the ratio of the energy cost to the investment cost.
Since typically ω has a value between 0 and 1, the preferred
solution is located at a lower exergy consumption level than
that for the economical consumption level.
Generally, economical and ecological viewpoints do not coin-

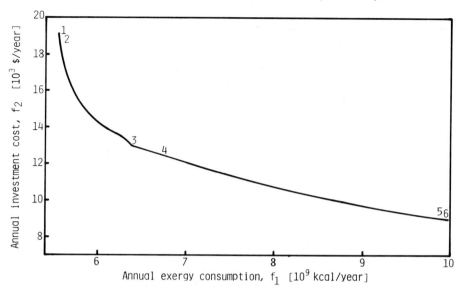

Figure 3.   Trade-off curve between the annual exergy
consumption and the annual investment cost.

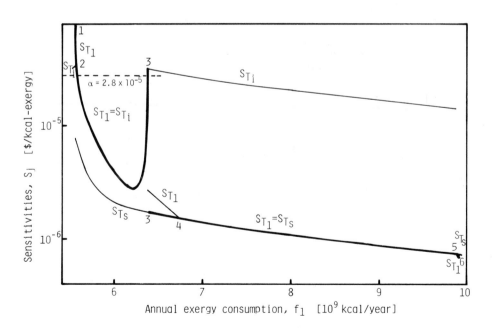

Figure 4.   Sensitivity profiles along the non-inferior
solution curve.

cide in the preferred solution. The conflict between the two
viewpoints is represented by a new two-objective optimization
problem. However this problem significantly reduces the size
of the non-inferior solution set from that of the former two-
objective problem (12). The discrepancy between economical and
ecological viewpoints is very small in this problem.

Conclusions

Energy and resource conservation in the process design was
evaluated based on the exergy consumption and the total investment cost, respectively. The conflict between the two objectives is obtained by solving the two-objective optimization
problem. The ε-constraint scalar optimization problem was solved by means of the max-sensitive method because it generates the
profile of the trade-off ratio as well as the trade-off curve.
The profile of the Lagrange multiplier along the non-inferior
solution curve is used to obtain the optimal solution when a unit
cost of exergy is specified. Using the two-objective analysis
we can obtain a much better understanding of the process design
under the present uncertain conditions with respect to energy.
A simple evaporation system for milk concentration was used
as an illustration.

Legend of Symbols

A_e	= heat transfer area of the evaporator	$[m^2]$
A_h	= heat transfer area of the heat exchanger	$[m^2]$
c_p	= specific heat capacity	$[kcal/kg/K]$
E	= boiling point rise	$[K]$
m	= mass flow rate	$[kg/s]$
P	= pressure	$[atm]$
T	= temperature	$[K]$
U	= overall heat transfer coefficient	$[kcal/m2/K/s]$
x	= weight fraction of solids in the solution	$[-]$
α	= unit cost of exergy	$[\$/kcal\text{-}exergy]$
β	= specific enthalpy	$[kcal/kg]$
ε	= specific exergy	$[kcal/kg]$
λ	= latent heat of vaporation	$[kcal/kg]$
μ	= Lagrange multiplier	$[-]$
ρ	= density of fluid	$[kg/m3]$

<Superscript>
0 = specific chemical quantity

<Subscript>
c = cooling water
f = feed
i = preheated liquid using the heat exchanger
ℓ = concentrated liquid from the evaporator

Continued on next page

s = steam
v = vapor from the evaporator
0 = dead state
1 = inlet to the multi-jet condenser

Literature Cited

1. Gaggioli, R.A.; Petit, R. J. Chemtech. 1977, 496-506.
2. Denbigh, K.G. Chem. Eng. Sci. 1956, 6, 1-9.
3. Riekert, L. Chem. Eng. Sci. 1974, 29, 1613-1620.
4. Fan, L.T.; Shieh, J.H. Energy 1980, 5, 955-966.
5. Cohon, J.L.; Marks, D.H. Water Resour. Res. 1975, 11, 208-220.
6. Shimizu, K. "System Optimization Theory (in Japanese)"; Corona-Sha: Tokyo, 1976; Chap. 4.
7. Takamatsu, T.; Hashimoto, I.; Nishitani H.; Tomita, S. Chem. Eng. Sci. 1976, 31, 705-717.
8. SCEJ, Ed. "Chemical Engineering Handbook (in Japanese)"; Maruzen: Tokyo, 3rd ed. 1968; p. 1298.
9. Saito, Y. "Cost Estimation Handbook for Chemical Engineers (in Japanese)"; Kogyo-Chosakai: Tokyo, 1977; Chap. 4.
10. Gaggioli, R.A.; Wepfer, W.J. Energy 1980, 5, 823-837.
11. Szargut J. Energy 1980,5, 709-718.
12. Nishitani, H.; Kungita, E. Computers & Chem. Eng. 1979, 3, 261-268 .

RECEIVED July 12, 1983

EXERGY PROPERTY EVALUATION

Energy and Exergy Estimation Using the Group Contribution Method

J. H. SHIEH and L. T. FAN

Department of Chemical Engineering, Kansas State University, Manhattan, KS 66506

A group contribution method is presented for the prediction of the specific chemical enthalpy and the specific chemical exergy (availability) for a gaseous or liquid material. These properties are essential in thermodynamic analysis and synthesis of a process system. Two sets of group contributions are developed, and the accuracy and range of applicability of each set are demonstrated by testing against 200 gaseous and liquid organic chemicals. These sets of the group contributions yield sufficiently accurate predictions.

A major role of a chemical engineer is to analyze and synthesize a process system. To incorporate thermodynamic bases into analysis and synthesis of a chemical or biochemical process system, the calorific thermodynamic functions of chemicals and substances involved in the process, such as enthalpy, entropy, Gibbs free energy and exergy (availability), need be known (1-3). Enormous amounts of data for these functions have been collected and/or correlated over the years, but the rapid advance of technology in discovering or synthesizing new materials seems always to create a significant gap between demand and availability of such data and correlations. In the light of this, a practical method is necessary not only for regenerating the existing data but also for predicting non-existing data.

In general, procedures for estimating physical and thermodynamic properties and functions can be divided into two categories, namely, group contribution methods and semi-empirical correlations. It is usually difficult, if not impossible, to employ a semi-empirical correlation for predicting the properties of a new material or those of an existing material at a condition different from that under which the available data were obtained. In contrast, the group contribution method, which is based on the assumption that the property of a material is contributed from

0097–6156/83/0235–0351$06.00/0

the properties of the constitutive atomic groups, provides a
convenient means for estimating the properties of both the
existing and non-existing materials.

The objective of this work is to develop a set of group
contributions for estimating the specific chemical enthalpy, β^0,
and the specific chemical exergy, ε^0, which are essential in
carrying out the thermodynamic analysis and synthesis of a
process system. This set of group contributions can be employed
not only for known organic compounds, but also for new organic
compounds which are yet to be synthesized or discovered.

Thermodynamic Framework

The conventional reference state for measuring thermodynamic
properties, which is collectively specified by all known elements
in their pure states, the temperature of 25°C (298.15 K) and the
pressure of 1 atm, has long been adopted as the basis to compute
the calorific thermodynamic functions (4,5). Adoption of a
different reference state will naturally give rise to different
values of these functions. It is possible that a compound or
substance will have a negative exergy content and that results of
the thermodynamic analysis of a process system will be erroneous
if an arbitrary reference state is selected. It is, therefore,
highly desirable to specify a reference state that will, in
theory, consistently generate non-negative exergy content for
every material species (substance or compound). With such a
reference state, in addition to the temperature and pressure of
the surroundings, each element contained in the material species
should have its own reference substance; in other words, a
reference substance is required for each known element. The
reference substance having neither energy nor exergy content at
the reference state is termed the datum level material (1,2,
6-12); this reference state is termed the dead state (1,2,6-11,
13,14).

Exergy (availability) is a property of a material system
that measures the maximum work which can be obtained when the
system is brought to the reference or dead state that is
thermally, mechanically and chemically in equilibrium with the
surroundings. As the dead state changes, so does the numerical
value of the exergy (availability). In other words, the value of
the exergy depends on the choice of the surroundings, which is
defined to be the dead state.

Chemical State

The chemical state is defined as the state where temperature is
the standard temperature, T^0, the pressure is the standard
pressure, P^0, and every material species is pure (1,2,8,14).
The room or environmental temperature and pressure are usually
adopted as the standard temperature and pressure, respectively.
For convenience, T^0 is often specified as 298.15 K and P^0 as 1
atm.

The datum level temperature, T_0, and the datum level pressure, P_0, are here considered to be identical to T^0 and P^0, respectively. However, different values can be specified when the system under examination is located in an environment whose temperature and pressure are persistently different from 298.15 K and 1 atm, respectively. An example of such a system is a heat pump located in the north or south pole or a spacecraft located on the moon surface. Regardless of the values specified for T_0 and P_0, they have to be consistent throughout an analysis. However, for any heat pump or a similar system, e.g., a heating or air-conditioning system, the temporal variation of T_0 may have to be taken into account; the use of average T_0's over extended time periods may lead to erroneous conclusions. On the other hand, for most systems, such as power plants and chemical processes, the use of an average T_0 suffices (12). The thermodynamic properties, e.g., enthalpy, entropy and exergy, of a material species at the chemical state are termed the chemical properties, e.g., chemical enthalpy, chemical entropy and chemical exergy (1,2,8,12,14).

Dead State (or Idealized Reference State)

The dead state is defined by specifying the mass (material species), chemical composition, and two thermostatic properties. In this work, T and P are chosen as these two properties. Thus, the dead state for a specific system is fixed when the intensive properties $(T_0, P_0, x_{A,0}, x_{B,0} \cdots)$ are specified where A, B, ... designate the reference material species at the dead state (10,15). Thus, the properties of each constituent of the system at the dead state are evaluated at the temperature, T_0, the total pressure, P_0, and the mole fraction, x_{i0}.

Conceptually, the dead state can be visualized from several different angles as follows: The universe contains a stable system which is composed of many stable materials existing in abundance and whose concentrations can be reasonably assumed to remain invariant (1,2,8,11,12,14,16,17). All the stable materials exist in thermodynamic equilibrium at the temperature, T_0, of 298.15 K, and under the total pressure, P_0, of 1 atm. This state is termed the "dead state" (1,2,8,16,18). The most stable materials, which are in the stable sector of the universe, i.e., the dead state, are termed "datum level materials" and have the availability (exergy) and energy (enthalpy relative to the dead state) of zero; the concentration of the datum level material is the datum level concentration. The values of availability (exergy) and energy (enthalpy relative to the dead state) of all materials that are in complete, stable equilibrium with the dead state are zero. The datum level materials and their concentrations that are used in this work to compute the specific chemical enthalpy, β^0, and the specific chemical exergy, ε^0, are listed in Table I.

Table I. β^0 and ε^0 of Material Species
Involved in the Examples

Material	Datum Level Material	Datum Level Concentration	β^0 kcal/g mole	ε^0 kcal/g mole
$H_2(g)$	H_2O (ℓ)	1	68.32	56.23
O_2 (g)	O_2 (g)	0.21	0	0.9263
N_2 (g)	N_2 (g)	0.78	0	0.147
CO_2 (g)	CO_2 (g)	0.000302	0	4.8002
C (s)	CO_2 (g)	0.000302	94.052	98.131
S (s)	$CaSO_4 \cdot 2H_2O$(s)	1	122.52	139.66
SO_2(g)	$CaSO_4 \cdot 2H_2O$(s)	1	51.58	68.85
F_2 (g)	FrF(s)	1	169.0	165.2
$C\ell_2$ (g)	$FrC\ell$ (s)	1	127.0	123.6
Br_2	FrBr(s)	1	109.0	112.57
I_2	FrI (s)	1	85.0	92.8
B	H_3BO_3 (s)	1	157.72	144.96
Ca	$Ca(NO_3)_2$ (s)	1	223.9	174.44
Si	SiO_2 (s)	1	217.7	203.78

Group Contribution Method

The group contribution method is based on the premise that many
properties and functions of a substance or a chemical can be
considered, at least roughly, as being made up of additive
contributions from individual atoms or bonds in the molecule of a
substance. Three approaches are available. The first approach
is based on the assumption that the properties of the atoms in a
molecule are linearly additive; this assumption is often called
the zero-order law (19,20). The second approach is based on the
assumption that properties of the bonds in a molecule are
additive; this assumption is often called the first-order law
(19,20). Unfortunately, accuracy of the zero-order and first-
order laws for estimating properties is relatively low, and their
applicability is somewhat limited. The third approach is based
on the assumption that properties of the atomic groups in a
molecule, each of which contains atoms and bonds, are additive.
This assumption is often called the second-order law. In
addition to the assumption of the additivity of group properties,
this method also assumes that the magnitude of the contribution
of a group remains invariant in different substances.

A group contribution method, specifically, the second-order
law, is adopted in this work for estimating the physical and
thermodynamic properties and functions of organic chemicals.
These properties and functions include the specific chemical
enthalpies of gas and liquid, β_g^0, and β_ℓ^0, respectively, and

the specific chemical exergies of gas and liquid, ε_g^0 and ε_ℓ^0,
respectively. These group contributions together with those for
the specific heat enable us to evaluate the thermal enthalpy, β_T,
and the thermal exergy, ε_T, at any temperature.

Different sets of groups have been used conventionally in
evaluating the different physical and thermodynamic properties
and functions. For example in predicting the critical properties, Lydersen (21) has generated 43 groups; Krevelen and
Chermin (22,23) have proposed a set of 92 groups in predicting the standard Gibbs free energy of formation, ΔG_f^0; Verma and
Doraiswamy (24) have suggested a set of 78 groups to calculate
the standard heat of formation, ΔH_f^0; Luria and Benson (25) have
computed the specific heat capacity of liquid material by means
of 33 groups; and Benson (20) has used a set of 431 groups to
estimate ΔH_f^0, ΔS_f^0 and c_p for vapor hydrocarbons. In contrast,
the present approach employs a single set of 127 groups to
evaluate β^0 and ε^0, which are summarized in Tables II and III.
Note that each group is coded for retrieval.

The set of 127 groups has been selected on the following
basis.
  (a) A group should be the simplest structure without losing
       the prediction accuracy.
  (b) A group should be a conventional functional group
       recognized in the field of chemistry.

Table II. Group Contributions for β^0 and ϵ^0 of
Gaseous Organic Chemicals

No.	Group	$\beta^0 \dfrac{kcal}{g\ mole}$	$\epsilon^0 \dfrac{kcal}{g\ mole}$
1	$-\overset{\shortmid}{\underset{\shortmid}{C}}-$	95.26	110.68
2	$-\overset{\shortmid}{\underset{\shortmid}{C}}H$	127.10	133.20
3	$-\overset{\shortmid}{C}H2$	157.49	156.31
4	$-CH3$	186.31	178.55
5	$CH4$	212.81	198.46
6	$=\overset{\shortmid}{C}-$	105.29	112.77
7	$=\overset{\shortmid}{C}H$	137.16	137.72
8	$=CH2$	168.33	162.10
9	$=C=$	129.97	132.54
10	$\equiv C-$	121.94	124.26
11	$\equiv CH$	154.73	150.62
12	$-\overset{\shortmid}{\underset{\shortmid}{C}}-$ (ring)	98.79	110.26
13	$-\overset{\shortmid}{\underset{\shortmid}{C}}H$ (ring)	130.21	134.15
14	$-\overset{\shortmid}{C}H2$ (ring)	160.87	158.17
15	$=\overset{\shortmid}{C}-$ (ring)	105.81	111.55

Note
Due to lack of space, only the first page of Table II is presented.
The table can be obtained in its entirety from the authors.

Table III. Group Contributions for β^0 and ε^0 of Liquid Organic Chemicals

No.	Group	$\beta^0 \dfrac{kcal}{g\ mole}$	$\varepsilon^0 \dfrac{kcal}{g\ mole}$
1	$-\overset{\shortmid}{\underset{\shortmid}{C}}-$	96.448	110.65
2	$-\overset{\shortmid}{C}H$	121.36	130.30
3	$-\overset{\shortmid}{C}H2$	155.69	155.58
4	$-CH3$	186.76	179.52
5	$CH4$	——	——
6	$=\overset{\shortmid}{C}-$	105.818	113.13
7	$=\overset{\shortmid}{C}H$	133.15	136.20
8	$=CH2$	173.11	161.37
9	$=C=$	128.89	133.75
10	$\equiv C-$	118.15	123.23
11	$\equiv CH$	154.37	151.59
12	$-\overset{\shortmid}{\underset{\shortmid}{C}}-$ (ring)	90.59	101.68
13	$-\overset{\shortmid}{C}H$ (ring)	117.30	129.77
14	$-\overset{\shortmid}{C}H2$ (ring)	157.31	156.10
15	$=\overset{\shortmid}{C}-$ (ring)	——	——

Note

Due to lack of space, only the first page of Table III is presented. The table can be obtained in its entirety from the authors.

Specific Chemical Enthalpies of Gaseous and Liquid Chemicals, β^0_g and β^0_ℓ

In performing an energy balance around any process system, the energy contents (enthalpies) associated with the material species involved in the process are needed. Values of the energy contents (enthalpies) have, conventionally, been evaluated with reference to the standard state (T = 298.15 K, P = 1 atm) and the pure elements at this state; however, the use of the standard state may yield a chemical enthalpy of a large negative value. This makes the evaluation of the thermodynamic analysis of the process system difficult or impossible. In view of this, the following definition has been introduced (8,14,26);

$$\beta^0 \equiv h^0 - h_0 \tag{1}$$

where β^0 is termed the specific chemical enthalpy, h^0 the specific enthalpy at the standard state, and h_0 the specific enthalpy at the so-called dead state. It is very difficult, if not impossible, to compute and store β^0 for a countless number of organic chemicals. A set of 127 groups is proposed in this work for estimating the specific chemical enthalpy by means of the group contribution method (see Tables II and III). The specific chemical enthalpy is thus computed as

$$\beta^0 = \sum_i a_i \beta^0_i \tag{2}$$

where β^0 stands for the specific chemical enthalpy, a_i for the number of group i in the molecule, and β^0_i for the specific chemical enthalpy of group i. Values of β^0_i obtained in the present work are listed in Tables I and II.

Specific Chemical Exergies of Gaseous and Liquid Chemicals, ε^0_g and ε^0_ℓ

Exergy is a property of a material system, which measures the maximum work obtainable when the system proceeds from its given state to the dead state while interacting only with the environment (1,8,14). The specific chemical exergy is defined as

$$\varepsilon^0 \equiv (h^0 - h_0) - T_0(s^0 - s_0)$$

$$= \beta^0 - T_0\gamma^0 \tag{3}$$

In the present scheme, the specific chemical exergy, ε^0, is computed as

$$\varepsilon^0 = \sum_i a_i \varepsilon_i^0 \qquad (4)$$

where a_i stands for the number of group i in the molecule, and ε_i^0 for the specific chemical exergy of group i. The values of ε_i^0 obtained here are tabulated in Tables II and III.

Specific Chemical Entropies of Gaseous and Liquid Chemicals, γ_g^0 and γ_ℓ^0

The specific chemical entropy, γ^0, is defined as (8,14)

$$\gamma^0 \equiv s^0 - s_0 \qquad (5)$$

where s^0 is the specific entropy at the standard state, and s_0 the specific entropy at the dead state. Note that once β^0 and ε^0 are given, γ^0 is fixed.

Examples

Two numerical examples are given in this section to illustrate the use of the present method. The values predicted by the group contribution method are compared with those obtained from the available experimental data, in the two monographs, one by Karapet'yants and Karapet'yants, (27); and the other by Barin and Knacke (28).

Example 1. What are the specific chemical enthalpy, β^0, and the specific chemical exergy, ε^0, of gaseous aniline, $C_6H_5NH_2$?
    The values of β^0 and ε^0 of the gaseous aniline can be predicted through its chemical structural information; aniline is composed of 3 groups, i.e., HC⤡, −C⤢, and −NH_2 (aromatic), which are listed as No. 21, No. 20, and No. 60 in Table II, respectively. According to Equations 2 and 4, its β^0 and ε^0 can be computed, respectively, as

$$\beta_{aniline}^0 = 5\beta_{HC}^0 + \beta_{-C}^0 + \beta_{-NH_2}^0 \text{(aromatic)} \qquad (6)$$

and

$$\varepsilon_{aniline}^0 = 5\varepsilon_{HC}^0 + \varepsilon_{-C}^0 + \varepsilon_{-NH_2}^0 \text{(aromatic)} \qquad (7)$$

where 5 stands for the number of group HC⤡. Substitution of values of β_{HC}^0, β_{-C}^0, $\beta_{-NH_2}^0$ (aromatic), ε_{HC}^0, ε_{-C}^0 and $\varepsilon_{-NH_2}^0$ (aromatic), tabulated in Table II, into Equations 6 and 7 gives rise to

$$\beta^0_{aniline} = 5(131.52) + 99.30 + 67.36$$

$$= 824.26 \text{ kcal/g mole}$$

and

$$\epsilon^0_{aniline} = 5(131.41) + 105.24 + 57.19$$

$$= 819.48 \text{ kcal/g mole,}$$

respectively.

The values of β^0 and ϵ^0 of aniline have also been calculated, respectively, as

$$\beta^0_{aniline} = 6\beta^0_C + \frac{7}{2}\beta^0_{H_2} + \frac{1}{2}\beta^0_{N_2} + \Delta H^0_r \tag{8}$$

and

$$\epsilon^0_{aniline} = 6\epsilon^0_C + \frac{7}{2}\epsilon^0_{H_2} + \frac{1}{2}\epsilon^0_{N_2} + \Delta G^0_r \tag{9}$$

By substituting the values of β^0_C, $\beta^0_{H_2}$, $\beta^0_{N_2}$, ϵ^0_C, $\epsilon^0_{H_2}$, and $\epsilon^0_{N_2}$ listed in Table I, and the available experimental values of ΔH^0_r and ΔG^0_r, i.e.,

$$\Delta H^0_r = 20.798 \text{ kcal/g mole}$$

$$\Delta G^0_r = 33.895 \text{ kcal/g mole}$$

into Equations 8 and 9, respectively, we have

$$\beta^0_{aniline} = 824.23 \text{ kcal/g mole}$$

and

$$\epsilon^0_{aniline} = 819.56 \text{ kcal/g mole}$$

The deviations of the predicted β^0 and ϵ^0 by means of the group contribution method from those estimated from the available experimental data are 0.03 kcal/g mole, and -0.08 kcal/g mole, respectively.

Example 2. What are the specific chemical enthalpy, β^0, and the specific chemical exergy, ϵ^0, of liquid methyl iso-butyl sulfide, $CH_3SC_4H_9$?

These quantities can be estimated from the chemical structural information of methyl isobutyl sulfide; it contains three $-CH_3$ groups, one $-CH_2$ group, one $-CH$ group, and one $-S-$ group. These four groups are listed as No. 4, No. 3, No. 2, and No. 70, respectively, in Table III. Accordingly, Equations 2 and 4 can be reduced to

$$\beta^0_{\text{iso-butyl sulfide}} = 3\beta^0_{-CH_3} + \beta^0_{-CH_2} + \beta^0_{-CH} + \beta^0_{-S-} \quad (10)$$

and

$$\varepsilon^0_{\text{iso-butyl sulfide}} = 3\varepsilon^0_{-CH_3} + \varepsilon^0_{-CH_2} + \varepsilon^0_{-CH} + \varepsilon^0_{-S-}, \quad (11)$$

respectively. Substituting the values of $\beta^0_{-CH_3}$, $\beta^0_{-CH_2}$, β^0_{-CH}, β^0_{-S-}, $\varepsilon^0_{-CH_3}$, $\varepsilon^0_{-CH_2}$, ε^0_{-CH}, and ε^0_{-S-} listed in Table III into Equations 10 and 11 gives rise to

$$\beta^0_{\text{iso-butyl sulfide}} = 3(186.76) + 155.69 + 121.36 + 126.42$$

$$= 963.75 \text{ kcal/g mole}$$

and

$$\varepsilon^0_{\text{iso-butyl sulfide}} = 3(179.52) + 155.58 + 130.30 + 147.48$$

$$= 971.92 \text{ kcal/g mole,}$$

respectively.

The values of β^0 and ε^0 for $CH_3SC_4H_9$ can also be calculated, respectively, as

$$\beta^0_{\text{iso-butyl sulfide}} = 5\beta^0_C + 6\beta^0_{H_2} + \beta^0_S + \Delta H^0_r \quad (12)$$

and

$$\varepsilon^0_{\text{iso-butyl sulfide}} = 5\varepsilon^0_C + 6\varepsilon^0_{H_2} + \varepsilon^0_S + \Delta G^0_r \quad (13)$$

Substituting the values of β^0_C, $\beta^0_{H_2}$, β^0_S, ε^0_C, $\varepsilon^0_{H_2}$ and ε^0_S listed in Table I and the available experimental values of ΔH^0_r and ΔG^0_r, i.e.,

$$\Delta H^0_r = -37.290 \text{ kcal/g mole}$$

$$\Delta G^0_r = 2.965 \text{ kcal/g mole}$$

into Equations 12 and 13, respectively, we obtain

$$\beta^0_{\text{iso-butyl sulfide}} = 965.21 \text{ kcal/g mole}$$

and

$$\varepsilon^0_{\text{iso-butyl sulfide}} = 970.66 \text{ kcal/g mole,}$$

Note that the deviations of the predicted β^0 and ε^0 from those estimated from available experimental data are -1.46 kcal/g mole and 1.26 kcal/g mole, respectively.

Results and Discussion

A set of groups numbering 127, listed in Tables II and III, has been employed to estimate the specific chemical enthalpies, β^0, and the specific chemical exergies, ε^0, of 200 gaseous organic chemicals and 200 liquid organic chemicals. The results for the gaseous organic chemicals are partially reproduced in Table IV and those for the liquid organic chemicals are partially reproduced in Table V; these results are compared with available experimental data.

The results from testing against 200 gaseous organic chemicals indicate that the use of Table I to predict ε^0 yields an average deviation of 0.65% from the experimental data, corresponding to an average deviation of 4.45 kcal/g mole and a maximum deviation of 4.80%. Similarly, the use of this table to predict β^0 gives rise to an average deviation of 2.18% from the experimental data, corresponding to an average deviation of 10.95 kcal/g mole, and a maximum deviation of 6.84%. For ε^0 of the liquid organic chemicals, the results from testing against 200 of them show an average deviation of 0.42% from the experimental data, corresponding to an average deviation of 4.02 kcal/g mole, and a maximum deviation of 4.60%. For β^0 of the liquid organic chemicals, the results show an average deviation of 0.70% from the experimental data, corresponding to an average deviation of 4.79 kcal/g mole, and a maximum deviation of 5.30%.

It is worth noting that the specific chemical enthalpy, β^0, of a combustible substance, which is composed of C, H, N, and O, is essentially equal to the higher heating value, H.H.V. One reason is that both β^0 and H.H.V. are computed based not only on the same set of reference materials but also on the same temperature and pressure. The conventionally employed reference materials are $CO_2(g)$ for C, $O_2(g)$ for O, $N_2(g)$ for N and $H_2O(\ell)$ for H. Another reason is that the pressure effect on β^0 is negligibly small around the conditions of the temperature at 298.15 K and the pressure at 1 atm. Notice that the complete combustion process of a compound containing C, H, N an O with excess oxygen yields $CO_2(g)$, $O_2(g)$, $N_2(g)$ and $H_2O(\ell)$. The difference between β^0 and H.H.V. can be substantial if a compound contains elements other than C, H, N and O.

In evaluating the thermodynamic properties, such as enthalpy, entropy, Gibbs free energy, specific chemical enthalpy and specific chemical exergy, it is essential that the reference state be specified; different reference states will inevitably generate different thermodynamic properties for material species. The definition of the reference state involves the specifications of two thermodynamic state functions, conventionally temperature and pressure, and material species and their compositions. The reference materials for determining the enthalpy, entropy and Gibbs free energy are the elements in their pure states; however, the reference materials for evaluating the higher heating value,

Table IV. β^0 and ϵ^0 of Gaseous Organic Chemicals

	$\beta^0 \frac{kcal}{g\ mole}$			$\epsilon^0 \frac{kcal}{g\ mole}$		
	Exp.	Pred.	Dev.*	Exp.	Pred.	Dev.*
CH_3Cl methyl chloride	239.37	241.65	2.28	229.33	231.85	2.52
CH_3Br methyl bromide	242.63	239.74	-2.89	232.56	231.37	-1.19
CH_3I methyl iodide	242.13	239.96	-2.17	232.41	232.85	0.45
CH_3SH methyl mercaptan	347.87	346.76	-1.12	348.02	345.54	-2.48
CH_3NO_2 nitromethane	178.67	176.20	-2.47	181.83	178.86	-2.97
C_2H_2 acetylene	310.61	309.46	-1.15	302.48	301.24	-1.24
C_2H_4 ethylene	337.24	336.66	-0.58	325.01	324.20	-0.81
C_2H_6 ethane	372.83	372.62	-0.21	375.09	357.10	0.01
C_2H_2O ketene	241.82	239.30	-2.52	238.76	236.05	-2.71
$HCOOCH_3$ methyl formate	243.74	241.05	-2.69	237.65	238.28	0.63
C_2H_5OH ethanol	336.89	335.52	-1.37	325.14	324.50	-0.64
$CH_2(OH)CH_2(OH)$ ethylene glycol	300.26	298.42	-1.84	294.36	291.90	-2.46
C_3H_6 cyclopropane	499.86	497.50	-2.36	488.02	486.23	-1.79
C_2H_5COOH propionic acid	378.37	378.40	0.03	375.74	374.98	-0.76

Continued on next page

Table IV. (Continued)

	$\beta^0 \dfrac{kcal}{g\ mole}$			$\varepsilon^0 \dfrac{kcal}{g\ mole}$		
	Exp.	Pred.	Dev.[*]	Exp.	Pred.	Dev.[*]
$C_3H_7NO_2$ 1-nitropropane	491.48	491.18	-0.30	492.28	491.48	-0.80
C_2H_6SO dimethylsulfoxide	479.62	479.63	0.01	485.60	485.60	0.0
$C_2H_6SO_2$ dimethylsulfone	426.88	426.89	0.01	440.34	440.35	0.01
C_3H_9N propyl amine	573.22	573.58	0.36	561.88	560.32	-1.56
C_3H_9B trimethyl boron	717.62	717.63	0.01	683.79	683.80	0.01
C_4H_6 butadiene-1,3	607.50	610.98	3.48	597.22	599.64	2.42
C_4H_4S thiophene	662.86	661.92	-0.94	679.42	675.63	-3.79
C_4H_4O furan	504.56	504.58	0.02	505.66	505.65	-0.01
C_4H_8S thiocyclopentane	763.93	763.17	-0.76	768.10	767.40	-0.70
C_4H_8O butanone	592.52	592.52	0.0	582.99	583.71	0.72
$CH_3COOC_2H_5$ ethyl acetate	546.05	545.81	-0.24	539.51	539.35	-0.16
$C_4H_8O_2$ p-dioxane	574.19	567.51	-6.68	575.16	571.60	-3.56
C_4H_9N pynolidine	682.83	681.27	-1.56	673.08	671.94	-1.14
C_3H_9OH n-butanol	650.89	650.50	-0.39	636.02	637.12	1.10
$C_2H_5OC_2H_5$ diethyl ether	658.11	660.93	2.82	646.05	648.41	2.36

Table IV. (Continued)

	$\beta^0 \frac{kcal}{g\ mole}$			$\varepsilon^0 \frac{kcal}{g\ mole}$		
	Exp.	Pred.	Dev.*	Exp.	Pred.	Dev.*
C_5H_8 pentyne-1	774.34	773.99	-0.35	761.98	761.93	-0.05
C_5H_8 cyclopentene	751.41	748.36	-3.05	742.06	739.47	-2.59
C_5H_5N pyridine	674.56	674.11	-0.45	676.77	676.56	-0.21

* % Deviation = $\dfrac{\text{Predicted} - \text{Experimental}}{\text{Experimental}}$ x 100

Table V. β^0 and ϵ^0 of Liquid Organic Chemicals

	$\beta^0 \dfrac{kcal}{g\ mole}$			$\epsilon^0 \dfrac{kcal}{g\ mole}$		
	Exp.	Pred.	Dev.*	Exp.	Pred.	Dev.*
CCl_4 carbon tetrachloride	314.75	315.76	1.0	329.35	332.37	3.02
$CHBr_3$ bromoform	284.91	286.42	1.51	293.93	293.89	-0.04
$CHCl_3$ chloroform	286.57	285.09	-1.48	294.35	294.34	-0.01
C_6H_5I iodobenzene	805.01	805.47	0.46	821.46	819.00	-1.86
C_6H_5Br bromobenzene	904.11	804.54	0.43	815.77	812.67	-3.10
C_8H_{10} ethyl benzene	1091.02	1091.97	0.95	1095.01	1093.24	-1.77
C_9H_{20} n-nonane	1463.81	1463.35	-0.46	1449.59	1448.10	-1.49
$C_{10}H_{22}$ n-decane	1620.09	1619.04	-1.05	1604.03	1603.68	-0.35
$C_{15}H_{32}$ n-pentadeane	2401.41	2397.49	-3.92	2382.28	2381.58	0.70
C_8H_{18} n-octane	1307.56	1307.66	0.10	1292.70	1292.52	-0.18
C_5H_{10} cyclopentane	786.56	786.55	0.01	780.51	780.50	-0.01
C_7H_{14} cycloheptane	1098.88	1101.77	2.29	1093.49	1092.70	-0.79
C_8H_{16} n-propyl-cyclopentane	1253.77	1254.68	0.91	1245.01	1244.85	0.16
C_9H_{18} n-propyl-cyclohexane	1404.37	1401.99	-2.83	1397.47	1400.95	3.48
C_8H_8 styrene	1050.51	1055.78	4.27	1058.53	1055.71	-2.82

Table V. (Continued)

	$\beta^0 \frac{kcal}{g\ mole}$			$\epsilon^0 \frac{kcal}{g\ mole}$		
	Exp.	Pred.	Dev.*	Exp.	Pred.	Dev.*
C_8H_{10} o-xylene	1088.18	1095.38	7.2	1092.57	1090.82	-1.75
$C_{12}H_{14}$ n-butylbenzene	1403.48	1403.35	-0.13	1405.91	1404.40	1.51
HCN hydrogen cyanide	154.23	154.37	0.14	156.18	156.59	0.41
CH_3CN acetonitrile	303.38	308.78	5.40	304.38	304.38	0.00
C_2H_5OH ethanol	326.70	327.44	0.74	323.64	322.84	-0.80
CH_3COCH_3 acetone	427.78	428.87	1.09	426.36	426.35	-0.01
$C_5H_{10}O$ cyclopentanone	740.16	740.25	0.09	737.02	737.51	0.49
CH_3OCH_3 dimethyl ether	343.04	342.17	-0.86	337.97	338.35	0.38
CH_3COOH acetic acid	209.04	209.04	0.00	216.55	216.55	0.0
C_3H_7COOH butyric acid	522.29	520.42	-1.87	528.47	527.71	-0.76
C_2H_4S cyclothiopropane	459.31	459.31	0.0	470.54	470.42	-0.1
$C_5H_{12}S$ methyl isobutyl sulfide	965.21	969.91	4.70	970.66	976.21	-5.5
CH_3SH methanathiol	342.13	344.00	1.87	348.40	348.40	0.0
C_4H_4O furan	497.95	497.94	-0.01	505.50	505.51	0.01
$CH_3COOC_2H_5$ ethyl acetate	535.0	537.79	2.79	538.85	538.85	0.0

Continued on next page

Table V. (Continued)

	$\beta^0 \dfrac{kcal}{g\ mole}$			$\epsilon^0 \dfrac{kcal}{g\ mole}$		
	Exp.	Pred.	Dev.*	Exp.	Pred.	Dev.*
C_7H_8 cyclokeptatuene	965.86	967.49	1.63	969.93	970.90	0.97
C_5H_5N pyridine	664.95	664.94	-0.01	673.63	673.65	0.02

* % Deviation $= \dfrac{\text{Predicted} - \text{Experimental}}{\text{Experimental}} \times 100$

H.H.V., are the products of complete combustion in their pure form at the standard state. In contrast, the so-called datum level materials are employed in computing the specific chemical enthalpy, β^0, and the specific chemical exergy, ε^0 ($\underline{1},\underline{2},\underline{8},\underline{12},\underline{14}$). As mentioned previously, the datum level materials are the most stable materials in the environment, which are specified to have zero exergy (availability) and enthalpy relative to the dead state.

Conclusion

Two sets of group contributions, one for predicting the specific chemical enthalpy, β^0, and the specific chemical exergy, ε^0, of gaseous organic chemicals, and the other for liquid organic chemicals, have been developed. These sets of the group contributions yield sufficiently accurate predictions. The estimated β^0 and ε^0 are useful in determining the thermodynamic first-law and second-law analyses. The present set icludes 127 groups; however, this number can be augmented once a new useful group is identified. The present approach can be easily implemented on the computer.

Notation

a_i = number of group i

c_p = specific heat, kcal/g mole, K

h_0 = specific enthalpy at the dead state, kcal/g mole

h^0 = specific enthalpy at the standard state, kcal/g mole

P_0 = pressure at the dead state, atm

P^0 = pressure at the standard state, atm

s_0 = specific entropy at the dead state, kcal/g mole, K

s^0 = specific entropy at the standard state, kcal/g mole, K

T_0 = temperature at the dead state, K

T^0 = temperature at the standard state, K

$x_{k,0}$ = concentration, in molar fraction, of material species k at the dead state, the so-called datum level concentration

ΔG_r^0 = Gibbs free energy change of reaction at the standard state, kcal/g mole

ΔH_f^0 = heat of formation at the standard state, kcal/g mole

Continued on next page

ΔH_r^0 = heat of reaction at the standard state, kcal/g mole

ΔS_f^0 = entropy of formation at the standard state, kcal/g mole, K

Greek Letters

β^0 = specific chemical enthalpy, kcal/g mole

γ^0 = specific chemical entropy, kcal/g mole, K

ϵ^0 = specific chemical exergy, kcal/g mole

Subscripts

k = material species

Literature Cited

1. Reikert, L., "The Efficiency of Energy-Utilization in Chemical Processes"; Chem. Engr. Sci. 1974, 29, 1613.
2. Gaggioli, R.A.; Petit, P.J., "Use the Second Law First"; Chemtech 1977, August, 496.
3. Fredrickson, A.G.; Stephanopoulos, G., "Microbial Competition"; Science 1981, 213, 28.
4. Hougen, O.A.; Watson, K.M., "Chemical Process Principles Part II, 2: Thermodynamics"; Wiley: New York, N.Y., 1957.
5. Himmelblau, D.M., "Basic Principles and Calculations in Chemical Engineering"; Prentice-Hall, Englewood Cliffs: N.J., 1974.
6. Bruges, E.A., "Available Energy and the Second Law Analysis"; Academic Press: London, 1959.
7. Baehr, von H.D.; Schmidt, E.F., "Definition und Berechnung von Brennstaffexergien"; BWK 1963, 15, 375.
8. Szargut, J.; Petela, R., "Egzergia"; Wydawnictwa Naukowo-Techniczne: Warezawa, Poland, 1965 (in Polish).
9. Evans, R.B., "A Proof that Exergy is the Only Consistent Measure of Potential Work"; Ph.D. Dissertation, Dartmouth College, Hanover, N.H., 1969.
10. Reistad, G. "Availability: Concepts and Application"; Ph.D. Dissertation, University of Wisconsin, Madison, Wisconsin, 1970.
11. Ahrendts, J., "Die Exergie chemishch reaktionfähiger System, VDI- Forschungsheft 579"; Verein Deutscher Ingenieuke, 1977.
12. Wepfer, W.J.; Gaggioli, R.A., in "Reference Datums for Available Energy"; Gaggioli, R.A., Ed.: ACS SYMPOSIUM SERIES NO. 122, American Chemical Society: Washington, D.C., 1980, p. 77.

13. Evans, R.B.; Tribus, M., "Thermo-Economics of Saline Water Conversion"; I&EC Process Design and Development 1965, 4, 195.

14. Fan, L.T.; Shieh, J.H., "Thermodynamically Based Analysis and Synthesis of Chemical Process Systems"; Energy 1980, 5, 955.

15. Gaggioli, R.A., "The Concept of Available Energy"; Chem. Engr. Sci. 1961, 16, 87.

16. Obert, E.F., "Thermodynamics"; McGraw Hill: New York, N.Y., 1948; pp. 389-395.

17. Rant, Z., "Bewertung und Praktische Verrechung Von Energien"; Allg. Wärmetechnik Bd. 1957, 8 (2), S 25/32.

18. Riekert, L., in "Flow and Loss of Available Energy in Chemical Processing System"; Koeteier, W.T., Ed.; Chemical Engineering in A Changing World, Elsevier Scientific: New York, N.Y., 1976.

19. Benson, S.W.; Buss, J.H., "Additivity Rules for the Estimation of Molecular Properties, Thermodynamic Properties"; J. Chem. Phys.958, 29, 546.

20. Benson, S.W., "Thermochemical Kinetics"; Wiley: New York, 1976; chapters 2 and 3.

21. Lydersen, A.L., "Estimation of Critical Properties of Organic Compounds"; Univ. Wisconsin Coll. Eng., Eng. Exp. Sta. Rep. 3: Madison, Wisconsin, April (1955).

22. Krevelen, D.W.; Chermin, H.A.G., "Estimation of the Free Enthalpy (Gibbs Free Energy) of Formation of Organic Compounds from Group Contributions"; Chem. Eng. Sci. 1951, 1, 66.

23. Krevelen, D.W.; Chermin, H.A.G., "Erratum: Estimation of the Free Enthalpy (Gibbs Free Energy) of Formation of Organic Compounds from Group Contributions"; Chem. Eng. Sci. 1952, 1, 238.

24. Verma, K.K.; Doraiswamy, L.K., "Estimation of Heats of Formation of Organic Compounds"; Ind. Eng. Chem. Fundam. 1965, 4, 389.

25. Luria, M.; Benson, S.W., "Heat Capacities of Liquid Hydrocarbons, Estimation of Heat Capacities at Constant Pressure as a Temperature Function Using Additivity Rules"; J. Chem. Eng. Data 1977, 22, 90.

26. Rodriguez, L., in "Calculation of Available Energy Quantities"; Gaggioli, R.A., Ed.: ACS SYMPOSIUM SERIES NO. 122, American Chemical Society: Washington, D.C.; 1980, p. 39.

27. Karapet'yants, M. Kh; Karapet'yants, M.L., "Thermodynamic Constants of Inorganic and Organic Compounds"; Ann Arbor-Humphrey Science Publishers: Ann Arbor, Michigan, 1970.

28. Barin, I.; Knacke, O., "Thermochemical Properties of Inorganic Substances," Springer-Verlag: New York, N.Y., 1973.

RECEIVED July 12, 1983

Thermodynamic Properties
of Coal and Coal-Derived Liquids
Evaluation and Application to the Exergy Analysis
of a Coal Liquefaction Process

MASARU ISHIDA and TAKAHIRO SUZUKI
Research Laboratory of Resources Utilization, Tokyo Institute of Technology,
4259 Nagatsuta-cho, Midori-ku, Yokohama, 227, Japan

NAONORI NISHIDA
Department of Management Science, Science University of Tokyo, 1-3 Kagurazaka,
Shinjuku-ku, Tokyo, 162, Japan

Methods to estimate heats of formation ΔH_f° and
absolute entropies S° for coal and coal-derived
liquids are proposed based on the group contribution method. Semiempirical formulas to estimate them are given on the basis of unit mole of
carbon in coal or coal-derived liquids. Necessary information on these formulas includes elemental composition data and normal boiling points
(only for coal-derived liquids). Using these
formulas and the Structured Process Energy-
Exergy-flow Diagram (SPEED), an exergy analysis
for the H-Coal process system for producing
synthetic fuels is performed.

The exergy (or availability) analysis has been applied to coal
conversion processes by various investigators. Most of them
have treated coal gasification processes (1,2), and, therefore,
the application of exergy analysis of liquefaction process has
not been made extensively. The COED process is the only direct
liquefaction process to which an exergy analysis has been performed (3-5). It has not been applied to other processes such
as the Exxon donor solvent, H-Coal, and SRC-II processes. One
reason why these processes have not been treated may be that
the necessary information such as pressures, temperatures, flow
rates, and compositions as to complete both the first law and
the second law analysis has been considered proprietary by their
developers. Another reason may be due to the lack of thermodynamic properties necessary to calculate both the enthalpy and
entropy of coal-derived liquids. Accordingly, Unruh et al. (5)
performed an exergy analysis on the COED process by assuming
that the exergy values for coal and coal-derived liquids are
equal to their standard heat of combustion.

Kidnay and his colleagues (6,7) have conducted experimental
enthalpy measurements on coal liquids, which were derived from

0097-6156/83/0235-0373$06.25/0
© 1983 American Chemical Society

various coal liquefaction processes. Experiments have been under-
taken for whole coal liquids, distillate samples from coal liquids,
and model compounds representative of coal-derived liquids. They
also have compared the observed enthalpies with those predicted
by correlations developed for petroleum fractions. They have
found that the thermodynamic properties of coal-derived liquids
differ considerably from those of petroleum liquids.

Methods to estimate the thermodynamic properties of coal
have been developed by previous investigators. Dulong (8) has
proposed a formula for the heat of combustion of coal from its
elemental composition data. Based on it, the enthalpy of formation of coal $\Delta H_f°$ may be determined. For the absolute entropy
of coal S°, Cheng et al. (9) have proposed a formula in terms of
elemental composition of coal. It has been derived by extrapo-
lating the known values of S° for low molecular weight hydrocarbons which are solid at room temperature. S° of coal may also
be determined backward from the Szargut-Stryska formula (10)
which estimates the exergy of coal from its elemental composition.

In view of the above survey, neither correlation nor estimation method has been developed to determine the enthalpy of
formation and the absolute entropy of coal-derived liquids.

In this paper, methods to estimate the heat of formation
$\Delta H_f°$ and the absolute entropy S° for coal and coal-derived liquids
are proposed based on the group contribution method. By applying
these methods and the Structured Process Energy-Exergy-flow
Diagram (SPEED, 11), an exergy analysis for the H-Coal process
is performed.

Thermodynamic Properties of Coal and Coal-Derived Liquids

Enthalpy of Formation and Absolute Entropy of Coal. We have applied the group-contribution method proposed by Benson and his
colleagues (12,13) to estimate the thermodynamic properties of
coal (14). In this method, the distribution of major atomic
groups which constitute coals was estimated by reviewing studies
on structural analyses of coals and by using the group contribution tables which Benson et al. developed. Then the heat of
formation and absolute entropy of coal were estimated by summing
up the contributions of major atomic groups. In the following,
we shall briefly outline that work.

For simplicity, coal is assumed to be constituted from
carbon, hydrogen and oxygen. This assumption may be valid since
nitrogen and sulfur contents are relatively small and their contribution to $\Delta H_f°$ and S° may be neglected.

Carbon atomic groups which constitute coal are classified
into eight groups indicated in Table I based on the structural
analyses of coals. The number of each group present in coal
denoted by CH_xO_y (MAF basis) is obtained as follows.

There are two types of carbon atom; aromatic (C_{ar}) and aliphatic (C_{al}). Figure 1 shows that good correlation can be ob-

tained for coals as well as for coal-derived liquids by plotting
the molar ratio of C_{ar} to total carbon C_{total} against x, i.e.,
the molar ratio of total hydrogen H_{total} to C_{total}. From this
correlation, carbon atoms in coal can be decomposed into aromatic
and aliphatic carbon atoms with the parameter H_{total}/C_{total}.

For coals whose carbon contents are less than 90 weight %,
it is reported (15,16) that the condensed aromatic rings are
mostly of the cata-condensed type with two through five rings.
This implies that about 30% of total aromatic carbon (C_{ar}) is
aromatic condensed carbon (C_{BF}) and hence we assume $C_{BF}/C_{ar} = 0.3$.

For aliphatic carbon, there are two types; one linked
directly to aromatic rings and the other linked to other aliphatic
carbon atoms. For example, CH_3 group may be classified into
C-(H)$_3$(C$_B$) and C-(H)$_3$(C); CH_2 group into C-(H)$_2$(C)$_2$, C-(H)$_2$(C)(C$_B$),
and C-(H)$_2$(C$_B$)$_2$; CH group into C-(H)(C)$_3$ and C-(H)(C)$_2$(C$_B$).
However, since the differences in the contributions of these
groups to $\Delta H_f°$ and $S°$ are insignificantly small, the average
values may be used, as summarized in Table I.

Major oxygen constituents present in coal are phenolic OH
or ethereal oxygen. Since no practical method to estimate their
precise distribution is available at the present stage, 60% of
total oxygen atoms are assumed to be phenolic and the rest ethereal. It is also assumed that all ethereal oxygen atoms have the
form of -O- linking two aromatic rings. With these assumptions,
the distribution of the C_B-(O) and C_B-(OH) groups can be determined.

Hydrogen atoms can be classified into aliphatic (H_{al}), aromatic (H_{ar}), and phenolic ones (H_{OH}). H_{al} may further be classified into H_{CH}, H_{CH2}, and H_{CH3}. Ladner et al. (17) obtained the
atomic ratio, ($H_{CH} + H_{CH3})/H_{CH2}$, for various coals by NMR analyses.
By denoting this ratio as α and assuming the atomic ratio, H_{al}/C_{al}
(= $[H_{CH} + H_{CH2} + H_{CH3}]/[H_{CH} + 1/2 \cdot H_{CH2} + 1/3 \cdot H_{CH3}])$, is equal to 2,
H_{CH3}/H_{al}, H_{CH2}/H_{al}, and H_{CH}/H_{al} are expressed as $0.75\alpha(\alpha+1)^{-1}$,
$(\alpha+1)^{-1}$, and $0.25\alpha(\alpha+1)^{-1}$, respectively. By referring to their
values of α, Figure 2 is obtained.

Since the number of H_{OH} is equal to the number of the C_B-(OH)
group, the number of H_{ar}, say the number of the group C_B-(H), can
be calculated by the relation $H_{total} = H_{al} + H_{ar} + H_{OH}$. Finally
the number of the C_B-(C) group can be determined by subtracting
the number of the groups C_B-(H), C_B-(O), C_B-(OH), and C_{BF}-(C")$_2$
from the total number of C_{ar}.

Using correlations shown in Figures 1 and 2, the distribution
of each atomic group per unit mole of carbon in coal was determined in terms of two parameters, x (= H_{total}/C_{total}) and y
(= O_{total}/C_{total}), and then $\Delta H_f°$ and $S°$ per unit mole of carbon in
coal CH_xO_y (MAF basis) were estimated by summing up the contribution of each atomic group. Based on the above structural analysis,
semiempirical formulas for estimating the heat of formation and
the absolute entropy per unit mole of carbon in coal were obtained
as follows:

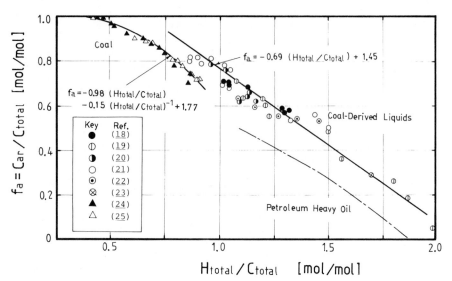

Figure 1. Relation between atomic ratio H_{total}/C_{total} and f_a of coal and coal liquids.

Table I. Group values of ΔH_f° and S° for coal

Group	ΔH_f° [kJ/mol]	S° [J/mol·K]
C --- (H)₃(C')	−42.68	127.2
C --- (H)₂(C')₂	−19.64	40.4
C --- (H)(C')₃	−6.02	−50.7
C_B --- (H)	13.81	48.2
C_B --- (C)	23.05	−32.2
C_B --- (O)	−47.90	−24.5
C_B --- (OH)	−162.30	79.1
C_{BF}--- (C'')₃	17.57	−20.9

C_B : C atm in a benzene ring

C_{BF}: C atm located at the border of fused rings

C' : C or C_B

C'' : C_B or C_{BF}

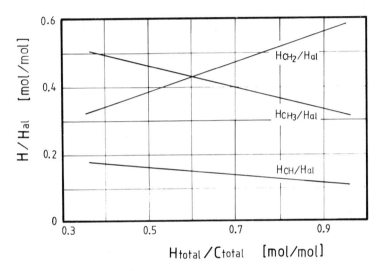

Figure 2. Relation between atomic ratio H_{total}/C_{total} and aliphatic hydrogen distribution of coal.

$$\Delta H_f^\circ \,[\text{kJ/mol-C}] = 1.312x^2 - 35.27x - 3.83/x - 162.3y + 40.97 \qquad (1)$$

and

$$S^\circ \,[\text{J/mol-C} \cdot \text{K}] = 4.414x^2 - 18.64x - 14.63/x + 24.7y + 47.53 \qquad (2)$$

The predicted values for ΔH_f° and S° for various coals are shown in Table II. In this Table, predicted and observed values of heat of combustion for the coals are also shown, where the former was calculated by the predicted value of ΔH_f°. It is found that predicted values of heat of combustion agree well with observed ones. For S°, on the other hand, methods of this work (14) and of Cheng et al. (9) yield comparable results.

Enthalpy of Formation and Absolute Entropy of Coal Liquids. Coal-derived liquids are of extremely complex compositions including highly aromatic groups. They are also characterized by their high contents of heteroatoms, such as oxygen, nitrogen and sulfur. Most of the oxygen atoms are contained in the form of alkyl phenols. Nitrogen atoms are mainly in the form of pirols, pyridine, or carbazole type compounds. Among them, the pyridine type compounds are major constituents of the nitrogen-containing hetero-cyclic ring compounds. Sulfur contents are relatively small. Therefore, its contribution to ΔH_f° and S° is neglected.

With this assumption, and utilizing a similar approach to the estimation of ΔH_f° and S° of coal, semiempirical formulas predicting ΔH_f° and S° of coal-derived liquids (26) are proposed as follows.

In order to deal with a coal-derived liquid as a mixture which has a statistically average chemical structure, we choose two measurable structural parameters, aromaticity, f_a ($=C_{ar}/C_{total}$), and the degree of substitution of the aromatic ring, σ. To identify major atomic groups of coal-derived liquids which contribute to ΔH_f° and S°, the following assumptions are made.
1) No double bonds exist except those in the aromatic rings.
2) All oxygen heteroatoms are present in phenolic OH form, and all nitrogen heteroatoms are in pyridine derivatives.
3) All aliphatic carbons are contained in aliphatic chains of aromatic rings and their average number of carbon atoms in chains is two.
4) Aromatic condensed rings are of cata-condensed type.

With these assumptions, six major atomic groups constituting coal liquids were derived, as shown in Table III. The group value for $C-(H)_n(C')_{4-n}$ in Table III is an average value for groups $C-(H)_3(C)$ and $C-(H)_2(C)(C_B)$, which are given by Benson et al. Similarly, the value for $C_{BF}-(C'')_3$ is an average value for groups $C_{BF}-(C_B)_2(C_{BF})$ and $C_{BF}-(C_B)(C_{BF})_2$.

Given a formula $CH_xO_yN_z$ of coal-derived liquid from its elemental analysis, moles of groups $C-(H)_n(C')_{4-n}$, $C_B-(OH)$, and $N_I-(C_B)$ are given as $1-f_a$, y, and z, respectively. Since the degree of oxygen substitution as well as of aliphatic substitution

Table II. Comparison of estimated and observed thermodynamic properties of coal

Coal type	C [wt% dafl]	H/C [mol/mol]	O/C [mol/mol]	ΔH°_f [kJ/mol-C] Eq. (1)	ΔH°_c [kJ/mol-C]		S° [J/mol-C·K]	
					Obsd.	Estd.	Eq. (2)	Cheng
Tempoku	72.7	0.836	0.208	-25.9	-482.5	-487.9	22.7	19.6
Western	74.8	0.898	0.174	-22.1	-496.3	-499.8	22.4	19.9
Kentucky No.9	78.4	0.796	0.109	-8.8	-500.5	-502.0	19.8	17.7
Pittsburgh	83.7	0.789	0.071	-2.4	-504.3	-506.0	18.8	17.0
Moura	87.6	0.720	0.044	3.8	-495.2	-500.6	17.2	15.8

Table III. Group values for coal-derived liquids

Group	ΔH_f° [kJ/mol]	S° [J/mol·K]
$C \text{---} (H)_n(C')_{4-n}$	-31.50	83.2
$C_B\text{---} (H)$	13.81	48.2
$C_B\text{---} (C)$	23.05	-32.2
$C_B\text{---} (OH)$	-162.30	79.1
$C_{BF}\text{---} (C'')_3$	17.57	-20.9
$N_I \text{---} (C_B)$	69.87	42.3

$N_I\text{-}(C_B)$: a pyridine nitrogen
n   : 2 or 3

determines the value of σ, the number of $C_B-(C)$ is obtained as $f_a\sigma-y$. By introducing the number of aromatic rings per molecule, R, the number of $C_{BF}-(C")_3$ is given as $2(R-1)$. By taking into consideration the ratio of total hydrogen to total carbon, R is found to be equal to $(4-f_a-x)/2$. Then the number of $C_{BF}-(C")_3$ becomes $2-f_a-x$. The rest of carbon is due to $C_B-(H)$. Hence its number is obtained as $f_a(2-\sigma)+x-2$ ($=1-[1-f_a]-[f_a\sigma-y]-[2-f_a-x]-[y]$). In this manner, the amounts of all groups are determined, giving rise to the following correlations for the standard heat of formation and the absolute entropy per unit mole of carbon in coal liquid ($CH_xO_yN_z$):

$$\Delta H_f° \text{[kJ/mol-C]} = -23.64+41.2f_a+9.24f_a\sigma-3.76x$$
$$-185.4y+69.87z-\Delta H_v \qquad (3)$$

and

$$S° \text{[J/mol-C·K]} = -54.4+33.5f_a-80.4f_a\sigma+69.1x$$
$$+111.3y+42.3z-\Delta S_v+\Delta S_{mix} \qquad (4)$$

where ΔH_v and ΔS_v, respectively, refer to the enthalpy and the entropy of vaporization. They are correction factors between the ideal gaseous state and the real liquid state. For the absolute entropy, it is necessary to take into account the entropy of mixing, ΔS_{mix}. The entropy of mixing was estimated to be 2 J/mol-C·K using the detailed component analyses on H-Coal liquids (27). For the enthalpy of vaporization, a correlation was derived from the measured data of SRC-II coal liquids (19) as:

$$\Delta H_v \text{[kJ/mol-C]} = T_b\Delta S_v = -5.0\times10^{-3}T_b + 6.55 \qquad (5)$$

where T_b [K] is the normal boiling point of coal liquid. To obtain the value of ΔH_v at 298 K more accurately, we must make a correction for the difference in the heat capacities between the liquid state and the ideal gaseous state. However, this information is not available, but, this correction will be small and for the sake of zeroth-order approximation for the exergy analysis of a coal liquefaction process system, it may be neglected.

Necessary data for estimating the standard heat of formation and the absolute entropy by Equations 3 and 4 are the elemental analysis, structural parameters, f_a and σ, and the normal boiling point. For a practical purpose, it will be more convenient if we could calculate $\Delta H_f°$ and $S°$ only from elemental analysis data and normal boiling point. The aromaticity f_a for coal liquids may be estimated by the correlation shown in Figure 1. On the other hand, the value of σ may be taken as 0.3 for its average value based on the reported data (18,20,21). Substitution of these relations into Equations 3 and 4 gives

$$\Delta H_f° \text{[kJ/mol-C]} = 40.1-34.1x-185.4y+69.9z-\Delta H_v \qquad (6)$$

and

$$S° \text{[J/mol-C·K]} = -38.8+62.6x+111.3y+42.3z-\Delta S_v \qquad (7)$$

Naphtha fractions require slightly different treatment. The major constituents of naphtha are reported to be cycloparafines. Since the contributions of CH_2 and CH substituents become significant, the values of ΔH_f° and S° for $C\text{-}(H)_n(C')_{4-n}$ in Table III are replaced by -24.25 kJ/mol and 57.8 J/mol·K. Hence, for naphtha fractions, we have

$$\Delta H_f^\circ [kJ/mol\text{-}C] = 37.1-29.3x-185.4y+69.9z-\Delta H_v \qquad (8)$$

and

$$S^\circ [J/mol\text{-}C·K] = -29.5+44.7x+111.3y+42.3z-\Delta S_v \qquad (9)$$

Proposed methods for predicting heats of formation and absolute entropies are tested on two fractions of synthetic crude oil obtained by the EDS process, one sample of H-Coal, one sample of Synthoil, two samples of Solvent Refined Coal, and five pure compounds found in coal liquefaction products. For these samples, the heats of combustion are calculated using predicted values of ΔH_f° and compared in Table IV with observed values. Note that Equations 8 and 9 were used to predict ΔH_f° and S° of the EDS heavy naphtha. Equations 6 and 7 are applied to other samples of coal-derived liquids, and Equations 3 and 4 to the pure compounds.

Exergy Analysis of H-Coal Process

A process flow diagram (PFD) of the H-Coal process for producing synthetic fuels from bituminous coal is shown in Figure 3. Raw coal is crushed and dried. The dried coal is then slurried with recycle oils which are recovered from subsequent processes. After that, it was mixed with recycle and make-up hydrogen. The coal-oil slurry and hydrogen gas are pumped to the preheating furnace, and then liquefied in direct contact with catalyst in an ebullated bed reactor. The reactor effluent is separated into recycle and net product streams by a series of flashes and by fractionation. Primary products include fuel gases, stabilized naphtha, turbine fuel, and distillate boiler fuel. Also, several by-products, such as ammonia, sulfur and phenols, are obtained from the plant. In addition to these primary products and by-products, the H-Coal liquefaction system produces the plant utility gases which are all consumed within the plant.

An exergy analysis of the conceptual design of the synclude mode H-Coal process for bituminous coal of Illinois No.6 has been carried out based on the data reported by the Fluor Engineers and Constructors, Inc., (31). The plant is designed to convert 14,448 tons (short ton) of coal per day.

The overall process has been analysed using a Structured Process Energy-Exergy-flow Diagram (SPEED) which has revealed the hierarchical structure of chemical process systems and facilitated computer-aided energy and exergy calculations (11). In this method, attention is paid on the changes in energy (enthalpy) and exergy in the processes rather than those for the streams. For

Table IV. Comparison of estimated and observed values for coal-derived liquids and pure compounds

Compound	Empirical formula	ΔH°_f [kJ/mol-C]	$-\Delta H^\circ_c$ [kJ/mol-C]			S° [J/mol-C·K]		
		Estd.	Obsd.	Estd.	%Error	Obsd.	Estd.	Δ
EDS Heavy Naphtha (70-200°C)	$C\,H_{1.517}O_{0.025}N_{0.002}S_{0.002}$	-16.07	597.24	595.82	0.24	–	31.6	–
EDS Fuel Oil (200-540°C)	$C\,H_{1.026}O_{0.015}N_{0.006}S_{0.002}$	-0.74	534.22	540.66	-1.21	–	22.3	–
H-Coal Oil (250-517°C)	$C\,H_{1.071}O_{0.016}N_{0.007}S_{0.002}$	-1.99	546.07	545.06	0.19	–	25.6	–
Synthoil Oil (172-477°C)	$C\,H_{1.092}O_{0.026}N_{0.010}S_{0.002}$	-4.53	538.03	545.51	-1.39	–	27.8	–
SRC Light Org.Liquid (83-294°C)	$C\,H_{1.407}O_{0.042}N_{0.002}S_{0.002}$	-19.71	567.54	575.34	-1.38	–	44.7	–
SRC Rec. Solvent (163-469°C)	$C\,H_{1.039}O_{0.036}N_{0.006}S_{0.001}$	-5.49	534.19	536.87	-0.50	–	23.1	–
i-Propylbenzene	$C\,H_{1.333}$	-4.17	579.49	579.89	-0.01	31.1	38.7	-7.6
Tetralin	$C\,H_{1.2}$	-2.40	558.19	558.82	-0.11	27.5	23.2	4.3
1-Methylnaphthalene	$C\,H_{0.909}$	7.30	528.55	530.74	-0.42	23.2	21.7	1.5
m-Cresol	$C\,H_{1.143}O_{0.143}$	-22.59	535.52	534.27	0.23	47.9	43.2	4.7
Quinoline	$C\,H_{0.778}N_{0.111}$	17.30	522.07	521.97	0.02	24.1	27.5	-3.4

%Error = (Obsd. - Estd.)/Obsd.

Δ = Obsd.- Estd.

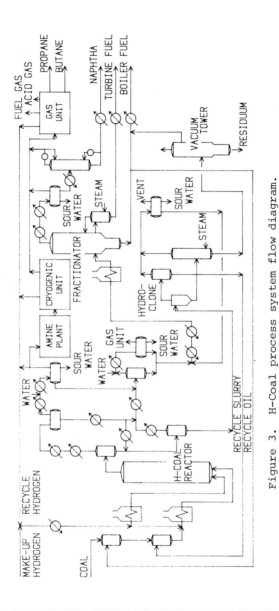

Figure 3. H-Coal process system flow diagram.

further detail on the SPEED, readers may refer to another paper in this book (32).

The standard heat of formation $\Delta H_f°$ and absolute entropy $S°$ of the substances appeared in this process have been estimated by the proposed methods and are tabulated in Table V, where $\Delta H_f°$ and $S°$ for each substance were evaluated for the value per unit mole of carbon in the substance. $\Delta H_f°$ and $S°$ for residuum were estimated by formulas for coal, i.e., by Equations 1 and 2. The heat capacity of the coal-derived liquids was estimated by refering the measured data by Lee and Bechtold (4).

Several processing conditions were, unfortunately, unavailable to calculate enthalpy and exergy due to the contractor's restrictions on proprietary data. In particular, pressures, temperatures and compositions of several streams associated with preheating and hydrogenation were not reported, so that several numerical values of their streams were estimated approximately.

In Figure 4, the exergy calculation for the H-Coal process by the SPEED is depicted. Several special symbols are used in the Figure. The sequence of the quotation mark implies that the reaction listed below it is the target process. The mark @ indicates the temperature and pressure of the inlet and outlet streams. For instance, @ 669, 25 stands for the condition at 669 K and 25 atm. The symbol { } indicates that streams within the parenthesis consist in the same phase and, therefore, they form a mixture. The sequence of the ripple mark $\sim$ denotes that the process below it is an exergy acceptor. D is the direction factor (32), which is defined as follows.

$$D = T_O \, \Delta S \, / \, \Delta H \qquad (10)$$

where the reference temperature T_O was chosen as 298.15 K. It is a dimensionless value denoting the entropy increase per unit enthalpy increase.

Calculations were carried out for unit mole of carbon in feed coal (MAF basis). Since the empirical formula of feed coal is denoted by $CH_{0.839}O_{0.096}N_{0.018}S_{0.019}$, its molecular weight per unit mole of carbon is 15.236. In the present exergy analysis, the exergy input to the liquefaction system is 528.71 kJ, which corresponds to the exergy value of 1 mol-C of feed coal, as shown in Table V.

The overall target reaction of the H-Coal liquefaction system is printed symbolically on the top of Figure 4. As shown in the Figure, it is a process in which 1 mol-C of coal, <COAL>, reacts with 0.448 mol-H_2 of make-up hydrogen, <MA.H_2>; eventually, they produce 0.045 mol-C of fuel gas, <GAS>, 0.535 mol-C of synthetic fuel oil, <OIL>, 0.001 mol-C of by-products, <BYPROD>, 0.314 mol-C of residuum, <RES>, and 0.111 mol-C of fuel gas as the utility source, <FUEL>. The <COAL>, <OIL>, <GAS>, and so on are defined so that their respective carbon mole number becomes unity, as shown in remarks of the Figure. Note that the synthetic

Table V. Estimated thermodynamic properties of coal and coal liquids for H-Coal system

Compound	Empirical formula	ΔH_f° [kJ/mol-C]	ΔH_c° [kJ/mol-C]	S° [J/mol-C·K]	ε° [kJ/mol-C]
Illinois No.6 Coal	$C\,H_{0.839}O_{0.096}N_{0.018}S_{0.019}$	-7.84	-511.22	19.9	528.71
Light Distillate (IBP/350°F)	$C\,H_{1.926}O_{0.005}N_{0.002}S_{0.0003}$	-24.72	-644.14	47.8	638.08
Middle Distillate (350/600°F)	$C\,H_{1.364}O_{0.008}N_{0.005}S_{0.0003}$	-11.61	-576.93	40.1	576.58
Heavy Distillate (500/800°F)	$C\,H_{1.141}O_{0.011}N_{0.006}S_{0.0003}$	-3.89	-552.78	28.5	557.29

```
" " " " " " " " " " " " " " " " " " " " " " "
<COAL> + 0.448<MA.H2> => 0.045<GAS> + 0.535<OIL> + 0.111<FUEL> + 0.314<RES>
+ 0.001<BYPROD> :
        " " " " " " " " " " " " " " " " " " " "
    ;Pretreating
    <COAL> + 1.699<RC.OIL> + 0.976<RC.SLUR> => 3.675<SLURRY> :
    ToΣΔSi= 7.926 kJ
    " " " " " " " " " " " " " " " " " " " "
    ;Preheating & Hydrogenation
    3.675<SLURRY> + 1.532<H2> => 3.694<CRUDE> :
    ToΣΔSi= 31.948 kJ
    " " " " " " " " " " " " " " " " " " "
    ;Separation
    3.694<CRUDE> => 0.045<GAS> + 0.535<OIL> + 0.111<FUEL> + 0.314<RES>
    + 0.001<BYPROD> + 1.699<RC.OIL> + 0.976<RC.SLUR> + 1.084<RC.H2> :
    ToΣΔSi= 15.955 kJ

;Accommodated system
        ;Oxygen production        ToΣΔSi= 4.860 kJ
        ;Hydrogen production       ToΣΔSi= 19.167 kJ
        ;Power production          ToΣΔSi= 28.145 kJ
        ;Utilities & Others        ToΣΔSi= 65.416 kJ

;Remarks
LIGHT       = C H1.926 0.005 N.002 S.0003
MIDDLE      = C H1.364 0.008 N.005 S.0003
HEAVY       = C H1.141 0.011 N.006 S.0003

<COAL>      = C H.839 0.096 N.018 S.019 + 1.982 ASH + 0.0195 H2O,L
<RC.OIL>    = {0.002 LIGHT + 0.442 MIDDLE + 0.556 HEAVY}
<RC.SLUR>   = {0.002 LIGHT + 0.151 MIDDLE + 0.267 HEAVY}
              + 0.580 C H.634 0.084 + 3.238 ASH
<SLURRY>    = [0.272<COAL> + 0.462<RC.OIL> + 0.266<RC.SLUR>]

<H2>        = 0.292<MA.H2> + 0.708<RC.H2>
<MA.H2>     = {H2 + 0.0068 N2 + 0.0226 CO + 0.0026 CO2 + 0.0055 CH4
              +0.0009 H2O}
<RC.H2>     = {H2 + 0.0034 N2 + 0.0046 CO}

<CRUDE>     = [0.012<GAS> + 0.145<OIL> + 0.030<FUEL> + 0.085<RES>
              + 0.0003<BYPROD> + 0.460<RC.OIL> + 0.264<RC.SLUR>
              + 0.293<RC.H2>]

<OIL>       = 0.285<NAPHTHA> + 0.441<TURBINE> + 0.273<BOILER>
<NAPHTHA>   = {0.003 C4H10 + 0.988 LIGHT}
<TURBINE>   = {0.190 LIGHT + 0.810 MIDDLE}
<BOILER>    = {0.030 LIGHT + 0.296 MIDDLE + 0.674 HEAVY}

<GAS>       = 0.128 C3H8 + 0.154 C4H10
<FUEL>      = {0.0381 CO + 0.3247 CH4 + 0.1433 C2H6 + 0.1166 C3H8
              + 0.0002 C4H10 + 0.1406 H2 + 0.0280 N2}
<RES>       = C H.698 0.077 + 1.982 ASH
<BYPROD>    = {CO2 + 13.4 H2S + 11.0 NH3} + 83.0 H2O,L
```

Figure 4. SPEED for H-Coal process system.

fuel oil, <OIL>, is composed of naphtha, <NAPHTHA>, turbine fuel, <TURBINE>, and boiler fuel, <BOILER>. Since the boiling ranges of light (denoted by LIGHT), middle (MIDDLE), and heavy (HEAVY) distillates of the synthetic liquids are defined as in Table V, naphtha, <NAPHTHA>, turbine fuel, <TURBINE>, and boiler fuel, <BOILER>, respectively, can be expressed as a mixture of these distillates and light hydrocarbons. For example, naphtha can be expressed as a mixture of 0.003 mole of butane and 0.988 mol-C of the LIGHT, as remarked in Figure 4. Molar fractions of these products in terms of the LIGHT, MIDDLE and HEAVY were determined by inspection of the material balance in the Fluor's report (31).

As shown in Figure 4, the overall target is decomposed into three subtargets using 1.699 mol-C of recycle oil, <RC.OIL>, and 0.976 mol-C of recycle slurry, <RC.SLUR>. They are pretreatment, preheating and hydrogenation, and separation.

In this way, we know that the coal liquefaction system constitutes a structure of multi-target system. In addition to these subtargets, the coal liquefaction system attaches an accomodate system which includes oxygen production, hydrogen production, power production, and utilities.

Figure 5 shows that the coal pretreatment subtarget is further decomposed into three sub-subtargets. The first sub-subtarget is coal crushing and drying. In this section, coal containing 10 weight % moisture is dried to 2% moisture. This is accomplished by donating 9.982 kJ of heat and 0.881 kJ of electric power to 1 mol-C of coal. Figure 5 also shows that 4.568 kJ of exergy $T_o\Sigma\Delta S_i$ is destructed in this stage.

In the second sub-subtarget system of coal pretreatment the dried coal, <COAL> @ 311, 2, is mixed with 1.699 mol-C of hot recycle oil, <RC.OIL> @ 567, 7. In this stage, power is considered as a coupled process. The most power is consumed in mixing pumps. The third sub-subtarget of coal pretreatment system is another mixing process in which the coal and oil mixture, [<COAL> +1.699<RC.OIL>] @ 444, 23 from the previous sub-subtarget process, is further mixed with the solids-containing hydroclone overflow stream, 0.976<RC.SLUR> @ 669, 26. Again, power required for mixing and slurry feed pumps is considered to be an exergy donor.

By summing up the exergy destruction of each sub-subtarget process system, the overall exergy destruction $T_o\Sigma\Delta S_i$ for the pretreating subsystem is given as 7.926 kJ, as listed in the SPEED in Figure 5. This value is also shown below the subtarget of pretreatment in Figure 4. Since there is not enough space for a complete drawing of SPEED for the overall process, Figures 4 and 5 are shown separately. However, if the content of Figure 5 is inserted into Figure 4 with an indent, the hierarchical structure of the liquefaction system will be depicted more clearly. When we compare this SPEED with the PFD of Figure 3, we learn that the SPEED plays a part of the PFD being attended with quantitative information.

The subtarget of preheating and hydrogenation shown in Figure 4 has been further subdivided into a sub-subtarget of preheating

```
" " " " " " " " " " " " " " " " " " " "
;Pretreating
<COAL> @312,2 + 1.699<RC.OIL> @567,7 + 0.976<RC.SLUR> @669,26
=> 3.675<SLURRY> @522,215 :
ΔH= -2.373 kJ,      Δε= -2.053 kJ,       D= 0.135
        " " " " " " " " " " " " " " " " " "
        ;Crushing & Drying
        <COAL> @298,1 => <COAL> @311,2 :
        ΔH= 0.289 kJ,      Δε= 0.006 kJ,       D= 0.979
                ------------------
                *ELECTRICITY SOURCE      ;For crushing
                ΔH= -0.881 kJ,     Δε= -0.881 kJ,      D= 0
                ------------------
                *HEAT SOURCE @473         ;For drying
                ΔH= -9.982 kJ,     Δε= -3.693 kJ,      D= 0.630
                ∿∿∿∿∿∿∿∿∿∿∿∿∿∿∿∿
                *HEAT SINK @298.15        ;Heat loss
                ΔH= 10.574 kJ,     Δε= 0 kJ,      D= 1
            ToΣΔSi= 4.568 kJ
        " " " " " " " " " " " " " " " " " "
        <COAL> @311,2 + 1.699<RC.OIL> @567,7
        => [<COAL> + 1.699<RC.OIL>] @444,23 :
        ΔH= -1.056 kJ,      Δε= -0.919 kJ,       D= 0.130
                ------------------
                *ELECTRICITY SOURCE
                ΔH= -0.155 kJ,     Δε= -0.155 kJ,      D=0
                ∿∿∿∿∿∿∿∿∿∿∿∿∿∿∿∿
                *HEAT SINK @298.15
                ΔH= 1.211 kJ,      Δε= 0 kJ,      D= 1
            ToΣΔSi= 1.074 kJ
        " " " " " " " " " " " " " " " " " "
        [<COAL> + 1.699<RC.OIL>] @444,23 + 0.976<RC.SLUR> @669,26
        => 3.675<SLURRY> @522,215 :
        ΔH= -1.317 kJ,      Δε= -1.134 kJ,       D= 0.139
                ------------------
                *ELECTRICITY SOURCE
                ΔH= -1.150 kJ,     Δε= -1.150 kJ,      D= 0
                ∿∿∿∿∿∿∿∿∿∿∿∿∿∿∿∿
                *HEAT SINK @298.15
                ΔH= 2.467 kJ,      Δε= 0 kJ,      D= 1
            ToΣΔSi= 2.284 kJ
    ToΣΔSi= 7.926 kJ
```

Figure 5.   SPEED for coal pretreatment subsystem.

and that of hydrogenation, as shown in Figure 6. Although it is obvious that some reactions proceed in the preheater, it is assumed here that no reaction takes place in the preheater. It is because compositions of the preheater effluent have not been reported. In the hydrogenation system, the reaction is exothermic with the heat of release 10.426 kJ. However, this value is calculated by considering temperature increase in reactants. When this heat of reaction is treated as a heat loss, 18.026 kJ of exergy is destructed at this stage.

If we assume that the reaction proceeds in an isothermal reactor at 740 K, the heat of reaction is obtained as 22.254 kJ, as shown in Figure 7. When this heat is absorbed by a heat sink at the same temperature, the exergy destruction caused by the reaction itself is given as 12.038 kJ.

In this way, the exergy loss of each subsystem can be calculated. The overall exergy efficiency η_ϵ of the process is defined here in the following way:

$$\eta_\epsilon = \frac{[\text{ all useful exergy outputs }]}{[\text{ all exergy inputs }]}$$
$$= \frac{[\text{ all exergy inputs }] - [\text{ all exergy losses }]}{[\text{ all exergy inputs }]}$$

where the exergy loss includes both exergy destruction and exergy wastes. Since the denominator of the above equation is given here by the exergy of 1 mol-C of feed coal, inspection of the SPEED yields

$$\eta_\epsilon = \frac{528.71 - (\ 7.936 + 31.95 + 15.95 + 117.59\ )}{528.71}$$
$$= \frac{355.29}{528.71}$$
$$= 67.2\%$$

Conclusion

The group contribution method is applied to estimate the heat of formation ΔH_f° and the absolute entropy S° of coal and coal-derived liquids. Semiempirical formulas are presented for estimating them. Using proposed formulas for the estimation of ΔH_f° of coal and coal-derived liquids, heats of combustion are predicted for various coals and coal-derived liquids. It is shown that the predicted values agree with the observed ones.

Thermodynamic analyses have been conducted for the various process steps in the H-Coal process system for producing synthetic fuels from bituminous coal. A Structured Process Energy-Exergy-flow Diagram (SPEED) for the H-Coal process is presented, which depicts the transformation of energy and exergy among the processes and the hierarchical structure of the process system with a compact format of the SPEED.

```
" " " " " " " " " " " " " " " " " " " "
;Preheating & Hydrogenation
3.675<SLURRY> @522,215 + 1.532<H2> @377 => 3.694<CRUDE> @740,208 :
ΔH= 20.302 kJ,      Δε= -4.330 kJ,      D= 1.213
       " " " " " " " " " " " " " " " " " " "
    ;Preheating
    3.675<SLURRY> @522,215 + 1.532<H2> @377
    => [3.675<SLURRY> + 0.736<H2>] @639,208 + 0.796<H2> @866
    ΔH= 30.728 kJ,      Δε= 14.489 kJ,      D= 0.528
           -----------------
           *HEAT SOURCE @700          ;Heat Exchange
           ΔH= -14.974 kJ,      Δε= -8.599 kJ,      D= 0.397
           -----------------
           *HEAT SOURCE @-2.08E05    ;Furnace
           ΔH= -21.267 kJ,      Δε= -21.297 kJ,      D= -0.0014
           ᜎᜎᜎᜎᜎᜎᜎᜎᜎᜎᜎᜎᜎᜎᜎ
           *HEAT RECOVERY          ;Steam generation
           ΔH= 3.379 kJ,      Δε= 1.485 kJ,      D= 0.561
           ᜎᜎᜎᜎᜎᜎᜎᜎᜎᜎᜎᜎᜎᜎ
           *HEAT SINK @298.15
           ΔH= 2.134 kJ,      Δε= 0 kJ,      D= 1
        ToΣΔSi= 13.922 kJ
    " " " " " " " " " " " " " " " " " "
    ;Hydrogenation
    [3.675<SLURRY> +0.736<H2>] @639,208 + 0.796<H2> @866
    => 3.694<CLUDE> @740 :
    ΔH= -10.426 kJ,      Δε= -17.895 kJ,      D= -0.716
           -----------------
           *ELECTRICITY SOURCE
           ΔH= -0.131 kJ,      Δε= -0.131 kJ,      D= 0
           ᜎᜎᜎᜎᜎᜎᜎᜎᜎᜎᜎᜎᜎᜎ
           *HEAT SINK @298.15
           ΔH= 10.557 kJ,      Δε= 0 kJ,      D= 1
        ToΣΔSi= 18.026 kJ
    ToΣΔSi= 31.948 kJ
```

Figure 6.   Preheating and hydrogenation subsystem.

```
" " " " " " " " " " " " " " " " " " " " "
3.675<SLURRY> @740,208 + 1.532<H2> => 3.694<CRUDE> :
ΔH= -22.254 kJ,      Δε= -25.330 kJ,      D= -0.138
       ᜎᜎᜎᜎᜎᜎᜎᜎᜎᜎᜎᜎᜎᜎ
       *HEAT SINK @740
       ΔH= 22.254 kJ,      Δε= 13.292 kJ,      D= 0.403
    ToΣΔSi= 12.038 kJ
```

Figure 7.   Hydrogenation under isothermal condition
at 740 K.

Literature Cited

1. Gaggioli, R. A., et al., "A Thermodynamic-Economic Analysis
    of the SYNTHANE Process," Report to U.S. Department of
    Energy, COO-4589-1, Nov. (1978).
2. Wen, C. Y., S. Ikumi, M. Onozaki, C. D. Luo, "Coal Gasifi-
    cation Availability Analysis," paper presented to the 2nd
    World Congress of a Chemical Engineering and World Chemical
    Exposition, Montreal, Canada, Oct. (1981).
3. Lin, C. Y., "Available Work Energy and Coal Conversion
    Processes," Ph.D. Dissertation, West Virginia University,
    Morgantown, W.Va. (1977).
4. Brainard, A. J., "The Application of an Availability
    Analysis to a Direct Liquefaction Process," paper
    presented to the 72nd Annual Meeting of the AIChE,
    San Francisco, Nov. (1979).
5. Unruh, T. L., and B. G. Kyle, "The Energetics of the COED
    Process for Coal Conversion from a Second Law Perspective,"
    paper presented to the 72nd Annual Meeting of the AIChE,
    San Francisco, Nov. (1979).
6. Kidnay, A. J., and V. F. Yesavage, "Enthalpy Measurement
    of Coal-Derived Liquids," Quarterly Technical Progress
    Reports, work done on DOE Contract No. EX-76-C-01-2035.
7. Sharma, R., et al., "The Measurement and Correlation of
    the Enthalpy of Coal-Derived Liquids," paper presented
    to the AIChE Spring 1981 National Meeting, Houston (1981).
8. Lowry, H. H.,(Ed.) "Chemistry of Coal Utilization,"
    Chap. 4, Wiley, New York (1945).
9. Cheng, W. B., et al., "Entropies of Coals and Reference
    States in Coal Gasification Availability Analysis,"
    paper presented to the 73rd AIChE Annual Meeting,
    Chicago (1980).
10. Szargut, J., and T. Styryska, Brennstoff-Warme-Kraft,
    16, 589 (1964).
11. Oaki, H., and M. Ishida, J. Japan Petrol. Inst., 24, 36
    (1981).
12. Benson, S. W., "Thermochemical Kinetics," 2nd ed., Chap.2,
    Wiley, New York (1976).
13. Shaw, R., et al., J. Physical Chemistry, 81, 1716 (1977).
14. Suzuki, T., and M. Ishida, "Estimation of the Enthalpy of
    Formation and the Absolute Entropy of Coal Based on the
    Group-Contribution Method," J. Fuel Society Japan, 61,
    250 (1982).
15. Hirsh, P. B., Proc. Roy. Soc., A226, 143 (1954).
16. Honda, H., Fuel, 36, 159 (1957).
17. Ladner, W. R., and A. E. Stacey, Fuel, 40, 295 (1961).
18. Kamiya, Y., J. Fuel Society Japan, 59, 229 (1980).
19. Gray, J. A., et al., "Selected Physical, Chemical, and
    Thermodynamic Properties of Narrow Boiling Range Coal
    Liquids," paper presented to the AIChE Fall 1981
    Annual Meeting, New Orleans, Nov. (1981).

20. Yokoyama, S., et al., Fuel, 60, 254 (1981).
21. Yoshida, T., and Y. Nakada, J. Fuel Society Japan, 60, 762 (1981).
22. Sakabe, T., Report of the Resources Research Institute of Japan, 49 (1961).
23. Annual Technical Progress Report 1977 SRC Process DOE, June (1978).
24. Sugimura, H., et al., J. Fuel Society Japan, 46, 911 (1967).
25. van Krevelen, D. W., "Coal", Elsevier, Amsterdam, p.323 (1961).
26. Suzuki, T., and M. Ishida, "Estimation of Enthalpies of Formation and Absolute Entropies of Coal-Derived Liquids Based on the Group-Contribution Method," J. Fuel Society of Japan, 61, 383 (1982).
27. Project H-Coal Report, HRI R&D No.26, OCR (1967).
28. Furlong, L. E., et al., Chem. Eng. Prog., August, 69 (1976).
29. Callen, R. B., et al., Ind. Eng. Chem., Prod. Res. Dev., 15, 222 (1976).
30. "Lange's Handbook of Chemistry," p.9-65, McGraw-Hill, New York (1979).
31. Buckingham, P. A., et al., "Engineering Evaluation of Conceptual Coal Conversion Plant Using the H-Coal Liquefaction Process," EPRI Report, AF-1297 (1979).
32. Ishida, M., "Analysis of Hierarchical Structure of a Process System Based on Energy and Exergy Transformation," in this book.

RECEIVED July 7, 1983

Thermodynamic Availability of Solar Radiation with Special Attention to Atmospheric Rayleigh Scattering

R. H. EDGERTON and J. A. PATTEN

School of Engineering, Oakland University, Rochester, MI 48063

This paper examines the available energy flux of scattered solar radiation in the atmosphere. It is shown that the available energy to energy flux ratio for Rayleigh scattered solar radiation is approximately 0.80 to 0.90. This is for clear sky conditions and with solar energy flux near the zenith and near the horizon. The available energy to energy flux of Rayleigh scattered radiation is shown to be nearly constant over the sky. From this analysis it can be implied that the solar radiation has high thermodynamic potential even if diffusely scattered over the sky.

In the evaluation of second law efficiencies of solar energy converters, a determination must be made of the available work function of the incident radiation. The author has derived this function for arbitrary spectral distributions and spacial or solid angle radiation inputs(1). The results have shown that the spectral distribution of typical solar radiation gives available energy inputs much higher than the inputs assuming a thermal equilibrium distribution with the same energy flux. The spacial effect on the available energy flux has also been examined theoretically with the available energy loss with scattering described. The spacial loss is apparent in the limited focusing capability of optical systems for diffuse radiation.

Both these effects require knowledge of the spectral and spacial radiation distributions of the radiation flux on a surface. The determination of both these distributions at a given location is a difficult instrumentation problem. In this paper, the effect of scattering processes in the atmosphere on the available energy of solar radiation on a surface is examined. Both the spectral and spacial effects of Rayleigh scattering are demonstrated.

In the establishment of a measure of the available energy of radiation energy inputs to processes, the question of the

0097–6156/83/0235–0395$06.00/0

angular distribution of the radiation becomes critical. The
author(1) and others (2-8) have attempted to resolve this problem by introduction of the angular or spacial dependence by
discussing the flux dependence on the solid angle of the
radiation input. In this approach the entropy and available
energy of a radiation input are all expressed as flux terms
rather than as thermodynamic properties of space.

If one examines the question of available energy of solar
radiation, two approaches appear attractive. The first is a
quantum perspective that the electromagnetic energy of the
photon stream from the sun is all convertible to thermodynamic
work. The second, based on the engineering perspective of
dealing with solar collectors which utilize the solar energy to
heat substances is that a thermodynamic limit is imposed by the
sun surface temperature as a radiating object at approximately
6000K.

For practical purposes both these answers are sufficient if
one is interested in second law analysis of energy conversion
devices operated outside the atmosphere. A "heat" source of
6000K has an available energy flux which is nearly equal to its
radiant energy flux.

The measurement of the spectral distribution of solar
radiation outside the atmosphere and the subsequent association
of this spectral distribution with the spectral distribution of
radiation in a blackbody cavity has, I believe, biased the
attempts to characterize the actual radiation in the atmosphere
to an undue extent. Figure 1 indicates typical spectral distributions of radiation in the atmosphere as compared to that of
solar radiation outside the atmosphere. Outside the atmosphere
m = 0 and if the flux is directly through m = 1. If slanted at
and angle from the zenith angle θ_o, then m is approximately
$1/\cos \theta_o$.

The processes of scattering and absorption of radiation in
the atmosphere so significantly alter the spectral distribution
that any similarity to extra terrestrial radiation is almost co-
incidental. Experiments with radiation between surfaces have
shown that blackbody radiation theory can be extended successfully to many radiation heat transfer situations. In these
situations the strict equilibrium requirements of the initial
model have so far not proved to be necessary for practical designs. Most importantly the concept of temperature has proved
useful in non-equilibrium radiation flux situations(3).

The spacial distribution of solar radiation is the other
important characteristic of terrestrial radiation which affects
its available energy. When an earth-sun system is examined in
a "universal" perspective, the sun can be visualized as a
spherical radiation emitter with the energy emitted from the
source expanding out into the universe (a divergent flux). The
radiation energy density of space decreasing with distance from
the sun. The energy flux per unit area decreasing but the

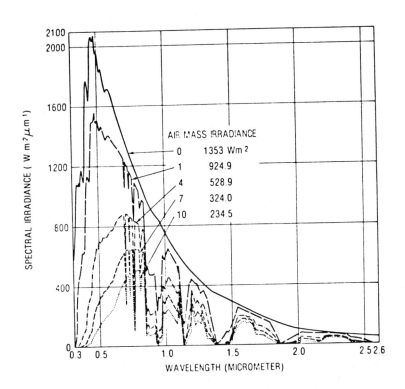

Figure 1. Spectral distribution of direct solar radiation through a clear atmosphere of different air masses.

intensity or flux per unit area per solid angle remaining constant. This constant intensity means that the temperature or
spectral distribution of the radiation remains constant even
though the flux per unit area is decreasing.

In an earth perspective however, the sun appears as a visible disk with radiation confined within a cone with convergence
of radiation within a solid angle of approximately 7×10^{-5}
steradians. This solid cone angle is critical in the development of focusing collectors used to increase the flux per unit
area or energy density of the electromagnetic radiation. This
in turn limits the maximum temperature obtainable with a passive
collector to the sun temperature of 6000K at a collector concentration ratio of approximately 40,000 to 1. Atmospheric scattering for the sun near the horizon can increase this solid
angle to 0.05 steradian.

In radiation through the atmosphere, the electromagnetic
energy is scattered and absorbed so that part of the radiation
is observed to be distributed over the entire sky. The radiation within the sun solid angle is usually considered as
direct radiation, but in addition, contains smaller amounts of
scattered radiation. The scattering and absorption processes
depend both spacially and spectrally upon the atmosphere composition and cloud distribution. It is useful to arbitrarily
separate the input radiation to a surface into the direct and
diffuse or scattered radiation. Figure 2 indicates the relative
spectral magnitudes of each of these components.

In addition, "sky radiation" indicating radiation from the
gases in the atmosphere are sometimes included. In a radiation
cooling situation this is the radiation flux which governs the
net flux from a surface at night. This sky radiation is dependent on the solar absorption heating of the atmosphere and has
been estimated as the order of magnitude of 20% of the extraterrestrial flux. The available energy of this flux is small
because the temperature is close to the environmental temperature. The sky has an equivalent emissivity of about 0.75. For
a sky energy flux of 300 W/m^2 the temperature is approximately
270 K.

In energy flux terms the input radiation appears to be
reasonably considered as a direct flux in the cone subtended by
the solar disk. The scattered radiation is distributed over the
sky in complicated geometrical ways depending on the position of
the sun. The two principal scattering mechanisms are molecular
or Rayleigh scattering, and Mie or particle scattering. Figure
3 shows a representative sky distribution of Rayleigh scattered
light.

To examine the available energy of solar radiation, let us
briefly examine the thermodynamic results for blackbody radiation within an insulated cavity. There is no net flux of

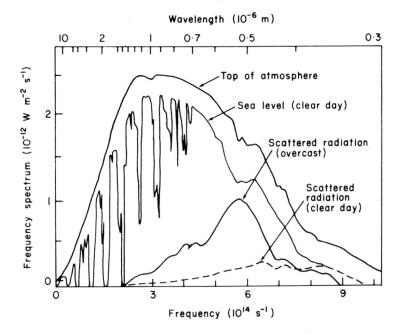

Figure 2. Spectral distribution of scattered solar radiation compared to extraterrestrial and direct solar radiation.

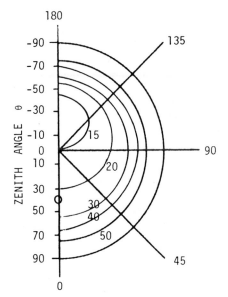

Figure 3. Distribution of Rayleigh scattered radiation over the sky for the sun at 45° from the zenith as shown. Intensity curves are normalized to 15 at the zenith.

energy at any point, and the radiation is uniformly distributed over a spherical angle of 4π. The expected energy in
the volume V at a temperature T can then be shown to be

$$E = 0.658 \ (2\pi kT/hc)^3 \ kTV$$

In this expression, h is Planck's constant, k is Boltzmann's
constant and c is the speed of light.

The entropy is

$$S = (4/3)E/T$$

The Gibbs'free energy is

$$G = E + PV - TS = 0$$

The closed system available energy is

$$A = E + P_0V - T_0S = E(1 + (T_0/T)^4/3 - (4/3)T_0/T)$$

The open system Available energy is

$$B = H - T_0S = (4/3)(1 - T_0/T)E = (1 - T_0/T)H$$

Note: T_0 is the environmental temperature and H is the enthalpy.

In dealing with problems of solar radiation, as opposed
to blackbody radiation, the effect of the solid angle in which
the radiation is confined has been examined (2-4) by considering
the volume density of photons to be reduced. Landsberg(6)
considers dilute radiation in the sense that the spectral distribution is retained but the radiation density is reduced. This
leads to defining the temperature of a spectral component as

$$T = h\nu/k \ \ln(1 + (h\nu^3\Omega/c^2e_\nu)$$

Here Ω is the solid angle of the radiation and e_ν is a
spectral energy flux per unit area per unit frequency.

Press (3) makes a similar approximation , comparing the closed
system available energy of a volume of radiation in a cone
compared to that in a full spherical space. His results trans-
lated into flux terms indicate that approximately 38% of the
available energy flux in a solar solid angle is retained if the
energy is uniformly scattered over a full 4π solid angle.

Scattered radiation has generally been considered as less available for energy conversion because of its diffuse spacial distribution. This diffuse spacial distribution imposes a collection limit on classical focusing systems. The capabilities of increasing the energy density or flux per unit area are reduced by scattering processes. The spacial flux characteristics are also important for flat plate collectors since the energy absorptivity of a surface is dependent on the incident angle of the radiation. The absorptivity for normal incidence being high compared to that for low incident angle radiation.

The radiation absorption and scattering of direct radiation by the atmosphere decreases the direct radiation flux at a surface. The scattering however increases the diffuse component of the energy flux (9). The Rayleigh and Mie scattering of light by the atmosphere are well understood. Scattering by clouds however presents an unsolved problem because of computational and measurement difficulties associated with the large variety of cloud characteristics.

Analysis

In this paper two basic conditions have been assumed.

1. In radiation problems, it is the available energy flux which is of concern, not the available energy density. This means that directional characteristics become more important and the steady flow available energy, $b = h - T_0 s$ is the more appropriate measure of thermodynamic work.

2. The spectral distribution of the radiation at the earth surface is only weakly related to the spectral distribution outside the atmosphere. This means that a measurement of the spectral distribution and spacial distribution will usually be required at a site to determine the available energy flux.

Calculation results are presented for the available energy flux at a horizontal surface due to Rayleigh scattered radiation in the atmosphere. Rayleigh scattering was chosen because of the good prediction of this component of energy flux both spectrally and spacially under clear sky conditions. It also provided a non uniform spacial distribution which could illustrate flux aspects of the available energy of solar radiation. The Rayleigh scattering model of Coulson (10) was used to calculate the available energy flux from the energy flux.

Determination of the available energy of sunlight if diffused uniformly over the sky have been previously examined by the author (1) with the results indicated in Figures 4 and 5. The principal result being that the available energy to energy flux ratio for uniform solar radiation is from 0.5 to 0.7.

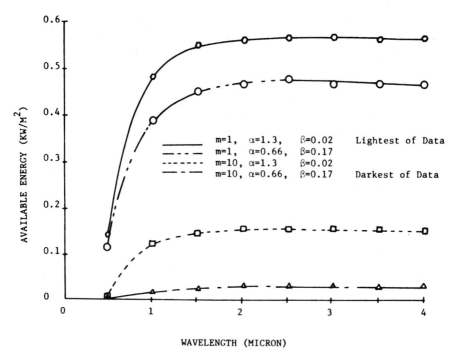

Figure 4. Available energy of solar radiation if scattered
uniformly over the sky as a function of the upper wavelength
received by a collector. Reproduced with permission from
Ref. 1, Copyright 1980, Pergamon Press Inc.

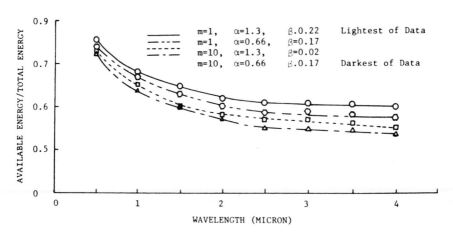

Figure 5. Available energy to energy flux ratio for solar
radiation if scattered uniformly over the sky. Reproduced
with permission from Ref.(1) Copyright 1980, Pergamon Press
Inc.

Assumptions and Procedures

In the calculations of the available energy flux reported in this paper for Rayleigh scattered solar radiation, the following assumptions and procedures have been utilized.

The Coulson-Dave-Sekera Tables (10) of diffuse radiation emerging from the bottom of a plane-parallel atmosphere due to Rayleigh scattering were used to calculate the spectral energy intensity $e_{\nu\omega}$. $e_{\nu\omega}$ is the energy flux per unit area, per unit frequency ν, per unit solid angle ω. This calculation procedure is outlined in the appendix. In this model, multiple scattering and a ground reflectivity of 0.25 is assumed. The input spectral radiation flux outside the atmosphere $F_{o\nu}$ is taken from the data of Thomas and Thekaekara (11). The spacial geometry utilized in this description is shown in Figure 6.

The basic equations used to evaluate the available energy flux are derived in reference(1). They are outlined below. The steady flow available energy flux per unit area is taken as

$$b = e + pv - T_0 s$$

This is found by integration of the spectral and spacial functions over the spectrum of the radiation and the solid angle over the hemisphere above the surface. The available energy intensity is

$$b_{\nu\omega} = e_{\nu\omega} + (2h\nu^3/c^2)\left(1-(\ln(D)/\ln(1-D))\right)$$

$$-kT_0\left((2\nu^2/c^2)\ln(1+(1/D))+(e_{\nu\omega}/h\nu)\ln(1+D)\right)$$

Note: $D = 2h\nu^3/c^2 e_{\nu\omega}$

The direction of the measurement of the energy flux is given by the zenith angle θ (or $\mu=\cos\theta$) and the azimuth angle ϕ of a vertical plane through the direction of observation. The sun is taken at an azimuth angle of zero. The direction of the sun is then noted as $\mu_0=\cos\theta_0$ and $\phi_0=0$.

The energy flux through a horizontal surface $e_{\nu\omega}$ is then found from the actual energy flux from a solid angle as

$$e_{\nu\omega} = (e_{p\nu\Omega}/\Omega)\cos\theta$$

The reason for this complication in the computation is that $e_{p\nu\Omega}$ is the spectral energy flux one would expect to measure with a spectral radiometer with an aperture which accepts radiation in a solid angle Ω. This energy flux is then measured on a surface perpendicular to the direction in which it is pointed. The surface for which the energy flux is to be found will be at a fixed angle and only the $\cos\theta$ component will be a flux through the surface.

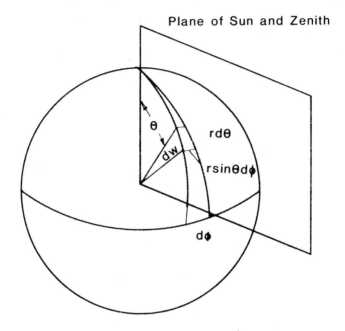

Figure 6. Geometry used for calculation of available energy flux and energy flux with spacially variable intensity.

Integrations over the spectrum and spherical angles were done numerically with azimuth angle increments $\Delta\phi = 30^\circ = 0.5236$ radians, and zenith angle increments $\Delta\mu = \Delta(\cos\theta) = 0.1$. The available energy flux per unit frequency is computed as

$$b_\nu = \sum_{\cos\theta} \sum_\phi b_{\nu\omega} \, \Delta(\cos\theta)\Delta\phi$$

The total available energy flux is then

$$b = \sum_\nu b_\nu \, \Delta\nu$$

The calculation of b was simplified by noting that for Rayleigh scattered solar radiation

$$D \gg 1$$

Thus, $\ln(1 + 1/D)$ is approximated by $1/D$. The available energy flux on a horizontal surface is then

$$b_{\nu\omega} = [e_{p\nu\Omega}/\Omega + 2h\nu^3/c^2(1 - (\ln(D)/(\ln(1+D))))$$

$$- kT_0(e_{p\nu\Omega}/h\nu\Omega)(1 + \ln(1+D))] \cos\theta$$

To the same level of approximation, the spectral temperature of a given spectral energy flux component is

$$T = (h\nu/k)\ln(D)$$

Calculated values of the separate components of the available energy flux are shown in Table III. These are tabulated as A,B, and C as follows

A   Energy Term : $e_{p\nu\Omega}/\Omega$

B   Pressure Term : $(2h\nu^3/c^2)[1 - (\ln(D)/(\ln(1+D)))]$

C   Entropy Term : $-kT_0(e_{p\nu\Omega}/h\nu\Omega)[1 + \ln(1+D)]$

Discussion of Results.

The results of the computation of the available energy of
Rayleigh scattering are presented in Tables I-V. In
discussing the results, it should be kept in mind that if all
the radiation could be converted to thermodynamic work then the
ratio b/e would be equal to 4/3. In the steady flow available
energy formulation the radiation pressure also is available for
producing useful work.

Let us begin by examining the overall results and then discuss the separate factors which account for the results. Table I
indicates that the ratio of available energy to energy flux on a
horizontal surface is 0.89 for the sun at the zenith $\mu_0=1$ and
0.82 for the sun at $\mu_0=0.1$ or approximately 6° above the
horizon. This is a surprising result when compared to statements
in the literature concerning the low quality of diffuse solar
energy. In order to explain this difference let us first
consider Table I which shows the variation of the available
energy to energy flux ratio as a function of the wavelength of
the scattered radiation. This table illustrates the expected
result that the shorter wavelength radiation has a higher
available energy than the long wavelength radiation as noted
before in the discussion of uniform radiation fields.

The available energy to energy ratio is also higher for the
sun at the zenith than at the horizon. This difference is not
significant and is within the computational error expected. The
magnitude however requires some discussion relative to the
result for direct solar radiation. Direct solar radiation of
energy flux 1.353 kW/m^2 outside the atmosphere in a solid angle
$\Omega = 6.78\times10^{-5}$ would give an available energy to energy flux
ratio of 1.26 thus the Rayleigh scattered radiation has reduced
the available energy from the direct solar energy. Computation
of available energy to energy flux ratios for actual direct
solar radiation, if uniformly scattered over the full sky
hemisphere (as noted in Figure 5), shows that the available
energy to energy flux ratio is smaller. The reason that the
Rayleigh scattered radiation has a higher available energy than
this uniformly distributed solar radiation, is that Rayleigh
scattering is predominantly a scattering of short wavelength
light as shown in Figure 7. Rayleigh scattering decreases as
λ^{-4}. Thus the spectrum of scattered radiation is shifted
toward the short wavelengths compared to direct solar radiation.
This shifting is also shown in Figure 2. The spectrum for the
Rayleigh scattered light is a high available energy spectrum.

Examination of Table II further clarifies the difference

Table I. Available Energy to Energy Flux Ratio for Rayleigh
Scattered Solar Radiation as a Function of Wavelength

Transmission τ	Wavelength $\lambda(\mu)$	Sun Near Horizon $\mu_o= 0.1$ (b_ν/e_ν)	Sun At Zenith $\mu_o= 1.0$ (b_ν/e_ν)
0.02	0.82	0.68	0.73
0.05	0.65	0.77	0.80
0.10	0.55	0.80	0.84
0.15	0.50	0.82	0.86
0.25	0.44	0.85	0.88
0.50	.37	0.88	0.89
1.00	.32	0.97	1.01

Overall Available Energy to Energy Flux Ratios.

Sun At Zenith   b/e = 0.89
Sun Near Horizon   b/e = 0.82

Table II. Available Energy to Energy Flux Ratio Per Unit
Frequency (b_ω/e_ω) as a Function of Azimuth Angle ϕ.

| Zenith Angle | Azimuth Angle | | | | | | |
	0°	30°	60°	90°	120°	150°	180°
6°	0.79	0.81	0.81	0.79	0.81	0.81	0.81
60°	0.82	0.82	0.84	0.82	0.82	0.82	0.82

Table III. Available Energy Flux Components $b_{\nu\omega}$ on a Horizontal Surface for the Sun at the Zenith. $e_{\nu\omega} = A \cos \theta$, $pv_{\nu\omega} = B \cos \theta$, $T_o S_{\nu\omega} = C \cos \theta$, $\mu = \cos \theta$, τ is a Transmission Parameter (See Table I). Units are $fW\text{-}s/m^2$

μ/τ		0.02	0.05	0.10	0.15	0.25	0.50	1.00
	A	4.2	8.8	11.75	13.5	12.1	6.3	2.9
0.1	B	0.252	0.530	0.698	0.800	0.694	0.343	0.151
	C	1.27	2.11	2.41	2.53	2.05	0.96	0.40
	$b_{\nu\omega}$	0.35	0.72	1.00	1.18	1.07	0.57	0.27
	A	1.6	3.6	5.5	7.2	7.8	5.4	3.1
0.3	B	0.089	0.210	0.315	0.408	0.436	0.290	0.162
	C	0.50	0.91	1.18	1.39	1.35	0.83	0.42
	$b_{\nu\omega}$	0.35	0.87	1.40	1.86	2.06	1.46	0.85
	A	1.0	2.5	3.8	5.1	5.9	4.6	3.0
0.5	B	0.058	0.137	0.214	0.287	0.328	0.245	0.155
	C	0.34	0.63	0.83	1.01	1.04	0.71	0.42
	$b_{\nu\omega}$	0.38	0.98	1.61	2.21	2.60	2.06	1.39
	A	0.8	2.0	3.2	4.3	5.0	4.1	3.0
0.7	B	0.045	0.110	0.176	0.238	0.275	0.219	0.151
	C	0.27	0.51	0.65	0.85	0.89	0.65	0.41
	$b_{\nu\omega}$	0.42	1.10	1.88	2.57	3.09	2.60	2.70
	A	0.7	1.7	2.9	3.9	4.6	3.9	2.9
1.0	B	0.040	0.096	0.153	0.214	0.250	0.208	0.048
	C	0.24	0.45	0.62	0.77	0.77	0.61	0.40
	$b_{\nu\omega}$	0.53	1.38	2.33	3.31	4.06	3.50	2.68

Table IV. Sample Available Energy Flux Per Unit Solid Angle
and Frequency Distributed Over the Sky for the Sun Near the
Horizon (μ= 0.1). Units are fW-s/m^2

WAVELENGTH   0.821 MICRONS

μ/ϕ	0	30	60	90	120	150	180
0.1	0.023	0.022	0.021	0.020	0.021	0.021	0.022
0.3	0.093	0.086	0.073	0.066	0.071	0.082	0.089
0.5	0.18	0.17	0.15	0.14	0.15	0.16	0.17
0.7	0.22	0.20	0.17	0.16	0.17	0.18	0.19
1.0	0.21	0.21	0.21	0.21	0.21	0.21	0.21

WAVELENGTH   0.395 MICRONS

μ/ϕ	0	30	60	90	120	150	180
0.1	0.70	0.62	0.46	0.37	0.45	0.61	0.69
0.3	0.80	0.71	0.53	0.44	0.50	0.66	0.78
0.5	0.76	0.68	0.53	0.45	0.49	0.64	0.70
0.7	0.69	0.64	0.53	0.44	0.47	0.55	0.61
1.0	0.43	0.43	0.43	0.43	0.43	0.43	0.43

Table V. Equivalent Temperature of Rayleigh Scattered Solar
Radiation for Clear Sky Conditions

Sun Position $\cos\theta_o$	Sky Direction $\cos\theta$	Wavelength microns	Temperature K
1.0	0.1	0.316	2290
1.0	1.0	0.316	2290
1.0	0.1	0.821	1050
1.0	1.0	0.821	950
0.6	0.6	0.316	2230
0.6	1.0	0.316	2160
0.6	0.6	0.821	1035
0.6	1.0	0.821	905
0.1	0.1	0.316	1870
0.1	1.0	0.316	1830
0.1	0.1	0.821	930
0.1	1.0	0.821	810

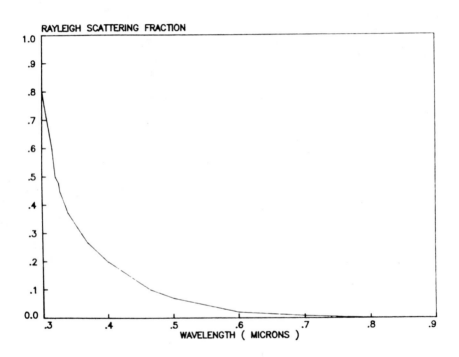

Figure 7. Fraction of direct solar radiation scattered by molecular constituents of the atmosphere (Rayleigh scattering) as a function of wavelength.

in the available energy as a function of wavelength.   In direct
solar radiation outside the atmosphere the pv term is approximately 0.33 of the energy flux and the T_0s term approximately
0.07 of the energy flux.   In the Rayleigh scattered radiation
the pv term is approximately 0.05 of the energy flux and is
essentially constant over wavelength and sky direction.   The
pv term reduction has a small practical significance since
technical methods to utilize this effect have so far not been
developed.   The T_0s term however ranges from 0.35 at long
wavelengths to 0.13 at short wavelengths and is important.   The
entropy effect is however nearly constant over the sky so that
horizon scattering does not have a strong effect on the
available energy flux.

It is of interest for assigning an equivalent temperature
to scattered radiation to examine Table III.   Equivalent temperatures calculated using Wein's distribution

$$T = (h\nu/k)\ln(2h\nu^3\Omega/c^2 e_{pv\Omega})$$

are tabulated for the cases of the sun near the horizon $\mu_0 =$
0.1 and at the zenith $\mu_0 = 1$.   Overall, these results indicate
that the equivalent temperature for Rayleigh scattered solar
radiation ranges from around 2300K to 800K.

For the sun at the zenith the horizon brightening is clear
with equivalent temperatures there higher than directly overhead.   The wavelength is the dominant factor however with short
wavelength temperatures of 2000K compared to long wavelenth
temperatures of 1000K.   It should be noted that these
temperatures are in the range of temperatures that Bosnjakovic
(12) assumes for his approximation for diffuse radiation in
calculating available energy fluxes.

Conclusions

The available energy flux of Rayleigh scattered solar radiation
in the atmosphere was examined in this paper.   It was demonstrated that the available energy to energy flux ratio on at
the earth surface is approximately 80% for molecularly
scattered solar energy.   This clarifies further the role of
spectral and special distribution effects on the available
energy of solar energy in the atmosphere.   It reinforces the
principal importance of the spectral distribution of the radiation in determining the quality of the radiation.   Scattering

processes in the atmosphere reduce the energy flux on surfaces but do not significantly reduce the available energy to energy flux ratio.

The results indicate that efforts to utilize scattered radiation should be directed toward research in spectral selective devices with less concern for the spacial distribution.

The collection of scattered radiation remains an engineering challenge. One need be less concerned about the quality but more at producing a high energy flux. One outcome of this research may be more effort being placed on the utilization of emitted radiation from collective surfaces which have so far been thought to have low quality because of their wide spacial distribution. It is hoped that this work will further efforts directed to using combined energy conversion systems where the transmitted, reflected and emitted radiation from a collection system are utilized.

Literature Cited

1. Edgerton, R. H. Energy, Int. J. 1980, 5, 693-707.
2. Press, W. H. Nature 1976, 264, 734-735.
3. Landsberg, P.T.; Tonge, G. J.Phys. A., 1979, 12, 551-562.
4. Leontovich, N. A. Sov. Phys-USP 1975, 6, 963-964.
5. Parrott, J. E. Solar Energy 1978, 21, 227.
6. Landsberg, P. J.; Tonge, G. J. Appl. Phys. 1980, 51, R1-R20.
7. Jeter, S. M. Solar Energy 1981, 26, 231-236.
8. Landsberg, P. T. "Thermodynamics and Statistical Mechanics"; Oxford University Press: Oxford, 1978.
9. Coulson, K. L. "Solar and Terrestrial Radiation"; Academic Press: New York, 1975.
10. Coulson, K.L.; Dave, J. V.; Sekera, Z., "Tables Related to Radiation Emerging from a Planetary Atmosphere with Rayleigh Scattering"; University of California, Berkeley, California, 1960.
11. Thomas, A. P.; Thekaekara, M. P., "Experimental and Theoretical Studies of Solar Energy for Energy Conversion"; NASA/Goddard Space Flight Center, 1973.
12. Bosnjakovic, F. M., "Studies in Heat Transfer"; Hartnett, J. et al., Ed.; Hemisphere Publishing Co: New York, 1979.

Appendix

The computational procedure for the calculation of energy flux per unit frequency per unit solid angle $e_{\nu\omega}$ due to Rayleigh scattered solar radiation is outlined in this section.

The geometry of Figure 6 is used to describe the calculations. The total energy flux through a surface due to a spectral radiation intensity I_ν is given by integrating the intensity over the hemispherical space and the frequency as

$$e = \int\int\int I_\nu \cos\theta \sin\theta \; d\theta \, d\phi \, d\nu$$

If I_ν is isotropic designated as I_ν' then the energy flux would be

$$e' = \pi \int I_\nu' d\nu$$

The energy flux can then be found relative to this isotropic energy flux as

$$e = (e'/\pi)\int\int\int (I_\nu/I_\nu') \cos\theta \sin\theta \, d\theta \, d\phi \, d\nu$$

I_ν/I_ν' is a relative intensity $I_{\nu R}$ tabulated for Rayleigh scattering by Coulson, Dave and Sekera (10).

Since the energy flux e' represents an energy flux outside the atmosphere then if the sun is at an angle θ_0 to the zenith this energy flux per unit frequency is represented by

$$e_\nu' = F_{0\nu} \cos\theta_0$$

$F_{0\nu}$ is the solar flux at the earth distance per unit frequency ν . Then for the solar energy the energy flux per solid angle ω per unit frequency can be written as

$$e_{\nu\omega} = ((F_{0\nu}\cos\theta_0)/\pi)I_{\nu R}$$

Then

$$e = \int\int\int ((F_{0\nu}\cos\theta_0)/\pi)I_{\nu R}\cos\theta \sin\theta \, d\theta \, d\phi \, d\nu$$

The values of $F_{0\nu}$ are found from reference(11) at the wavelengths or frequencies at which $I_{\nu R}$ values are tabulated. These values are summarized in Table A.

Table A. Parameter Values Used To Calculate Energy Flux.

τ	$\lambda(\mu)$	ν(PHz)	$\Delta\nu$(PHz)	$F_{0\nu}$(PW-s/m^2)
0.02	0.821	0.365	0.096	2.25
0.05	0.654	0.456	0.090	2.16
0.10	0.551	0.545	0.0715	1.75
0.15	0.498	0.602	0.0685	1.61
0.25	0.440	0.682	0.100	1.17
0.50	0.374	0.802	0.1335	0.54
1.00	0.316	0.949	0.150	0.25

Note that the scattering as a function of wavelength is tabulated in terms of τ an optical thickness for corresponding wavelengths noted in Table A. In tables of results then, increments of τ are used as well as wavelength λ to show wavelength dependence of scattering.

The integral for the energy flux is summed over the spectrum

$$e = \sum_{\nu} e_{\nu} \Delta\nu$$

where

$$e_{\nu} = \left((F_{o\nu} \cos\theta_o)/\pi\right) \sum_{\cos\theta} \sum_{\phi} I_{\nu R} \cos\theta \Delta\mu \Delta\phi$$

and $\Delta\mu = 0.1$; $\Delta\phi = 0.5236$ radians.

RECEIVED July 12, 1983

BIBLIOGRAPHY

Y. A. LIU

Department of Chemical Engineering, Virginia Polytechnic Institute and State University, Blacksburg, VA 24061

W. J. WEPFER

School of Mechanical Engineering, Georgia Institute of Technology, Atlanta, GA 30332

This bibliography contains most of the English literature on second law analysis (SLA) published by early 1983. It also includes most of the European literature which appeared after early 1977. A related bibliography, compiled by Wepfer in appendix A.1 of reference 236, covers 404 European publications available before early 1977.

The specific publications are listed according to the following topics.

A. Selected textbooks and general articles for background on thermodynamics (1 to 7).

B. Selected textbooks and published notes on SLA (8 to 14).

C. Selected conference proceedings on SLA (15 to 18).

D. General publications on the theory and applications of SLA (19 to 70).

E. General publications on SLA and energy utilization (71 to 104).

F. Basic principles and applications of thermoeconomics and available energy costing (105 to 115).

G. Applications of SLA to the design, evaluation and optimization of cryogenic equipment/processes (116 to 140).

H. Desalination processes (141 to 150).

I. Fluid flow and heat transfer (thermal) equipment/processes, including solar energy systems (151 to 207).

J. Power plants and complex energy systems, including electrochemical, geothermal, nuclear and hydrogen energy systems (208 to 238).

K. Mass transfer and separation equipment/processes (239 to 267).

L. Combustion and chemical reaction processes (268 to 295).

M. Fuel conversion processes (296 to 321).

N. Applications of SLA to the optimal design and systematic synthesis of energy-efficient chemical processes (322 to 356).

0097-6156/83/0235-0415$08.75/0
© 1983 American Chemical Society

BIBLIOGRAPHY

A. Selected Textbooks and General Articles for Background on
    Thermodynamics

1. Baehr, H. D., Thermodynamics; An Introduction to Fundamental
    Principles and Engineering Applications, 2nd Edition, Spring-
    Verlag, Berlin (1966).

2. Bett, K. E., Rowlinson, J. S. and Saville, G., Thermodynamics
    for Chemical Engineers, MIT Press, Cambridge, MA (1975).

3. Hatsopolous, G. N. and Keenan, J. H., Principles of General
    Thermodynamics, Wiley, New York (1965).

4. Keenan, J. H., Thermodynamics, Wiley, New York (1944).

5. Gaggioli, R. A., "Principles of Thermodynamics," in Thermo-
    dynamics: Second Law Analysis, ACS Symp. Ser., No. 122,
    R. A. Gaggioli, Editor, pp. 3-14, ACS, Washington, D.C.
    (1980).

6. Obert, E. F. and Gaggioli, R. A., Thermodynamics, 2nd
    Edition, McGraw-Hill, New York (1963).

7. Sussman, M. V., Elementary General Thermodynamics, Addison-
    Wesley, Reading, MA (1972).

B. Selected Textbooks and Published Notes on Second Law Analysis

8. Ahern, J. E., The Exergy Method of Energy System Analysis,
    Wiley, New York (1980).

9. Bejan, A., Entropy Generation Through Heat and Fluid Flow,
    Wiley, New York (1982).

10. Bruges, E. A., Available Energy and the Second Law Analysis,
    Academic Press, London (1959).

11. Moran, M. J., Availability Analysis - A Guide to Efficient
    Energy Use, Prentice-Hall, Englewood Cliffs, NJ (1982).

12. Rathore, R. N. S. and Kenney, W. F., Thermodynamic Analysis
    for Improved Energy Efficiency, AIChE Today Series, AIChE,
    New York, (1980).

13. Seader, J. D., Thermodynamic Efficiency of Chemical
    Processes, MIT Press, Cambridge, MA (1982).

14. Sussman, M. V., "Availability (Exergy) Analysis: A Self-Instruction Manual," M. V. Sussman, Publisher, Department of Chemical Engineering, Tufts University, Medford, MA (1980).

C. Selected Conference Proceedings on Second Law Analysis and Its Applications

15. Cambel, A. B., Heffernan, G. A., Cutler, D. W. and Ghamarian, A., Editors, Proceedings of a Workshop on Second Law Analysis of Energy Devices and Processes, George Washington University, Washington, D.C., Aug./Sept., 1979; published in Energy (Oxford), 5, 665-1012, Aug./Sept. (1980).

16. Fazzolare, R. A. and Smith, C. B., Editors, Energy Use Management (Proceedings of the International Conference, Tuscon, Arizona), Vol. II, pp. 1-106, Pergamon Press, New York (1978).

17. Gaggioli, R. A., Editor, Thermodynamics: Second Law Analysis, ACS Symp. Ser., No. 122, ACS, Washington, D.C. (1980).

18. Symposium on Second Law Analysis and Applications, in Vol. II, pp. 238-301 and 385-425, Proceedings of the 2nd World Congress of Chemical Engineering, Montreal, Canada, Published by the Canadian Society of Chemical Engineers, October (1981).

D. General Publications on the Theory and Applications of Second Law Analysis

19. Ahrendts, J., "Reference States," Energy (Oxford), 5, 667 (1980).

20. Borel, L., "General Theory of Exergy and Practical Applications: 1. Energy Balance, Coenergy and coenthalpy," Entropie, 15, No. 85, 3 (1979).

21. Borel, L., "General Theory of Energy and Practical Applications: 2. Heat Exergy, Transformation Exergy and Exergetic Losses," Entropie, 15, No. 86, 3 (1979).

22. Chandrashekar, M. and Wong, F. C., "Thermodynamics Systems Analysis - 1. A Graphic-Theoretic Approach," Energy (Oxford), 7, 539 (1982).

23. Chen, H. T. and Tucker, W. H., "Availability Analysis - An Academic Viewpoint," ASHRAE Symposium, Paper No. LO-73-4 (1973).

24. Chiogioli, M. H., Industrial Energy Conservation, Chapter 3, "Thermodynamic Availability Analysis," Marcel Dekker, Inc., New York (1979).

25. Dealy, J. M., and Weber, M. E., "Thermodynamic Analysis of Process Efficiency," Appl. Energy, 6, 177 (1980).

26. Denbigh, K. G., "The Second-Law Efficiency of Chemical Processes," Chem. Eng. Sci., 6, 1 (1956).

27. De Nevers, N., "Two Fundamental Approaches to Second-Law Analysis," in Foundations of Computer-Aided Chemical Process Design, Vol. II, pp. 501-536, R. S. H. Mah and W. D. Seider, Editors, Engineering Foundation, New York, NY (1981).

28. De Nevers, N., and Seader, J. D., "Thermodynamic Lost Work and Thermodynamic Efficiencies of Processes," AIChE National Meeting, Houston, TX, April (1979).

29. De Nevers, N. and Seader, J. D., "Lost Work: A Measure of Thermodynamic Efficiency," Energy (Oxford), 5, 757 (1980).

30. El-Sayed, Y. M. and Tribus, M., "Strategic Use of Thermo-economic Analysis for Process Improvement," AIChE National Meeting, Detroit, MI, Aug. (1981).

31. Evans, R. B., Crellin, G. L. and Tribus, M., "Basic Relation-ships among Entropy, Exergy, Energy and Availability," in Principles of Desalination, K. S. Spiegler, Editor, pp. 44-66, Chapter 2, Appendix A, Academic Press, New York (1966).

32. Evans, R. B., "A Proof That Essergy Is the Only Consistent Measure of Potential Work," Ph.D. Dissertation, Thayer School of Engineering, Dartmouth College, N.H. (1969).

33. Evans, R. B., "Thermoeconomic Isolation and Essergy Analysis," Energy (oxford), 5, 805 (1980).

34. Gaensslen, H., "Thermal Efficiency and Economic Optimization in Chemical Plants," In reference 18, p. 281 (1981).

35. Gaggioli, R. A., "The Concept of Available Energy," Chem. Eng. Sci., 16, 87 (1961).

36. Gaggioli, R. A., "The Concepts of Thermodynamic Friction, Thermal Available Energy, Chemical Available Energy and Thermal Energy," Chem. Eng. Sci., 17, 523 (1962).

37. Gaggioli, R. A. and Wepfer, W. J., "Available-Energy Accounting," AIChE National Meeting, Philadelphia, PA, June (1978).

38. Gaggioli, R. A., "More on Generalizing the Definitions of Heat and Entropy," Int. J. of Heat and Mass Transfer, 12, 656 (1969).

39. Gaggioli, R. A., "Second Law Analysis for Process and Energy Engineering," AIChE Annual Meeting, Chicago, IL, Nov. (1980).

40. Grant, C. D. and Anozie, A. N., "The Use of Exergy Analysis for Process Plant Improvements," In reference 18, p. 392 (1981).

41. Gyftopoulos, E. P. and Widmer, T. F., "Availability Analysis: The Combined Energy and Entropy Balances," In reference 17, p. 61 (1980).

42. Haywood, R. W., "A Critical Review of the Theorems of Thermodynamic Availability, with Concise Formulations," J. Mech. Eng. Sci., 16, No. 3, 160 and 258 (1974).

43. Hevert, H. W. and Hevert, S. C., "Second-Law Analysis: An Alternative Indicator of System Efficiency," Energy (Oxford), 5, 865 (1980).

44. Kameyama, H., Yoshida, K., Yamauchi, S. and Fueki, K., "Evaluation of Reference Exergies for the Elements," Appl. Energy, 11, No. 1, 69 (1982).

45. Keenan, J. H., "Availability and Irreversibility in Thermodynamics," Brit. J. Appl. Phys., 2, 183 (1951).

46. Kestin, J., "Availability: The Concept and Associated Terminology," Energy (Oxford), 5, 679 (1980).

47. Mansoori, G. and Gomez, A. L., "Thermodynamic Efficiencies Revisited: Upper and Lower Bounds to the Efficiency and Coefficient of Performance Based on 2nd Law of Thermodynamics," In reference 18, p. 241 (1981).

48. Naimpally, A., "New Thermodynamic Function, Fugergy, to Facilitate Computation of the Thermodynamic Availability Function (Exergy)," Proc. 16th Intersociety Energy Convers. Eng. Conf., Atlanta, GA, Vol. 1, p. 9, Aug. (1981).

49. Parker, A. L., "Availability Energy Analysis Within a Chemical Simulator," In reference 18, p. 289 (1981).

50. Petit, P. J. and Gaggioli, R. A., "Second Law Procedures for Evaluating Processes," In reference 17, p. 15 (1980).

51. Pittas, A. C., "Energy, Its Quality and Efficiency of Conversion. The First and Second Law Analysis of Engineering Systems." Australian Institute of Refrigeration, Air-Conditioning and Heat (AIRAH) Conference, Hobart, Tasmania, April (1975).

52. Reistad, G., "Availability: Concepts and Applications," Ph.D. Dissertation, University of Wisconsin, Madison, WI (1970).

53. Rodriguez, L. S. J., "Calculation of Available-Energy Quantities," In reference 17, p. 39 (1980).

54. Rubin, M. H., Andersen, B. and Berry, R. S., "Finite-Time Constraints and Availability," AIChE National Meeting, Detroit, MI, Aug. (1981).

55. Sato, M., "Presentation of a New Formulation of Negetropy: 2. Description of Work by Negentropy," Bull. Japanese Soc. Mech. Eng., 25, 1551 (1982).

56. Schiff, D., "Entropy (Available Energy) Analysis, Energy and Environmental Control Applications," Technical Report, Mitre Corporation, McLean, VA, NTIS PC A05/MF A01, National Technical Information Service (1978).

57. Shieh, J. H. and Fan, L. T., "Estimation of Energy (Enthalpy) and Exergy (Availability) Contents in Structurally Complicated Materials," Energy Sources, 6, 1 (1982).

58. Standart, G., and Lockett, M. J., "The Available Energy Balance in Heterogeneous Flow Systems," Chem. Eng. J., 2, 143 (1971).

59. Sussman, M. V., "Availability Analysis," In reference 16, p. 57 (1978).

60. Sussman, M. V., "Standard Chemical Availability," Chem. Eng. Progr., 76, No. 1, 37 (1980).

61. Sussman, M. V., "Steady-State Availability and the Standard Chemical Availability," Energy (Oxford), 5, 793 (1980).

62. Sussman, M. V., "Geometric Aspects of the Availability Functions," In reference 18, p. 286 (1981).

63. Sussman, M. V., "Second Law Efficiencies and Reference States for Exergy Analysis," In reference 18, p. 420 (1981).

64. Szargut, J. and C. Dziedziniewicz,"Utilizable Energy of Inorganic Chemical Substances," Entropie, 40, 14 (1971).

65. Szargut, J., "International Progress in Second Law Analysis," Energy (Oxford), 5, 709 (1980).

66. Tapia, C. and Moran, M. J., "A Generalized Property Chart to Evaluate Exergy," In reference 18, p. 385 (1981).

67. Tolman, R. C. and P. C. Fine, "On Irreversible Production of Entropy," Rev. Mod. Phys., 10, No. 1 (1948).

68. Tribus, M. "Generalizing the Meaning of 'Heat'." Int. J. Heat and Mass Transfer, 11, 9 (1968).

69. Wepfer, W. J. and Gaggioli, R. A., "Reference Datums for Available Energy," In reference 17, p. 77 (1980).

70. Wepfer, W. J., "Applications of Available-Energy Accounting," In reference 17, p. 161 (1980).

E. General Publications on Second Law Analysis and Energy Utilization

71. Baloh, T., "Methods for Energy Investigations in Sugar Factories," Zuckerindustrie (Berlin) 106, No. 1, 29 (1981).

72. Berg, C. A., "A Technical Basis for Energy Conservation," Technology Review, 76, February (1974); also in Mech. Eng., 96, No. 5, 30 (1974).

73. Berg, C. A., "Process Integration and The Second Law of Thermodynamics: Future Possibilities," Energy (Oxford), 5, 733 (1980).

74. Boberg, R., "The Preparation of Thermodynamic Balances for Plants for Energy Utilization," Proc. 8th World Energy Conf., 7, 148 (1971).

75. Calm, J. M., "Energy Effectiveness Measurement for Integrated Energy Systems," In reference 16, p. 23 (1978).

76. Cirrito, A. J., "The Energy Reformation," in Proceedings of the 14th Intersociety Energy Conversion Engineering Conference, Vol. 2, pp. 1758-1761, ACS, Washington, D.C. (1979).

77. Cozzi, C. "Thermodynamics and Energy Accountancy in
     Industrial Processes." Energy Sources, 2, No. 2, 165 (1975).

78. Deroutte, J. J., "Application of the Second Law of Thermo-
     dynamics of Basic Industrial Processes," pp. 303-311, in
     Proceedings of an International Seminar on New Ways to Save
     Energy, Brussels, Belgian, Oct. 1979; D. Reidel Publ. Co.,
     Dordrecht, Halland and Boston, Mass. (1980).

79. Edgerton, R. H., "Measurements of Energy Effectiveness of
     Interacting Resources Processing Systems," Energy (Oxford),
     4, 1151 (1979).

80. Evans, R. B., "A New Approach for Deciding upon Constraints
     in the Maximum Entropy Formalism," pp. 169-203, in The
     Maximum Entropy Formalism, MIT Press, Cambridge, MA (1978).

81. Ford, K. W., Rochlin, G. J. and Socolow, R. H., Part I in
     "Efficient Use of Energy," API Conference Proceedings, No.
     25, American Institute of Physics, New York (1975).

82. Gyftopoulos, E. P., Lazaridia, L. and Widmer, T., Potential
     Fuel Effectiveness in Industry, Ballinger Publ. Co.,
     Cambridge, MA (1974).

83. Gypftopoulos, E. P. and Widmer, T. F., "Effective Energy End-
     Use: Opportunities and Barriers," In reference 16, p. 45
     (1978).

84. Hall, E. and Hanna, W., "Evaluation of the Theoretical
     Potential for Energy Conservation in Seven Basic Industries,"
     Report from Battelle Columbus Laboratories to the Federal
     Energy Administration, Report No. FEA/D-75/CE1 (1975).

85. Hamel, B. B. and Brown, H. L., "Measures of Thermal Energy
     Utilization," NBS Publications No. 403, pp. 57-64, National
     Bureau of Standards, Washington, D.C., June (1976).

86. Hanna, W. T. and Frederick, W. J., "Theoretical Potential
     for Energy Conservation in the Pulp and Paper Industry,"
     in AIChE Symp. Series, Vol. 74, No. 177, Energy and Environ-
     mental Concerns in the Forest Products Industry, Edited by
     W. T. McKean, AIChE, New York (1978).

87. Hedman, B. A., Brown, H. L. and Hamel, B. B., "Second Law
     Analysis of Industrial Processes," Energy (Oxford), 5, 931,
     (1980).

88. Hendrix, W. A., "Practical Application of Useful Energy Analysis for Industrial Energy Conservation." Proceedings of Conference on Industrial Energy Conservation Technology, Houston, TX, April (1980).

89. Jorgensen, S. E., "Exergy as Key Function in Econological Models: Energy and Ecological Modeling," pp. 587-590, in Proc. Intern. Symp. on Energy and Ecological Modeling, Louisville, KY, Elesvier, Amsterdam (1981).

90. Keenan, J. H., Gyftopoulos, E. P. and Hatsopoulos, G. N., "The Fuel Shortage and Thermodynamics: The Entropy Crisis," Proceedings of MIT Energy Conference, M. Macrakis, Editor, MIT Press, Cambridge, MA (1972).

91. Lewis III, J. H., "Propulsive Efficiency from an Energy Utilization Standpoint," J. Aircr., 13, 299 (1976).

92. Linnhoff, B. and Carpenter, K. J., "Energy Conservation by Exergy Analysis: The Quick and Simply Way," In reference 18, p. 248 (1981).

93. Michaelides, E. E., "The Concepts of Available Work as Applied to the Conservation of Fuel Resources," in Proceedings of the 14th Intersociety Energy Conversion Engineering Conference, Vol. 2, pp. 1762-1766, American Chemical Society, Washington, D.C. (1979).

94. Naghdi, P. M., "On the Role of the Second Law of Thermodynamics in Mechanics of Materials," Energy (Oxford), 5, 771 (1980).

95. Reistad, G. M., "Available Energy Conversion and Utilization in the United States," Trans. ASME: J. Eng. Power, 97, 429 (1975).

96. Reistad, G. M., "Available-Energy Utilization in the U.S.," In reference 17, p. 93 (1980).

97. Shapiro, H. N. and Kuehn, T. H., "Second Law Analysis of the Ames Solid Waste Recovery System," Energy (Oxford), 5, 985 (1980).

98. Silver, R. S., "Considerations Regarding Potential Convertibility of Heat into Work," J. Heat Recovery Syst., 1, 205 (1981).

99. Streich, M. and Bolkart, A., "Utilization of Waste Heat from Process Plants," In reference 18, p. 265 (1981).

100. Timmerhaus, K. D. and Flynn, T. M., "Energy Conservation Through Use of Second Law Analysis," Erdoel Kohle, Erdgas, Petrochem. Brennst-Chem., 35, No. 5, 208 (1980).

101. Traegardh, C., "Energy and Exergy Analysis in Some Food Processing Industries," Lebensm-Wiss Technol., 14, No. 4, 213 (1981).

102. Tribus, M., "The Case for Essergy," J. Mech. Eng., p. 75, April (1975).

103. Urdaneta, A. and Schmidt, P. S., "Evaluation of Energy Utilization Analysis Methods for Industrial Processes," In reference 16, p. 65 (1978).

104. Van Gool, W., "Thermodynamic Aspects of Energy Conservation," Energy (Oxford), 5, 783 (1980).

  F. Basic Principles and Applications of Thermoeconomics and Available Energy Costing

105. El-Sayed, Y. M. and Tribus, M., "A Specific Strategy for the Improvement of Process Economics through Thermoeconomic Analysis," In reference 18, p. 278 (1981).

106. Evans, R. B., "A Contribution to the Theory of Thermo-Economics." Master's Thesis, University of California, Los Angeles, CA (1961).

107. Gaggioli, R. A., "Proper Evaluation and Pricing of Energy," In reference 16, p. 57 (1978).

108. Gaggioli, R. A. and Wepfer, W. J., "The Composition of Thermoeconomic Flow Diagrams," In reference 18, p. 295 (1981).

109. Ocallaghan, P. W. and Probert, S. D., "Exergy and Economics," Appl. Energy, 8, 227 (1981).

110. Reistad, G. M. and Gaggioli, R. A., "Available-Energy Costing," In reference 17, p. 143 (1980).

111. Soma, J., and Morris, H. N., "Exergy Management: The Seminal Synergism of Thermodynamics and Economics," Energy Econ. and Policy Manage., 1, No. 4, 6 (1982).

112. Tribus, M. and Evans, R. B., "Thermoeconomics," University of California, Department of Engineering, Los Angeles, Report No. 52-63 (1962).

113. Tribus, M. and Evans, R. B., "Thermoeconomic Design under
      Conditions of Variable Price Structure," International
      Symposium on Water Desalination, Washington, D.C., Oct.
      (1965).

114. Tribus, M., Evans, R. B., and Crellin, G. L., "Thermo-
      economics," Chapter 3, in Principles of Desalination,
      K. Spiegler, Editor, Academic Press, New York (1966).

  G. Applications of Second Law Analysis to the Design, Evalu-
      ation and Optimization of Cryogenic Equipment/Processes

115. Ahern, J. E., "Applications of the Second Law of Thermo-
      dynamics to Cryogenics - A Review," Energy (Oxford), 5,
      891 (1980).

116. Ball, W., "From Cascade Refrigerator for Liquid Air Plant
      Precooler," in Advances in Cryogenic Engineering, Vol. 3,
      K. D. Timmerhaus, Editor, Plenum Press, New York (1960).

117. Barclay, J. A., "Analysis of Liquefaction of Helium Using
      Magnetic Refrigerators," Technical Report, Los Alamos
      National Lab., NTIS PC AO3/MF AO1, National Technical
      Information Service, Dec. (1981).

118. Barclay, J. A., "Can Magnetic Refrigerators Liquefy Hydrogen
      at High Efficiency?" NTIS PC AO2/MF AO1, National Technical
      Information Service, Aug. (1981).

119. Briggs, S. W., "Second-Law Analysis of Absorption Refriger-
      ation," A.G.A/I.G.T. Conference on Natural Gas Research and
      Technology, Chicago, IL, (1971).

120. Brodyanskii, V. M., "Thermodynamic Analysis of Gas-
      Liquefaction Processes: Part I. Basic Methods of Analysis
      and Part II. Analysis of Air Liquefaction by Linde
      Process," Inzhnerno-Fizicheskii Zhurnal, 6, No. 7 and
      No. 10 (1963); English translation appears in reference 8,
      pp. 217-236 (1980).

121. Chiu, C. H. and Newton, C. L., "Second Law Analysis in
      Cryogenic Processes," Energy (Oxford), t, 899 (1980).

122. Chiu, C. H., "Exergy Analysis for Cryogenic Process and
      Equipment Optimizations," In reference 18, p. 269 (1981).

123. Chiu, C. H., "Exergy Analysis Aids Equipment Design for
      Cryogenic Process," Oil Gas J., 80, No. 3, 88, Jan. 18
      (1982).

124. Daly, W. O. and Harris, J. B., "Construction and Use of Exergy Diagrams," in Advances in Cryogenic Engineering, Vol. 25, K. D. Timmerhaus, Editor, Plenum Press, New York (1980).

125. Doering, R., "Thermophysical Properties of Ammonia (R717) and of Fluorinated Refrigerants R11 and R113," pp. 574-588, in Proceedings of the 7th Symp. on Thermophys. Prop., National Bureau of Standards, Gaithersburg, MD, published by ASME, NY, May (1977).

126. Evseev, V. S., "Analysis of a Two-Loop Gas-Liquid Refrigerating Power Plant," Power Eng., 19, No. 3, 150 (1981).

127. Gardner, J. B. and Smith, K. C., "Power Consumption and Thermodynamic Reversibility in Low Temperature Refrigeration and Separation Processes," in Advances in Cryogenic Engineering, Vol. 3, K. D. Timmerhaus, Editor, paper A-4, Plenum Press, New York (1960).

128. Guenther, B., "Process-Integrated Refrigeration Systems Save Driving Energy," Verfahrenstechnik (Mainz)., 15, 751 (1981).

129. Gutowski, H. and Wanner, M., "Liquefying Hydrogen on a Large Scale," pp. 60-69, in Proc. HDT Meeting on Hydrogen-Energy Source of the Future, Essen, Germany, March (1981).

130. Kimenov, G. and Schalapatora, E., "Thermodynamic Analysis of a Vapor Jet Refrigeration Plant," Ki Klima Kaelte Heiz, 10, 187 (1982).

131 Kun, L. C. and Ranov, T., "Efficiency of Low Temperature Expansion Machines," in Advances in Cryogenic Engineering, Vol. 10, K. D. Timmerhaus, Editor, Plenum Press, New York (1965).

132. Leidenfrost, W., Lee, K. H. and Korenic, B., "Conservation of Energy Estimated by Second Law Analysis of a Power-Consuming Process," Energy (Oxford), 5, 47 (1980).

133. Martnovskii, V. S., Cheilyakh, V. T., and Shnaid, T. N., "Thermodynamic Effectiveness of a Cooled Shield in Vacuum Low-Temperature Insulation," Isvestiya Akadami Nauk SSR, Energitika; Transport, No. 2 (1971); English translation appears in reference 8, p. 263 (1980).

134. Matsubara, Y., Kaneko, M., Hiresaki, Y., and Yasukochi, K., "Exergetic Analysis of Multi-Staged Claud Cyclic Helium Refrigerator," pp. 131-134, in Cryogenic Processes and Equipment in Energy System, ASME, NY (1980).

135. Meltser, L. H., "Method of Thermodynamic Evaluation of Theoretical and Actual Cycles of Cooling Machines," Kholodilnaya Teknike i Teknologiya, No. 6, Kiev (1968); English translation appears in reference 8, p. 242 (1980).

136. Shen, L. S., "Essergy Analysis of Basic Vapor Compression Refrigeration Systems," Master's Thesis, Ga. Inst. of Technology, Atlanta, GA (1979).

137. Swers, R., Patel, Y. P., Steward, R. B., "Thermodynamic Analysis of Compression Refrigeration Systems," ASHRAE Semi-Annual Meeting, New Orleans, LA, Jan. (1972).

138. Thirumaleshwar, M., "Exergy Method of Analysis and Its Application to Helium Cryorefrigeration," Cryogenics, 19, 355 (1979).

139. Trepp, C., "Refrigeration Systems for Temperatures Below 25°K with Turbo Expanders," in Advances in Cryogenic Engineering, Vol. 7, K. D. Timmerhaus, Editor, Plenum Press, New York (1961).

140. Tripp, W., "Second-Law Analysis of Compression Refrigeration," ASHRAE J., p. 49, January (1966).

H. Applications of Second Law Analysis to the Design, Evaluation and Optimization of Desalination Processes

141. El-Sayed, Y. M. and Aplenc, A. J., "Application of the Thermoeconomic Approach to the Analysis and Optimization of Vapor-Compression Desalting System," Trans. ASME: J. Eng. Power, 92, 17 (1970).

142. Evans, R. B., "Thermodynamic Availability as a Resource and a Tool for System Optimization," (1958). Appendix II of the report by Tribus, M. et al., Thermodynamic and Economic Considerations in the Preparation of Fresh Water From the Sea, Revised September, 1960. University of Calif., Department of Engineering, Los Angeles, Report No. 59-34 (1960).

143. Evans, R. B. and Tribus, M., "Thermo-Economics of Saline Water Conversion," Ind. Eng. Chem. Proc. Des. Dev., 4, 195 (1965).

144. Evans, R. B., Crellin, G. L., and Tribus, M., "Thermo-
      economic Consideration of Sea Water Demineralization,"
      Chapter 2, in Principles of Desalination, K. S. Spiegler,
      Editor, Academic Press, New York (1966).

145. Kvajic, G., "Solar Desalination by Freezing and Distill-
      ation," Proc. 2nd Intern. Conf. on Alternative Energy
      Sources, Vol. 3, pp. 1009-1046, Miami Beach, FL (1979).

146. Nishimoto, W., "Loss of Available Energy in Desalination
      Plant," pp. 185-194, in Vol. 1, pp. 185-194, Proceedings
      of the 5th Intern. Symp. on Fresh Water from the Sea,
      European Federation of Chem. Engrs., Alghero, Italy,
      May (1976).

147. Slesarenkov, V., and Shtim, A., "Comparative Analysis of
      the Efficiency of Thermal Desalination Plants,"
      Desalination, 37, 269 (1981).

148. Tribus, M., and Evans, R. B., "Economic and Thermodynamic
      Aspects of Sea Water Conversion," Proceedings: Conference
      on Water Research at the University of California, May
      1960, at Davis, California. University of California
      Water Resources Center, Progress Report No. 2, November
      (1960).

149. Tribus, M., and Evans, R. b., "Thermo-Economic Consider-
      ations in the Preparation of Fresh-Water from Sea-Water,"
      (A paper read before the European Symposium, "Fresh-Water
      From the Sea," Sponsored by the European Federation of
      Chemical Engineering at Athens, Greece, June 1962), Dechema
      Monographien, NR. 781-834 BAND 47. Verlag Chemie, GMBH,
      Weinheim/Bergstrasse (1962).

150. Tribus, M. and Evans, R. B., "Optimum-Energy Technique
      for Determining Costs of Saline-Water Conversion," J. Amer.
      Water Works Asso., 54, 1473 (1962).

   I. Applications of Second Law Analysis to the Design, Evalu-
       ation and Optimization of Fluid Flow and Heat Transfer
       (Thermal) Equipment/Processes, Including Solar Energy
       Systems

151. Akau, R. L. and Schoenhals, R. J., "Second Law Efficiency
      of a Heat Pump System," Energy (Oxford), 5, 853 (1980).

152. Bejan, A., "The Concept of Irreversibility in Heat Exchanger
      Design: Counter-flow Heat Exchangers for Gas-to-Gas
      Applications," Trans. ASME: J. Heat Transfer, 99, 374
      (1977).

153. Bejan, A., "General Criterion for Rating Heat-Exchanger
      Performance, Int. J. of Heat and Mass Transfer, 21, 655
      (1978).

154. Bejan, A., "Two Thermodynamic Optima in the Design of
      Sensible Heat Units for Energy Storage," Trans. ASME:
      J. Heat Transfer, 100, 708 (1978).

155. Bejan, A., "Second Law Analysis in Heat Transfer," Energy
      (Oxford), 5, 721 (1980).

156. Bejan, A., "Extraction of Exergy from Solar Collectors
      Under Time-Varying Conditions," Int. J. Heat and Fluid
      Flow, 3, No. 2, 67 (1982).

157. Bejan, A., and Pfister, P. A., Jr., "Evaluation of Heat
      Transfer Augmentation Techniques Based on Their Impact
      on Entropy Generation," Letters on Heat and Mass Transfer,
      7, 97 (1980).

158. Bejan, A., Kearney, D. W. and Kreith, F., "Second Law
      Analysis and Synthesis of Solar Collector Systems," Trans.
      ASME: J. Heat Transfer, 103, 23, Feb. (1981).

159. Belousov, V. S. and Yasnikov, G. P., "Analysis of Exergetic
      Losses in Heat Conductivity Processes," Izu. Wyssh. Uchebn
      Zared Energ., No. 2, 80, Feb. (1978).

160. Bosnjakovic, E. H. F., "Solar Collectors as Energy
      Converters," pp. 331-381, in Studies in Heat Transfer,
      Hemisphere Publ. Corp., Washington, D.C. (1979).

161. Bosnjakovic, E. H. F., "Thermodynamics of Solar Collectors,"
      Fortschr Ber VDI Z. Reiche 6, No. 89, 57 (1981).

162. Boteler, Kevin, Essergy Analysis of Fuel Fired Boilers,
      Master's Thesis, Ga. Inst. of Tech., Atlanta, GA (1981).

163. Brunklaus, J. H., "Pros and Cons of Exergy Considerations
      in the Design of Furnaces," Gas Waerme Int., 26, No. 1,
      19 (1977).

164. Buchet, E., "Heat Exchangers: An Energy Viewpoint
      Approach," Entropie, 17, 124 (1981).

165. Bullock, C. E., "The Application of Availability Analysis
      to Psychrometric Processes," ASHRAE Symposium, Paper No.
      LO-73-4 (1973).

166. Edgerton, R. H., "Second Law and Radiation," Energy (Oxford), 5, 693 (1980).

167. Edgerton, R. H., "Thermodynamic Availability Analysis Applied to Systems with Solar Energy Inputs," AIChE National Meeting, Detroit, MI, Aug. (1981).

168. El-Sayed, Y. M. and Evans, R. B., "Thermoeconomics and the Design of Heat Systems," Engineering School Report, Dartmouth College, Hanover, NH, Feb. (1969).

169. El-Sayed, Y. M. and Evans, R. B., "Thermoeconomics and the Design of Heat Systems," Trans. ASME: J. Eng. Power, 92, 27 (1970).

170. Evans, R. B., and Hendrix, W. A., "Second Law Analysis for Optimum Design of Feedwater Heaters," AIChE Annual Meeting, San Francisco, CA, Nov. (1979).

171. Evans, R. B., Hendrix, W. A., and Kadaba, P. V., "Thermo-dynamic Availability Analysis for Complex Thermal Systems Design and Synthesis," AIChE Annual Meeting, Chicago, IL, Nov. (1980).

172. Fehring, T. and Gaggioli, R. A., "Economics of Feedwater Heater Replacement," Trans. ASME: J. Eng. Power, 99, 482 (1977).

173. Fewell, M. E., Reid, R. L., Murphy, L. M., and Ward, D. S., "First and Second Law Analysis of Steam Steadily Flowing through Constant-Diameter Pipes," Proc. 3rd Annual Conference on Systems Simulation, Economic Analysis/Solar Heating and Cooling Operational Results, pp. 712-718, Reno, NV, April, 1981, ASME, NY (1981).

174. Fiala, W., "Exergetic Analysis of Various Heat Accumulators at Changing Ambient Temperatures," Brennst-Waerme-Kraft, 33, 482 (1982).

175. Gaggioli, R. A., Wepfer, W. J. and Chen, H., "A Heat-Recovery System for Process Steam Industries," Trans. ASME: J. Eng. Power, 110, 511 (1978).

176. Gaggioli, R. A., Wepfer, W. J., and Elkouh, A. F., "Avail-able Energy Analysis for HVAC. I. Inefficiencies in a Dual-Duct System," Energy Conservation and Building Heating and Air-Conditioning Systems, ASME Symp. Vol., No. H00116, pp. 1-20 (1978).

177. Gaggioli, R. A., Wepfer, W. J. and Elkouh, A. F., "Available Energy Analysis for HVAC. II. Comparison and Recommended Improvements," ibid., pp. 21-30 (1978).

178. Gaggioli, R. A., and Wepfer, W. J., "Second Law Analysis of Building Systems," Energy Conver. Manage., 21, No. 1, 65 (1981).

179. Geskin, E. S. and Foster, J., "Exergy Analysis of Fuel Utilization in Heating Furnaces," AIChE National Meeting, Detroit, MI, Aug. (1981).

180. Harrison, R. F. and Dean, R. B., "Availability Ratio for Performance of Pipeline Components in Two-Phase Flow," Trans. ASME: J. Fluids Eng., 100, 350 (1978).

181. Hendrix, W. A., "Essergy Optimization of Regenerative Feedwater Heaters," Master's Thesis, GA Inst. of Technology, Atlanta, GA (1978).

182. Hussein, M., Wood, R. J., Ocallaghan, P. W. and Probert, S. D., "Efficiencies of Exergy Transductions," Appl. Energy, 6, 371 (1980).

183. Johnson, D. H., "Exergy of the Ocean Thermal Resource and the Second-Law Efficiency of Idealized Ocean Thermal Energy Conversion: Power Cycles," Solar Energy Research Institute, Golden, CO, Report No. SERI/TR-252-1420R, Available as NTIS PCA03/MF 01, National Technical Information Service (1982).

184. Knoche, K. F. and Stehmeier, D., "Exergetic Criteria for the Development of Absorption Heat Pumps," AIChE National Meeting, Detroit, MI, Aug. (1981).

185. Kotas, T. J., "Exergy Concepts for Thermal Plants," Int. J. Heat Fluid Flow, 2, No. 3, 105 (1980).

186. Kotas, T. J., "Exergy Criteria of Performance for Thermal Plant," Int. J. Heat Fluid Flow, 2, No. 4, 147 (1980).

187. Kreider, J. F., "Second-Law Analysis of Solar-Thermal Processes," Int. J. Energy Res., 3, 325 (1979).

188. Kreith, F., Kearney, D. and Bejan, A., "End-Use Matching of Solar Energy Systems," Energy (Oxford), 5, 875 (1980).

189. Lavan, Z. and Wovek, W. M., "Cooled-Bed Solar-Powdered Desiccant Air Conditioning," Proc. 16th Intersociety Energy Conv. Eng. Conf., Vol. 2, pp. 1654-1661, Aug. (1981).

190. Lavan, Z., Monnier, J. B. and Worek, W. M., "Second Law
      Analysis of Desiccant Cooling Systems," Trans. ASME: J.
      Solar Energy Eng., 104, 229 (1982).

191. Lay, J. E., "Second Law Assessment of Solar Heating and
      Cooling Systems," Solar Cooling and Heating Forum, Miami,
      FL, Dec. (1976).

192. Levshakov, A. M., "Exergetic Balance for Polydispersive
      Fluxes of Gaseous Suspensions," Izr. Vyssh. Uchebn Zaved
      Energ., No. 1, 123, Jan. (1979).

193. Loewer, H., "Sorption Heat Pumps as Heating Machines,"
      Klim Kaelte Ing., 5, 447 (1977).

194. Lu, P. C., "On Optimum Disposal of Waste Heat," Energy
      (Oxford), 5, 993 (1980).

195. Munroe, M. and Shepherd, W., "Assessment of Solar Energy
      Availability in Different Regions of the Solar Spectrum,"
      Solar Energy, 26, 41 (1981).

196. Nishitani, H. and Kunugita, E., "Multiobjective Analysis of
      Energy and Resource Conservation in an Evaporator System,"
      In reference 18, p. 273 (1981).

197. Oulette, W. R. and Bejan, A., "Conservation of Available
      Work (Exergy) by Using Promoters of Swirl Flow in Forced
      Convection Heat Transfer," Energy (Oxford), 5, 587 (1980).

198. Poulikakos, D., and Bejan, A., "Fin Geometry for Minimum
      Entropy Generation in Forced Convection," Trans. ASME:
      J. Heat Transfer, 104, 616 (1982.

199. Preisegger, E., "Theoretical Fundamentals of Heat Pumps,"
      Feuerungstechnik (Stuttgart), 18, No. 3, 41 (1980).

200. Reistad, G. M., "Availability Analysis of the Heating Process
      and a Heat Pump System," ASHRAE Symp. Paper No. LO-73-74
      (1973).

201. Sama, D. A., "Economic Optimum LMTD at Heat Exchangers,"
      AIChE National Meeting, Houston, TX, March (1983).

202. San, J. Y., Lavan, Z., Worek, W. M., Monnier, J. B., Franta,
      G. E., Haggard, K., Glenn, B. H., Kolar, W. A., and Howell,
      J. R., "Exergy Analysis of Solar Powdered Desiccant Cooling
      Systems," pp. 567-572, in Proc. of the American Section of
      the Intern. Solar Energy Society, Houston, TX (1982).

203. Stoecker, W. F., Design of Thermal Systems, 2nd edition, McGraw-Hill, NY (1980).

204. Wepfer, W. J., Gaggioli, R. A., and Obert, E. F., "Proper Evaluation of Available Energy for HVAC," Trans. ASHRAE, 85, No. 1, 214 (1979).

205. Wepfer, W. J., Gaggioli, R. A., and Obert, E. F., "Economic Sizing of Steam Piping and Insulation," Trans. ASME: J. Engineering for Industry, 101, 427 (1979).

206. Wepfer, W. J. and Gaggioli, R. A., "An Instructional Experiment for First and Second Law Analysis of a Gas-Fired Air Heater," Intern. J. Mech. Eng. Education, 9, No. 4, 283 (1981).

207. Zschernig, J. and Dittmann, A., "Evaluation of Heat Transformation Processes," Energietechnik, 31, 451 (1981).

J. Applications of Second Law Analysis to the Design, Evaluation and Optimization of Power Plants and Complex Energy Systems, Including Electrochemical, Geothermal, Nuclear and Hydrogen Energy Systems

208. Bloomster, C. H., and Fassbender, L. A., "Role of Second Law Analysis in Geothermal Economics," Energy (Oxford), 5, 839 (1980).

209. Davis, M. E. and Conger, W. L., "An Entropy Production and Efficiency Analysis of the Bunsen Reaction in the General Atomic Thermochemical Hydorgen Production Cycle," Int. J. Hydorgen Energy, 5, 475 (1980).

210. De Vries, B., and Nieuwlaar, E., "Dynamic Cost-Exergy Evaluation of Steam and Power Generation," Resour. Energy, 3, 359 (1981).

211. Fehring, T., "Application of the Second Law of Thermodynamics to Power Plant Problems," Master's Thesis, Marquette University, Milwaukee, Wisconsin (1975).

212. Fratzscher, W., and Eckert, F., "Experience Gained by the Introduction of the Exergy Concept for the Standardized Evaluation of Power and Technological Processes in an Industrial Chemical Complex," Proc. 9th World Energy Conf., 7, 85 (1974).

213. Funk, J. E. and Eisermann, W., "Exergetic/Energetic/
     Economic Analysis of Three Hydrogen Production Processes:
     Electrolytic, Hybrid and Thermochemical," Proc. 2nd Intern.
     Conf. on Alternative Energy Sources, Vol. 8, pp. 3285-3320,
     Miami Beach, FL, Dec. (1979).

214. Gaggioli, R. A., Yoon, J. J., Patulski, S. A., Latus, A. J.,
     and Obert, E. F., "Pinpointing the Real Inefficiencies in
     Power Plants and Energy Systems," Proc. Amer. Power Conf.,
     37, 671 (1975).

215. Gaggioli, R. A. and Fehring, T., "Economics of Boiler Feed
     Pump Drive Alternative," Combustion, 49, No. 7, 35 (1978).

216. Gaggioli, R. A. and Wepfer, W. J., "Available Energy
     Accounting--A Cogeneration Case Study," AIChE National
     Meeting, Philadelphia, PA, June (1978).

217. Gaggioli, R. A. and Wepfer, W. J., "Second Law Analysis of
     Energy Devices and Processes," Energy (Oxford), 5, 823
     (1980).

218. Ileri, A., Reistad, G. M. and Schmisseur, W. E., "Urban
     Utilization of Waste Energy from Thermal-Electric Plants,"
     Trans. ASME; J. Eng. Power, 98, 309 (1976).

219. Keller, A., "The Evaluation of Steam Power Plant Losses by
     Means of the Entropy Balance Diagram," Trans. ASME, 72, 949,
     (1959).

220. Kestin, J., "Available Work in Geothermal Energy," Report
     No. CATMEC/20, Division of Engineering, Brown University,
     Providence, RI, July (1978).

221. Khalifa, H. E., "Towards a Systematic Definition of the
     Thermodynamic Efficiency of Energy Conversion Systems,"
     In reference 18, p. 260 (1981).

222. Khalifa, H. E., "Economic Implications of the Exergy and
     Thermal Efficiencies of Energy Conversion Systems,"
     Proc. 16th Intersociety Energy Conversion Eng. Conf.,
     Vol. 2, pp. 3-8, Aug. (1981).

223. Li, K. W., Duckwitz, N. R., "Cycle Analysis of Air-Storage
     Power Plants," ASME Paper No. 76-GT-41, ASME, NY (1976).

224. Li, K. W., "Second-Law Analysis of the Air-Storage Gas
     Turbine System," ASME Paper No. 76-JPGC-GT-2, ASME, NY
     (1976).

225. Li, K. W., "A Parametric Study of Hot Water Storage for Peak Power Generation." Trans. ASME: J. Eng. Power, 100, 229 (1978).

226. Meyer, C. A., Silvestri, G. I., and Martin, J. A., "Availability Balance of Steam Power Plants," Trans. ASME, Ser. A., 81, 81 (1959).

227. Nesterov, B. P., Korovin, N. V., Brodyanskii, V. M., and Trerodokhlebov, E. S., "Principles of Exergetic Analysis of Electrochemical Energy Generators," Izv. Vyssh, Vchebn. Zaved. Energ., No. 3, 61, March (1977).

228. Obert, E. and Birnie, C., "Evaluation and Location of Losses in a 60 MW Power Station, Proc. Midwest Power Conf. 187 (1949).

229. Reistad, G. M. and Ileri, A., "Performance of Heating and Cooling Systems Coupled to Thermal-Electric Power Plants," ASME Paper No. 74-WA/PID-17, ASME, NY (1974).

230. Reistad, G. M., Yao, B. and Gunderson, M., "Thermodynamic Study of Heating with Geothermal Energy," Trans. ASME: J. Eng. Power, 100, 503, (1978).

231. Roth, J. R. and Miley, G. H., "Implications of the Second Law for Future Directions in Controlled Fusion Research," Energy (Oxford), 5, 967 (1980).

232. Smith, M. S., "Efforts of Condenser Design Upon Boiler Feedwater Essergy Costs in Power Plants," Master's Thesis, GA Inst. of Technology, Atlanta, GA (1981).

233. Tabi, R. and Mesko, J. E., "First and Second Law Analysis of an Advanced Steam Cycle with Fluidized-Bed Heat Input," in Proceedings of the 14th Intersociety Energy Conversion Engineering Conference, Vol. 2, pp. 1767-1773, ACS, Washington, D.C. (1979).

234. Tsujikawa, Y., and Sawada, T., "Analysis of a Gas Turbine and Steam Turbine Combined Cycle with Liquefied Hydrogen as Fuel," Int. J. Hydorgen Energy, 7, 499 (1982).

235. Voigt, H., "Evaluation of Energy Processes Through Entropy and Exergy," RM-78-60, International Institute for Applied Systems Analysis, 2361 Laxenburg, Austria, November (1978).

236. Wepfer, W. J., "Application of the Second Law to the
     Analysis of Energy Systems," PhD Dissertation, Univ. of
     Wisconsin, Madison, WI, Available from the University
     Microfilms, Ann Arbor, MI, Order No. 79-28,679 (1979).

237. Wepfer, W. J. and Crutcher, B. G., "Comparison of Costing
     Methods for Cogenerated Process Steam and Electricity,"
     Proc. Amer. Power Conf., 43, 1070 (1981).

238. Wyatt, J. L., Jr., Chua, L. O., Gannett, J. W., Goeknar,
     I. C., and Green, D. N., "Energy Concepts in the State-
     Space Theory of Nonlinear N-Ports: 2. Losslessness,"
     IEEE Trans. Circuits Syst., 29, 417 (1982).

  K. Applications of Second Law Analysis to the Design, Evalu-
     ation and Optimization of Mass Transfer and Separation
     Equipment/Processes

239. Abrams, H., "Energy Reduction in Distillation," pp. 295-306,
     in Alternatives to Distillation, Inst. Chem. Engrs., Symp.
     Series No. 54, London (1978).

240. Baloh, A., "Energy Consumption in Boil-Down and Drying
     Processes," pp. 1-21, in Proc. of Meeting: Can the Amount
     of Energy Used in Vaporization and Drying Be Reduced,
     Wuerzburg, Germany, F. R., Sept. (1979).

241. Benedict, M. and Gyftopoulos, E. P., "Economic Selection
     of the Components of an Air Separation Process," In
     reference 17, p. 195 (1980).

242. Fitzmorris, R. E. and Mah, R. S. H., "Improving Distillation
     Column Design Using Thermodynamic Availability Analysis,"
     AIChE J., 26, 265 (1980).

243. Flower, J. R. and Jackson, R., "Energy Requirements in
     the Separation of Mixtures by Distillation," Trans. Inst.
     Chem. Engr., 42, T249 (1964).

244. Fonyo-Z, R. E., "The Thermodynamic Efficiency and Energy
     Conservation of Industrial Distillation Systems," In
     reference 18, p. 298 (1981).

245. Fonyo-Z., R. E., "General Interpretation of the Thermo-
     dynamic Efficiency for Separation Processes," Hungarian
     J. Ind. Chem., 10, 89 (1982).

246. Freshwater, D. C., "Thermal Economy in Distillation,"
     Trans. Inst. Chem. Engr., 29, 149 (1951).

247. Freshwater, D. C., "The Heat Pump in Multicomponent Distillation," Brit. Chem. Eng., 6, 388 (1961).

248. Hauer, C. R., "Coupled Transport Membranes for Ore Beneficiation," Energy (Oxford), 5, 937 (1980).

249. Henley, E. J. and Seader, J. D., Equilibrium-Stage Separation Processes, Chapter 17, "Energy Consumption and Thermodynamic Efficiency," Wiley, NY (1981).

250. Itoh, J., Niida, K., Shiroko, K. and Umeda, T., "Analysis of the Available Energy of a Distillation System," Intern. Chem. Eng., 20, 379 (1980).

251. Kato, K., "Energy Savings in Grain Drying - A Thermodynamic Evaluation," Energy Dev. Japan, 4, 153, Oct. (1981).

252. Kayihan, F., "Optimum Distribution of Heat Load in Distillation Columns Using Intermediate Condensers and Reboilers," in Recent Advances in Separation Techniques-II, Norman N. Li, Editor, AIChE Symp. Ser., Vol. 76, No. 192, pp. 1-5 (1980).

253. King, C. J., Separation Processes, 2nd Edition, Chap. 13, "Energy Requirements of Separation Processes," McGraw-Hill, New York (1980).

254. Krishna, R., "A Thermodynamic Approach to the Choice of Alternatives to Distillation," pp. 185-214, in Alternatives to Distillation, Inst. Chem. Engrs. Symp. Series No. 54, London (1978).

255. Mah, R. S. H., Nicholas, J. J., Jr. and Wodnik, R. B., "Distillation with Secondary Reflux and Vaporization: A Comparative Evaluation," AIChE J., 23, 651 (1977).

256. Mah, R. S. H. and Wodnik, R. B., "On Binary Distillation and Their Idealizations," Chem. Eng. Comm., 3, 59 (1979).

257. Mah, R. S. H. and Fitzmorris, R. E., "Approaches to Reversible Multi-Component Distillation," AIChE Annual Meeting, Chicago, IL, Nov. (1980).

258. Meckler, M., "Use Peltier Heat Pumps to Improve Process Separation Availability," in Proceedings of the 14th Intersociety Energy Conversion Conference, Vol. 2, pp. 1780-1787, ACS, Washington, D.C. (1979).

259. Naka, Y., Terashita, M., Hayashiguchi, S. and Takamatsu, T., "Intermediate Heating and Cooling Method for a Distillation Column," J. Chem. Eng. Japan, 13, 123 (1980).

260. Null, H. R., "Energy Economy in Separation Processes," Chem.
     Eng. Progr., 76, No. 8, 42 (1980).

261. Petit, P. J., "Economic Selection of a Venturi Scrubber,"
     In reference 17 , p. 187 (1980).

262. Platonov, V. M. and Zhvanetskii, I. B., "Minimum Work of
     Separating Solutions by Rectification," Theor. Found. Chem.
     Eng., 14, No. 1, 1 (1980).

263. Pratt, H. R. C., Countercurrent Separation Processes,
     Elsevier Publishing Company, New York (1967), pp. 16-23,
     159-171, 238-241, 296, 317-318 and 333.

264. Shinskey, F. G., Distillation Control for Productivity and
     Energy Conservation, Chapters 6 and 7, McGraw-Hill, NY (1977).

265. Stephenson, R. M. and Anderson, T. F., "Energy Conservation
     in Distillation," Chem. Eng. Prog., 76, No. 8, 68 (1980).

266. Vruggink, R. S. and Collins, T. F., "Apply Thermo Laws with
     Care," Hydrocarbon Processing, 61, No. 7, 129 (1982).

  L. Applications of Second Law Analysis to the Design, Evalu-
     ation and Optimization of Combustion and Chemical Reaction
     Processes

267. Appelbaum, B. and Lannus, A., "Available Energy Analysis
     of a Dry Process Cement Plant," AIChE National Meeting,
     Philadelphia, PA, June (1978).

268. Auerswald, O., "Exergetic Analysis of a Sugar Industry with
     Vapor Compression Using the Example of Aarberg Sugar
     Factory (Switzerland)," Zuckevindustrie (Berlin), 106, 804
     (1980).

269. Bidard, R. A., "Energy Conservation in Chemical Reactions:
     Some Thermodynamic Aspects," In reference 16, p. 11 (1978).

270. Brzustowski, T. A. and Brena, A., "Exergy Analysis of
     Combustion for Process Heat," In reference 18, p. 389 (1981).

271. Brzustowski, T. A., "Toward a Second-Law Texonomy of Com-
     bustion Processes," Energy (Oxford), 5, 743 (1980).

272. Cambell, A. B. and Ghamarin, A., "Second Law (Exergy)
     Analysis of Industrial Processes: Application to
     Pressurized Fluidized Bed Combustor with Steam and Gas
     Turbine Generators," Paper 10, Intern. Conference on Co-
     generation, Washington, D.C., Oct. (1981).

273. Clarke, J. M., Horlock, J. H., "Availability and Propulsion," J. Mech. Eng. Sci., 17, 223 (1975).

274. Cremer, H., "Thermodynamic Balance and Analysis of a Synthesis Gas and Ammonia Plant," In reference 17, p. 111 (1980).

275. England, C. and Funk, J. E., "Reduced Product Yield in Chemical Processes by Second Law Effects," Energy (Oxford), 5, 941 (1980).

276. Funk, J. E., and Knoche, K. F., "Irreversibilities, Heat Penalties and Economics for the Methanol/Sulfuric Acid Process," Proc. 12th Intersociety Energy Conversion Conference, 1, 933, ANS, LaGrange Park, IL (1977).

277. Geskin, E. S., "Second Law Analysis of Fuel Consumption in Furnaces," Energy (Oxford), 5, 949 (1980).

278. Kaiser, V., "Energy Optimization," Chem. Eng., 88, No. 4, 62, Feb. 23 (1980).

279. Kapner, R. S. and Lannus, A., "Thermodynamic Analysis of Energy Efficiency in Catalytic Reforming," Energy (Oxford), 5, 915 (1980).

280. Lazovskaya, V. V., Atamanchuk, L. I. and Tyutyunik, L. N., "Evaluating the Efficiency of a Coke Dry Quencing Chamber as a Thermodynamic System," Coke Chem. (USSR), No. 6, 34 (1981).

281. Lewis, J. H., III, "Propulsive Efficiency from an Energy Utilization Standpoint," J. Aircr., 13, 299 (1976).

282. Maloney, D. P., and Burton, J. R., "Using Second Law Analysis for Energy Conservation Studies in the Petro-chemical Industry," Energy (Oxford), 5, 925 (1980).

283. Ravindranath, K. and Thiyagarajan, S., "Available Energy Analysis of a Sulfuric Acid Plants", AIChE National Meeting, Detroit, MI, Aug. (1981).

284. Richter, H. J. and Knoche, K. F., "Revasibility of Combustion Processes," AIChE National Meeting, Detroit, MI, Aug. (1981).

285. Riekert, L., "The Efficiency of Energy-Utilization in Chemical Processes," Chem. Eng. Sci., 29, 1613 (1974).

286. Riekert, L., "The Conversion of Energy in Chemical
      Reactions," Energy Conversion, 15, 81 (1976).

287. Riekert, L., "Flow and Loss of Available Energy in Chemical
      Processing Systems," pp. 483-494, in Chemical Engineering
      in a Changing World: Proceedings of the Plenary Sessions
      of the First World Congress on Chemical Engineering, Edited
      by W. T. Koetsier, Elsevier, Amsterdam (1976).

288. Riekert, L., "Flow and Conversion of Energy in Chemical
      Processing Networks," pp. 35-44, in Proceedings of 4th
      Intern. Symp. on Large Chemical Plants, Antwerp, Belgian,
      Oct. 1979, Elsevier, Amsterdam (1979).

289. Rickert, L., "Energy Storage in Chemical Operations," Ber
      Bunsenges Phys. Chem., 84, 964 (1980).

290. Shieh, J. H. and Fan, L. T., "Thermodynamic Analysis of the
      Portland Cement Production Process," In reference 18, p.
      396 (1981).

291. Smith, S. V., Sweeney, J. C., Brown, H. L., Hamel, B. B.,
      and Grossmann, E. D., "A Thermodynamic Analysis of a
      Refinery Process," Energy Institute Report No. 75-3, Drexel
      University, Philadelphia, PA, June (1975).

292. Sweeney, J. C., Smith, S. V., Brown, H. L., Hamel, B. B.,
      and Grossman, E. D., "A Fundamental Approach to the Thermo-
      dynamic Configurational Analysis of Process Flowsheets as
      Applied to a Refinery Unit," Energy Institute Report No.
      75-4, Drexel University, Philadelphia, PA, June (1975).

293. Vakil, H. B., "Thermodynamic Analysis of Chemical Energy
      Transport," AIChE Annual Meeting, Chicago, IL, Nov. (1980).

294. Vakil, H. B., "Thermodynamic Analysis of Gas-Turbine Cycles
      with Chemical Reactions," In reference 18, p. 538 (1981).

  M. Applications of Second Law Analysis to the Design, Evalu-
      ation and Optimization of Fuel Conversion Processes

295. Baehr, H. D., "The Exergy of Fuels," Brennst-Waerme-Kroft,
      31, 292 (1979).

296. Cheng, W. B., Ikumi, S. and Wen, C. Y., "Entropies of Coals
      and Reference States in Coal Gasification Availability
      Analysis," AIChE Annual Meeting, Chicago, IL, Nov. (1980).

297. Gaggioli, R. A. and Petit, P. J., "Second Law Analysis for
      Pinpointing the True Efficiencies in Fuel Conversion
      Systems," Amer. Chem. Soc., Fuel Chem. Division, Preprints,
      21, No. 2, 56 (1976).

298. Gaggioli, R. A. and Petit, Peter J., "Use the Second Law
      First," CHEMTECH, 496, August (1977).

299. Gaggioli, R. A., Roddriquez, L. S. J., and Wepfer, W. J.,
      "Thermodynamic-Economic Analysis of the Synthane Process
      Using Available Energy Concepts," NTIS PC A06/MF A01,
      National Technical Information Service, Nov. (1978).

300. Gaggioli, R. A. and Wepfer, W. J., "Second-Law Costing
      Applied to Coal Gasification," AIChE Technical Manual,
      Coal Processing Technology, Vol. 6, 140, AIChE, NY (1980).

301. Ghamarian, S. and Cambel, A. B., "Exergy Analysis of
      Illinois No. 6 Coal," Energy (Oxford), 7, 483 (1982).

302. Grossman, E. D., Smith, S. V., and Sweeney, J. C.,
      "Calculation of the Availability of Petroleum Fractions,"
      AIChE National Meeting, Detroit, MI, Aug. (1981).

303. Ishida, M. and Nishida, N., "Evaluation of Coal Conversion
      Processes from an Energy Efficient Use Viewpoint (II):
      Energy and Exergy Analysis of a Process System," Fuel
      Soc. J. (Japan), 60, 952 (1981).

304. Ishida, M. and Nishida, N., "Evaluation of Coal Conversion
      Processes from an Energy Efficient Use Viewpoint (III):
      Energy and Exergy Analysis of a Gasification Process,"
      Fuel Soc. J. (Japan), 61, 82 (1982).

305. Johnson, P. J., "Computer Simulation, Second-Law Analysis,
      and Economics of Coal Gasification Processes," Ph.D.
      Dissertation, Univ. of Kentucky, Lexington, KY, Available
      from the University Mocrofilms, Ann Arbor, MI, Order No.
      80-27,985 (1980).

306. Johnson, P. J., and Conger, W. L., "Availability (Exergetic)
      Analysis of Coal Gasification Processes - 1. Theoretical
      Considerations," Fuel Proc. Technol., 5, No. 1-2, 141 (1981).

307. Klose, E. and Heschel, W., "Calculation of Energy in
      Chemical Engineering Processes with Particular Consideration
      of Fuel Engineering: 1. Definitions and Exergy Calculation
      Rules," Energietechnick, 30, No. 8, 295 (1980).

308. Klose, E. and Heschel, W., "Calculating Exergy in
     Chemical Engineering Processes with Particular Consider-
     ation of Fuel Engineering: 2. Calculation Examples and
     Exergy Charts," Energietechnik, 30, No. 12, 471 (1980).

309. Lin, C. Y., "Available Work Energy and Coal Conversion
     Processes," Ph.D. dissertation, West Virginia University,
     Morgantown, WVA (1977).

310. Nishida, N. and Ishida, M., "Evaluation of Coal Conversion
     Processes from an Energy Efficient Use Viewpoint (I),"
     Fuel Soc. J. (Japan), 60, 806 (1981).

311. Nishida, N. and Ishida, M., "Evaluation of Coal Conversion
     Processes from an Energy Efficient Use Viewpoint (IV):
     Energy and Exergy Analysis of Liquefaction Process,"
     Fuel Soc. J. (Japan), 61, 291 (1982).

312. Nishida, N. and Ishida, M., "Evaluation of Coal Conversion
     Processes from an Energy Efficient Use Viewpoint (V): The
     H-Coal Process," Fuel Soc. J. (Japan), 61, 728 (1982).

313. Peters, W. C., Ruppel, T. C. and Mulvihill, J. W., "The
     Role of Thermodynamic Effectiveness in Evaluating Coal
     Conversion RD&D," In reference 16, p. 75 (1978).

314. Purcupile, J. C. and Stas, J. D.,"Energy Conservation in
     Coal Conversion and Energy Conservation Potential in Heat
     Recovery Techniques: A Case Study," In reference 16,
     p. 601 (1978).

315. Rodriquez, S. J. L. and Gaggioli, R. A., "Second Law of
     a Coal Gasification Process," Can. J. Chem. Eng., 58,
     376 (1980).

316. Singh, S. P., Weil, S. A. and Babu, S. P., "Thermodynamic
     Analysis of Coal Gasification Processes," Energy (Oxford),
     5, 905 (1980).

317. Tabi, R. and Mesko, J. E., "Combined Gas-Steam Turbine
     Cycle Using Coal-Derived Liquid Fuel - A Viable Alternative
     to Direct Combustion of Coal," Proc. 2nd Intern. Conf.
     on Alternative Energy Sources, Vol. 7, pp. 2837-2846,
     Miami Beach, FL, Dec. (1979).

318. Tsatsaronis, G., Schuster, P. and Rortgen, H., "Thermo-
     dynamic Analysis of a Coal Hydrogasification Process for
     SNG Production by Using Heat from a High-Temperature
     Nuclear Reactor," In reference 18, p. 401 (1981).

319. Unruh, T. L. and Kyle, B. G., "The Energetics of the COED
      Process for Coal Conversion from a Second Law Prospective,"
      AIChE Annual Meeting, San Francisco, CA, Nov. (1979).

320. Wen, C. Y., Ikumi, S., Onozaki, M. and Luo, C. D., "Coal
      Gasification Availability Analysis," In reference 18, p.
      256 (1981).

  N. Applications of Second Law Analysis to the Optimal Design
      and Systematic Synthesis of Energy-Efficient Chemical
      Processes

321. Andrecovich, M. J. and Westerberg, A. W., "A Simple
      Synthesis Method Based on Utility Bounding for Heat
      Integrated Distillation Sequences," AIChE National Meeting,
      Houston, TX, March (1983).

322. Baehr, H. D., "Exergy - A Useful Tool for Chemical
      Engineers," In reference 18, p. 238 (1981).

323. Barnes, F. J. and King, C. J., "Synthesis of Cascade
      Refrigeration and Liquefaction Systems," Ind. Eng. Chem.
      Process Des. Develop., 13, (4), 421 (1974).

324. Cheng, W. B. and Mah, R. S. H., "Interactive Synthesis of
      Cascade Refrigeration Systems," Ind. Eng. Chem. Process Des.
      Dev., 19, 410 (1980).

325. Evans, R. B., Hendrix, W. A., Kadaba, P. V. and Wepfer,
      W. J., "Exergetic Functional Analysis for Process Design
      and Synthesis," AIChE National Meeting, Detroit, MI, Aug.
      (1981).

326. Fan, L. T. and Shieh, J. H., "Thermodynamically-Based
      Analysis and Synthesis of Process Systems," Energy (Oxford),
      5, 955 (1980).

327. Fan, L. T., Shieh, J. H., Shimizu, Y. and Chiu, S. Y.,
      "Thermodynamically-Oriented Analysis and Synthesis of a
      Process System - HYGAS Process," AIChE Annual Meeting,
      San Francisco, CA, Nov. (1979).

328. Flower, J. R. and Linnhoff, B., "Thermodynamic Analysis
      in the Design of Process Networks," pp. 472-486, in
      Proceedings of the 12th Symp. on Computer Appl. in
      Chem. Eng., Montreaux, Switzerland, April (1979).

329. Gaggioli, R. A., "Second Law Analysis for Process and
      Energy Engineering," AIChE National Meeting, Chicago, IL,
      Nov. (1980).

330. Hohmann, E. C., "Optimal Networks for Heat Exchange," Ph.D.
      dissertation, University of Southern California, Los
      Angeles, CA (1971).

331. Hohmann, E. C. and Sander, M. T., "A New Approach to the
      Synthesis of Multicomponent Separation Sequences," AIChE
      Annual Meeting, Chicago, IL, Nov. (1980).

332. Ishida, M. and Oaki, H., "Chemical Process Design Based on
      the Structured Process Energy-Exergy-Flow Diagram (SPEED),"
      AIChE National Meeting, Detroit, MI, Aug. (1981).

333. Ishida, M., and Kawamura, K., "Energy and Exergy Analysis
      of a Chemical Process System with Distributed Parameters
      Based on the Enthalph-Directed Factor Diagram," Ind. Eng.
      Chem. Process Des. Dev., 21, 690 (1982).

334. Ishida, M. and Tunaka, H., "Computer-Aided Reaction System
      Synthesis Based on Structured Process Energy-Exergy-Flow
      Diagram," Computer in Chem. Eng., 6, 295 (1982).

335. King, C. J., Gantz, D. W. and Barnes, F. J., "Systematic
      Evolutionary Process Synthesis," Ind. Eng. Chem. Process
      Des. Develop., 11, No. 2, 271 (1972).

336. Linnhoff, B., "Thermodynamic Analysis in the Design of
      Process Networks," Ph.D. Dissertation, Univ. of Leeds,
      United Kingdom (1979).

337. Linnhoff, B., "Entropy in Practical Process Design,"
      pp. 537-572, in Foundations of Computer-Aided Chemical
      Process Design, Vol. II, Edited by R. S. H. Mah and W. D.
      Seider, Engineering Foundation, New York (1981).

338. Linnhoff, B., Townsend, D. W., Boland, B., Hewitt, G. F.,
      Thomas, B. E. A., Guy, A. R. and Marsland, R. H., A User
      Guide on Process Integration for the Efficient Use of
      Energy, Inst. of Chem. Engrs., London (1982).

339. Liu, Y. A., "Thermodynamic Availability Analysis in Process
      Design, Evaluation and Synthesis: Review and Extensions,"
      AIChE Annual Meeting, Chicago, IL, Nov. (1980).

340. Liu, Y. A., "Recent Advances Toward the Systematic Multi-
      objective Synthesis of Heat Exchanger Networks," AIChE
      National Meeting, Orlando, FL, Feb. (1982).

341. Liu, Y. A., "A Practical Approach to the Multiobjective
      Synthesis and Optimizing Control of Resilient Heat Exchanger
      Networks," Proc. Amer. Control Conf., 3, 1115-1126,
      Arlington, VA, IEEE, NY, June (1982).

342. Liu, Y. A. and Williams, D. C., "Optimal Synthesis of Heat-Integrated Multicomponent Separation Systems and Their Control System Configurations," Proc. of the 1981 Summer Computer Simulation Conf., Washington, D.C., pp. 238-248, published by ISA, Research Triangle Park, NC, July (1981).

343. Liu, Y. A., Pehler, F. A. and Cahela, D. R., "Studies in Chemical Process Design and Synthesis: VII. Systematic Synthesis of Multipass Heat Exchanger Networks," AIChE J., in press (1983).

344. Naka, Y., Terashita, M. and Takamatsu, T., "Thermodynamic Approach to Multicomponent Distillation System Synthesis," AIChE J., 28, 812 (1982).

345. Nishio, M., Itoh, J., Shiroko, K. and Umeda, T., "A Thermodynamic Approach to Steam-Power System Design," in Proceedings of the 14th Intersociety Energy Conversion Conference, Vol. 2, pp. 1751-1757, American Chemical Society, Washington, D.C. (1979).

346. Pehler, F. A. and Liu, Y. A., "Thermodynamic Availability Analysis in the Synthesis of Energy-Optimum and Minimum-Cost Heat Exchanger Networks," AIChE National Meeting, Detroit, MI, Aug. (1981).

347. Pehler, F. A. and Liu, Y. A., "Studies in Chemical Process Design and Synthesis: VI. A Thermoeconomic Approach to the Evolutionary Synthesis of Heat Exchanger Networks," Chem. Eng. Commu., in press (1983).

348. Petlyuk, F. B., Platonou, V. M. and Slavinskii, D. M., "Thermodynamically Optimal Method for Separating Multi-component Mixtures," Intern. Chem. Eng., 5, 555 (1965).

349. Shieh, J. H. and Fan, L. T., "Multiobjective Optimal Synthesis of Methanation Process," AIChE Annual Meeting, Chicago, IL, Nov. (1980).

350. Sophos, A., Rotstein, E., and Stephanopoulos, G., "Thermo-dynamic Bounds and the Selectivity of Technologies in the Petrochemical Industry," Chem. Eng. Sci., 35, 1049 (1980).

351. Takamatsu, T. and Naka, Y., "Design Method of Chemical Processes for Energy Saving," Energy Develop., Japan, 5, 149 (1982).

352. Townsend, D. W., "Second Law Analysis in Practice," Chem. Eng., (London), 361, 628 (1980).

353. Umeda, T., Itoh, J. and Shiroko, K., "Heat Exchange System Synthesis by Thermodynamic Approach," Chem. Eng. Progr., 74, No. 7, 70 (1978).

354. Umeda, T., Harada, T. and Shiroko, K., "Thermodynamic Approach to the Synthesis of Heat Integration Systems in Chemical Processes," Computers in Chem. Eng., 3, 273 (1979).

355. Umeda, T., Niida, K. and Shiroko, K., "A Thermodynamic Approach to Heat Integration in Distillation Systems," AIChE J., 25, 423 (1979).

356. Zudkevitch, D. and Wenzel, L. A., "The Economic Role of Entropy Minimization in Chemical Process Design," AIChE Annual Meeting, Chicago, IL, Nov. (1980).

RECEIVED July 12, 1983

INDEXES

Author Index

Subject Index

449

Thermodynamics: Second Law Analysis

Richard A. Gaggioli, Editor

Contents

461

Book production by Anne Riesberg
Indexing by Florence Edwards
Jacket design by Anne Bigler

Elements typeset by Hot Type Ltd., Washington, DC
Printed and bound by Maple Press Co., York, PA

Recent ACS Books

"Xenobiotics in Foods and Feeds"
Edited by John W. Finley and Daniel E. Schwass
ACS SYMPOSIUM SERIES 234; 432 pp.; ISBN 0-8412-0809-3

"Nonlinear Optical Properties of Organic and Polymeric Materials"
Edited by David J. Williams
ACS SYMPOSIUM SERIES 233; 244 pp.; ISBN 0-8412-0802-6

"Rings, Clusters, and Polymers of the Main Group Elements"
Edited by Alan H. Cowley
ACS SYMPOSIUM SERIES 232; 182 pp.; ISBN 0-8412-0801-8

"Bacterial Lipopolysaccharides: Structure, Synthesis, and Activities"
Edited by Laurens Anderson and Frank M. Unger
ACS SYMPOSIUM SERIES 231; 326 pp.; ISBN 0-8412-0800-X

"Geochemistry and Chemistry of Oil Shales"
Edited by Francis P. Miknis and John F. Mckay
ACS SYMPOSIUM SERIES 230; 557 pp.; ISBN 0-8412-0799-2

"The Effects of Hostile Environments on Plastics"
Edited by David P. Garner and G. Allan Stahl
ACS SYMPOSIUM SERIES 229; 330 pp.; ISBN 0-8142-0798-4

"Chemistry and Modern Society"
Edited by John Parascandola and James C. Whorton
ACS SYMPOSIUM SERIES 228; 204 pp.; ISBN 0-8412-0795-X

"Chemorheology of Thermosetting Polymers"
Edited by Clayton A. May
ACS SYMPOSIUM SERIES 227; 338 pp.; ISBN 0-8412-0794-1

"Chemical Reaction Engineering--Plenary Lectures"
Edited by James Wei and Christos Georgakis
ACS SYMPOSIUM SERIES 226; 202 pp.; ISBN 0-8412-0793-3

"Fate of Chemicals in the Environment"
Edited by Robert L. Swann and Alan Eschenroeder
ACS SYMPOSIUM SERIES 225; 336pp.; ISBN 0-8412-0792-5

"Dopamine Receptors"
Edited by Carl Kaiser and John W. Kebabian
ACS SYMPOSIUM SERIES 224; 290 pp.; ISBN 0-8412-0781-X

"Molecular-Based Study of Fluids"
Edited by J. M. Haile and G. A. Mansoori
ADVANCES IN CHEMISTRY 204; 524 pp.; ISBN 0-8412-0720-8

"Polymer Characterization--Spectroscopic, Chromatographic
and Physical Instrumental Methods"
Edited by Clara D. Craver
ADVANCES IN CHEMISTRY 203; 792 pp.; ISBN 0-8412-0700-3